9 **Chemistry of Plant Protection**

Pesticides in Ground and Surface Water

Volume Editor: H. Börner

With contributions by
H. Beitz, D. W. Bewick, C. N. Guyot,
M. Häfner, F. Herzel, M. O. James,
G. Mattheß, H. Schmidt

With 28 Figures and 59 Tables

Springer-Verlag
Berlin Heidelberg NewYork
London Paris Tokyo
HongKong Barcelona Budapest

Volume Editor:

Prof. Dr. *H. Börner*
Institut für Phytopathologie
der Universität Kiel
Hermann-Rodewald-Str. 9
24118 Kiel, FRG

This series continues the handbook
"Chemie der Pflanzenschutz- und Schädlingsbekämpfungsmittel"
edited by R. Wegler

ISBN-13:978-3-642-79106-2 e-ISBN-13:978-3-642-79104-8
DOI: 10.1007/978-3-642-79104-8

Library of Congress Cataloging-in-Publication Data
Pesticides in ground and surface water / volume editor, H. Börner with contributions by H. Beitz ... [et al.]. (Chemistry of plant protection : 9)
Includes index.
ISBN-13:978-3-642-79106-2
1. Pesticides--Environmental aspects. 2. Groundwater--Pollution. 3. Water--Pollution.
I. Börner, Horst, 1926- . II. Beitz, Horst. III. Series.
TD427.P35P468 1994 628.1'6842--dc 20 94-23050 CIP

Softcover reprint of the hardcover 1st edition 1994

Typesetting: Macmillan India Ltd., Bangalore-25
SPIN: 10077350 52/3020 - 5 4 3 2 1 0 - Printed on acid-free paper

Managing Editors

Editor-in-Chief
Prof. Dr. W. Ebing
Biologische Bundesanstalt für Land- und Forstwirtschaft
Königin-Luise-Str.
14195 Berlin/FRG

Prof. Dr. H. Börner
Institut für Phytopathologie der Universität Kiel
Hermann-Rodewald-Str. 9
24118 Kiel/FRG

Prof. Dr. D. Martin
Biologische Zentralanstalt Berlin
Institut für Phytopharmakologie
Stahnsdorfer Damm 81
14532 Kleinmachnow/FRG

Dr. V. Sjut
Schering AG, Pflanzenschutzforschung
Gollanczstr. 57-101
13465 Berlin/FRG

Prof. Dr. H.-J. Stan
Technische Universität Berlin
Gustav-Meyer-Allee 25
13355 Berlin/FRG

Dr. J. Stetter
ZF - FWI, Bayer AG, Q 18
51368 Leverkusen, Bayerwerk/FRG

Editorial Board

Editorial

The managing editors and I are proud that ten volumes of the book series "Chemistry of Plant Protection" have been published before a decade has passed since the start of this project in 1986.

The objectives of this enterprise are still the same: To supply readers with reviews about scientific fields of modern interest; the reviews are directed to new developments in research of all questions regarding chemical plant protection in its broadest sense. This includes synthesis and properties of new biochemically active substance classes, their trace analysis in environmental specimens, the environmental influence and behaviour of pesticides as well as compounds attributed with properties generating and controlling biological processes which may protect nutritional plants against diseases or other influences detrimental to growth.

In order to reach a wide readership all papers are written in English. The publisher and the board endeavour to engage highly specialized scientists for the respective articles and it is self-evident that only high qualified contributions will be accepted.

The contributions are written to a fairly advanced level and should be of interest to graduate students, research fellows and practising scientists. I am sure that the subject matter treated is of high interest to scientists in industry, universities, as well as in governmental and regulatory bodies. In addition, natural products chemists, biologists, ecologists, and organic chemists will benefit from several of the chapters within this book series.

It is with great satisfaction that I thank all authors and editors involved for the time they have devoted to this effort.

Berlin, October 1994

W. Ebing
Editor-in-Chief

Table of Contents

Occurrence, Toxicological and Ecotoxicological Significance of Pesticides in Groundwater and Surface Water

Horst Beitz[1], Heinz Schmidt[2] and Fritz Herzel[3]

[1] Umesa – Umweltsanierung Boden und Grundwasser, Max-Eyth-Allee 54, D-14469 Potsdam, FRG, formerly[2]
[2] Biologische Bundesanstalt für Landwirtschaft und Forstwirtschaft in Berlin und Braunschweig, Institut für Ökotoxikologie, Außenstelle Kleinmachnow, Stahnsdorfer Damm 81, D-14532 Kleinmachnow, FRG
[3] Bundesgesundheitsamt, Max von Pettenkofer Institut, Thielallee, D-14195 Berlin, FRG

The contamination of groundwater and surface water is a consequence of the use of pesticides in the agriculture and out of the agriculture. Their significance for human health and the environment largely depends on the quantity applied annually within an area, the toxicological and ecotoxicological properties of the pesticides and the persistence in the environment. The regulations and guidelines for maximum allowable concentrations in drinking water are described for the EC, WHO, Germany, the former Soviet Union and the USA. An overview of the contamination of groundwater and drinking water in West and East Germany, Western Europe and the USA is given. With regard to the evaluation of the quantity of contaminated samples the results of various monitoring programmes in the mentioned countries are analyzed. The contamination of water bodies is described by selected data from Western and Eastern Europe. The ecotoxicological evaluation of the contamination of water bodies with pesticides is difficult because international harmonized guidelines.

Chemistry of Plant Protection, Vol. 9

1 Application of Pesticides – A Source of Environmental Contamination

The use of pesticides is associated with their entering ecosystems. For this reason, if possible, resulting hazards have to be cleared up before their registration. It has to be taken into account that only a small part of the applied pesticides reaches its final destination, the biological targets, and is toxicologically active. The greater part gets lost during "transport" and must be degraded in the ecosystems to prevent accumulation or even contamination. Thus, pesticides are an essential part of xenobiotics brought into the environment by human activities. Their significance for human health and the environment largely depends on

- the quantity annually applied in a country or within an area,
- the toxicological and ecotoxicological properties of the pesticides, and
- the persistence in the environment, especially in soil and water.

These criteria indicate that the coincidence of high persistence in soil and water (half-life >1 month) and a good lipophility (log $P_{OW} > 3.0$) can cause an accumulation in soil or in sensitive organisms which in turn causes the threat of chronical toxicity or delayed damage (human beings) or reproductive toxicity (field organisms). Therefore, the FAO recommended all countries to introduce a statutory obligation to protect human health and the environment from unwanted damage and to create a register or pesticides [1]. For this purpose the manufacturer has to submit documentation [2, 3] on the active ingredient and its formulations. It includes

- a toxicological characterization,
- the ecotoxicological properties,
- the chemical behaviour and fate in the environment,
- analytical methods to detect the active ingredient and its by-products, metabolites and impurities. They serve together for the detection of pesticides on sprayed plants and above all in substrates of the environment and the evaluation of analysed residues and concentrations in ground- and surface water.

Present documentation of registered pesticides reveal that these substances are the most thoroughly analyzed compounds apart from drugs, however, particularly for the active ingredients developed in the last 10 to 15 years. For this reason biological and chemical monitoring of ecosystems, especially for older formulations, is very important. This becomes obvious when considering the multiple interactions between pesticides and ecosystems (Fig. 1).

The registration procedure evaluates biological effectiveness and the side effects of a pesticide and possible risks in the sense of a benefit-risk analysis. According to the Plant Protection Act of the Federal Republic of Germany

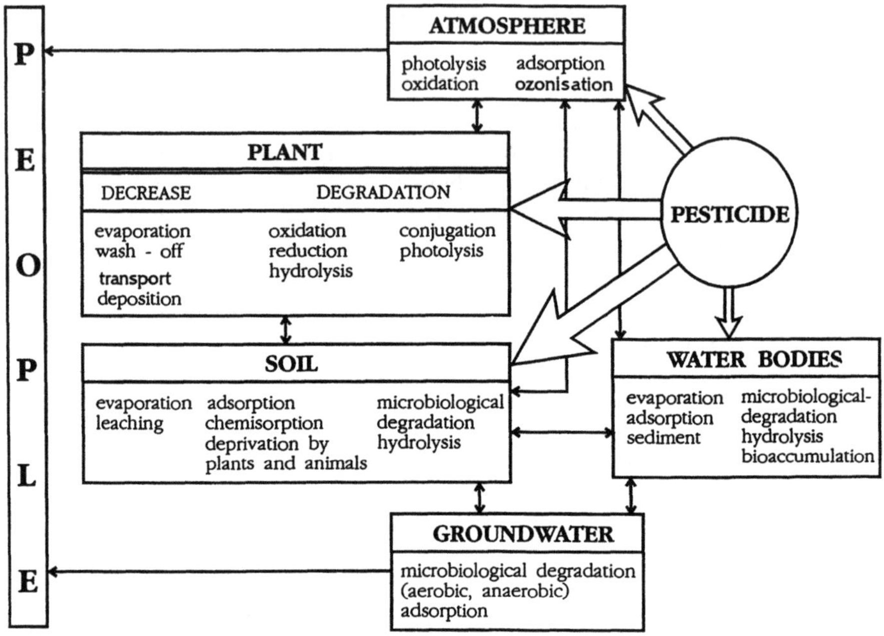

Fig. 1. Interactions between pesticides and ecosystems

Article 15, water, especially groundwater, is expressive laid down as a commodity to be protected. This is understandable as in Germany groundwater supplies about 70% of the drinking water. As drinking water is one of the basic necessities, which, moreover, cannot be replaced, especially high purity and sanitary requirements have to be met. It is urgently necessary to reduce the amounts of pesticides entering groundwater and surface water. Such measures can differ from region to region and mainly depend on the amounts of pesticide applied and climatic factors. This is also evident from the market shares of the individual regions in the estimated world-wide sales of 36.5 thousand millions german Marks [5] in 1991:

Western Europe	31%
Eastern Europe	4%
North America	26%
Latin America	11%
Asia	15%
Japan	9%
Africa	4%

The Situation in Germany has changed considerably since reunification (Table 1). The enormous decline in the sales of active ingredients in 1991 can be

Table 1. Active ingredients provided to German agriculture in 1989 and 1991 (metric tons) [4, 5]

Group of pesticides	1989 FRG	1989 GDR	1991 FRG (East and West)
Herbicides	16539	21458	17633
Fungicides	10241	5507	8483
Insecticides	1575	1352	1264
Other	2329	2774	2300
Total	30684	31091	29680

attributed to the decline of East German agriculture, but also to a decreased chemical plant protection measures and land set-aside programmes. An increase of arable land by 50% in comparison to 1989, for example, is in contrast to an increase in domestic net sales of pesticides of only 21% [5]. A dramatic decline as a consequence of economic and political problems has also taken place in East-European countries where the market share was reduced by 3% in 1991 alone. Also in Japan the market share was reduced by 1% to 9%, i.e. relatively by 10%. Such dramatic changes have an effect on the environment and thus on the contamination of groundwater and surface water.

2 Pathways of Pesticide Pollution of Groundwater and Surface Water

The contamination of groundwater and surface water with pesticides can on the one hand be attributed to improper handling of these environmentally significant chemicals and accidents in agriculture, industry and trade. On the other hand pesticides can also reach groundwater and surface water even with good agricultural practice. From the viewpoint of agriculture the following pathways are of special importance [6]:

- transport in the soil profile with leaching, by rainwater through the unsaturated zone to the groundwater,
- drain flow into water bodies and probably reservoirs, and
- runoffs from treated crops and soils entering water bodies and probably reservoirs. Furthermore, attention has to be given to
- direct contamination of surface water by drift and
- deposition of pesticides by rainfall and other kinds of immissions from the atmosphere.

Taking into consideration all sources of contamination of drinking water, the following facts are of occasional importance:

- The use of bank-filtrates of contaminated surface water bodies and
- the feeding of contaminated surface water into the groundwater, but also
- contaminated waste water resulting from pesticide production and
- of leftovers of spray liquids and waste water from cleaning spray equipment entering water bodies and groundwater.

2.1 Pollution in Consequence of Plant Protection Measures

Since the beginning of the 1960s, it has been known that the application of pesticides can cause groundwater contamination. Beran and Guth [7], for example, reported on ground water contamination with organic insecticides, especially organochlorine insecticides (DDT, lindane, aldrin, dieldrin) within the German speaking countries. They found out that the leaching of pesticides into groundwater mainly due to extreme hydrogeological conditions. This was confirmed by Leh [8], Heinisch et al. [9] and other authors. Studies on accidental groundwater contamination by 2,4-D, MCPA, DNOC and chlorate [10], MCPA [11], and dimethoate [12] emphasized this.

On the other hand, pesticides were also found in drinking water from shallow groundwater. Engst and Knoll [13], for instance, found up to 250 $mg\,l^{-1}$ DDT is bank-filtrate and Vrochinskij [14] up to 300 $mg\,l^{-1}$ DDT is covered artesian aquifers and up to 500 $mg\,l^{-1}$ in shallow groundwater.

Despite the then prevalent view that a "normal use", i.e. complying with "good agricultural practice" recommended by the WHO/FAO, does not contaminate groundwater there are a series of publications disproving this.

Althaus [15], for instance, pointed out the possibility of groundwater contamination by pesticides and the resulting hazards such as accumulation in the food chain and contamination of drinking water, because conventional water treatment procedures remove pesticides insufficiently. For this reason the Althaus rejects limits for drinking water as it must be absolutely free from such substances. Furthermore, lack of possibilities for analysis and the absence of a monitoring system are criticized.

Hellmann [16] gives an overview of literature dealing especially with pesticides residues in the environment and their identification. It reveals that pesticides in connection with environment protection problems gained ever growing importance in the middle of the 1960s and at the beginning of the 1970s (Table 2). It is interesting that, according to the author, only after the fish poisoning in the river Rhine caused by endosulfan in 1969 were long and short-term scientific research programmes on the monitoring of water bodies for pesticide residues were initiated in Europe. The Netherlands, which are dependent on the river Rhine as source of drinking water, at first had a dominating position. This trend advanced also in the following years, especially because of people's growing environmental awareness. Such activities could not only be observed in Western countries, but also in the former socialist countries research in this field increased.

Table 2. Overview of publications on environmental problems in connection with pesticides in the 1960s and 1970s [16]

	Publications in the years								
Subject	1964	1965	1966	1967	1968	1969	1970	1971	1972
Analysis		1	1	1	7	3	2	3	8
Toxicity		1	1	4	5	2	10	7	6
Fate in soil					2	1		2	
Fate in water	1			1	6	1	5	4	4
Limits			1	2		3	2	2	
Biological degradation		1	2	1	4	1	7	4	6

Important impulses to investigate the leaching and deposition of agrochemicals and their fate in the environment by monitoring programmes came in the 1980s from the USA where long-term studies were carried out by order of the EPA and ASAE [17–21].

In general it has been proven that, especially in case of shallow groundwater layers, absolute protection from contamination cannot be guaranteed. The active ingredients of pesticides in the soil are subject to specific conditions which strongly differ according to soil texture. Therefore, their fate in the environment can only be described by complex long-term monitoring.

These few examples show that pesticides from very different sources can reach water bodies and groundwater in the many different ways.

Groundwater contamination is mainly caused by agricultural practices, but also by the treatment of paths, squares, industrial and sports facilities and the use of pesticides in household gardens and domestic areas. Especially hazardous is weed removal from railway tracks.

The leaching of pesticides and possibly their metabolites to groundwater normally takes place through the soil. On moving downwards from the soil surface to an aquifer a number of purification processes is induced. Thus, all soils are able to buffer environmental impacts on groundwater. Apart from these natural site conditions, properties of pesticides and application practice play an important part in the evaluation of groundwater vulnerability to contamination.

Helling and Gish [22] classify the individual factors influencing leaching behaviour as follows:

I. Classification of soil properties
   1. Physical composition
      - soil texture (% sand, silt, clay; gravel)
      - soil organic matter content
   2. Chemical composition
      - clay mineralogy
      - organic matter type

3. Physical properties
   - bulk density
   - field moisture capacity
   - hydraulic conductivity
   - pore size distribution, macropores; tendency to crack on drying
4. Chemical properties
   - pH
   - cation-exchange capacity; anion exchange-capacity
   - % base saturation
   - redox potential
5. Transient soil properties
   - soil moisture content
   - soil temperature
6. Biological/biochemical properties
   - number and type of microorganisms
   - activity of specific enzymes

II. Classification of macro soil properties
1. Surface
   - relief
   - slope
2. Subsurface
   - profile changes (type, depth, and areal homogeneity)
   - restricting layers
   - depth to groundwater (perched and unconfined aquifer)

III. Classification of non-soil factors potentially affecting transport to groundwater
1. Climate
   - rainfall (temporal distribution, intensity)
   - temperature
   - evapotranspiration
2. Pesticide properties
   - soil adsorption coefficient (K_{OC})
   - water solubility
   - octanol: water partition coefficient (P_{OW})
   - ionization constant (pK_a, pK_p)
   - chemical and biological stability (persistence in soil)
   - volatility
3. Pesticide application
   - formulation
   - method of application (foliar, soil surface, soil incorporation)
   - rate
   - timing
   - history of pesticide use (accelerated degradation, build-up)

4. Agricultural measurements
   - cropland
     - conventional tillage
     - conservation tillage
     - irrigation
   - non-cropland
     - fallow
     - rangeland, forest, etc.
   - soil amendments

To protect the ground water from pesticides or to minimize their leaching the large number of factors influencing leaching behaviour require a complex solution. Either the complete prohibition of chemical pesticides at least in water protection areas and the transition of crop farming to ecological arable farming or the reduction of the amount of pesticides used in water protection areas on the basis of their leaching behaviour and the hydrogeological conditions of the field with the help of computer-aided consultation systems (models, expert systems) and the step by step extension of this procedure and combination of these measures mentioned. Approaches to realize such a concept on the territory of the former GDR have been described by Beitz and Schmidt [23] and a group of authors [24] which are based on

- general application restrictions in drinking water protection zones
- reduction of the maximum application rate, and
- exclusion of sites vulnerable to contamination with low retention capacity of the soil.

As experimental work on the leaching behaviour of pesticides is very labour-, material- and time-consuming without being able to take into consideration all necessary hydrogeological, climatic and agricultural conditions and each pesticide in all its respect, there have been intensive efforts to use mathematical modelling and computer simulation for their evaluation over the last 10 to 15 years. Only some procedures will be mentioned here.

Aller et al. [25] elaborated the numerical method DRASTIC on the basis of hydrogeological parameters, but without taking into account pesticide properties. Rao et al. [26] recommend the determination of indices characterizing groundwater vulnerability on the basis of the estimated retention time in soil and the biochemical half-life. Blume and Brümmer [27] assign numerical values to the various cause variables, like pesticide properties, soil cohesion, soil type, etc. to evaluate qualitatively the degree of groundwater vulnerability. The model of Jury et al. [28] allows a qualitative and quantitative assessment of groundwater vulnerability taking into account a limit, e.g. drinking water limit, which must not be exceeded. This model underlies the model TERRA [29] with the help of which various scenarios can be chosen describing the fate of active ingredients in the aeration zone and thus their immediate or delayed leaching into aquifers.

2.2 Atmospheric Deposition

Pesticide contents in the atmosphere are a consequence of their application. Because of the drift and evaporation of active ingredients at first amounts of pesticides get into the atmosphere. The quantities depend on the application method, chemico-physical properties of the active ingredient and weather conditions (especially temperature and wind). Thereafter the most important sources of atmosphere contamination are, according to Boehncke et al. [30] and Oberwalder et al. [31], volatilization from treated soil and plant surfaces and drift of soil particles. Finally, the active ingredients and their metabolites in the atmosphere reach soil and water bodies with wet and dry deposits. These atmospheric deposits are of minor significance in comparison to the contamination of arable land resulting from the purposeful application of pesticides. However, they are the only source of contamination of distant ecosystems and of protected ecosystems, like water bodies for drinking water supply in water protection areas and aquatic ecosystems in nature reserves. Therefore, the magnitude and the toxicological or ecotoxicological potential of atmospheric pesticide deposition have to be evaluated.

At the beginning of the 1960s, monitoring of wet and dry atmospheric deposits of pesticides was restricted to organochlorine compounds. Meanwhile, monitoring data have also been published for other groups of active ingredients [32, 33], PAH [34, 35], PCB [36] and volatile organic compounds [37]. The measured concentrations of selected pesticides in rainwater are given in Table 3.

The individual active ingredients occur in the atmosphere bound to particles, droplets or in gaseous state and are transported in this form [38–40]. In general

Table 3. Concentrations of selected pesticides analyzed in rainwater

Active ingredient of pesticide	Concentration ($\mu g\,l^{-1}$)	Author(s)
Atrazine	1.0	Oberwalder et al. [31]
	0.24	Siebers et al. [42]
	0.4	Herterich [44]
	0.6	Buser [43]
Simazine	0.2	Oberwalder et al. [31]
	0.1	Buser [43]
2,4-D	0.1	Oberwalder et al. [31]
Mecoprop	1.0	Oberwalder et al. [31]
Dichlorprop	1.0	Oberwalder et al. [31]
Lindane	0.55	Oberwalder et al. [31]
	0.52	Siebers et al. [42]
	0.01	Braun et al. [51]

the investigations of Braun et al. [41], Oberwalder et al. [31] and Siebers et al. [42] have shown that the highest concentrations were found during the main application periods.

However, there have been no systematic studies for Germany like in other countries [43, 32, 33]. For this reason, an ecological and ecotoxicological evaluation of the deposition and impact of contaminants transmitted by air on individual ecosystems has not been made so far. In this connection Schleyer et al. [45] have drawn the attention to possible effects of organic air contaminants on the quality of groundwater and surface water and soil. The authors stated that the large-scale leaching of pesticides can even lead to the contamination of hardly protected aquifers. Thus, phthalates have already been ubiquitously detected.

The RIVM study [46] describes possible deposition rates for pesticides. On the assumption of a roughly estimated volatilization rate of about 50% of the application rate Baart and Van Diederen [47] calculated for 12 pesticides deposition rates of 85 mg ha^{-1} yr^{-1} to 11 g ha^{-1} yr^{-1}.

For the most important pesticides, atrazine and lindane, they state 8.1 g ha^{-1} yr^{-1} and 1.9 g ha^{-1} yr^{-1}, respectively. Taking into account the inaccuracies of the calculation, these figures are comparable with measured values obtained by various authors (Table 4).

In connection with the above-mentioned pesticide concentrations the computed deposition rates cannot lead to an exceeding of the EC drinking water limit in groundwater and storage reservoirs. Only in extremely shallow basins can the EC limits be theoretically exceeded because of pesticide leaching, the analyzed maximum deposition rate for lindane in surface water with a depth of 1 m gave a concentration of 0.08 μg l^{-1} per year. For this reason the conclusion of Schleyer et al. [45] is exaggerated under normal conditions of pesticide use and good agricultural practice.

2.3 Pollution from Contaminated Sites

The soil and groundwater vulnerability has to be judged as especially critical in the case of a locally limited leaching of pesticides in large quantities. In addition to environmental risks arising from agricultural practices attention has to be given to the illegal dumping of pesticides and their packings on dumping sites or that someone buries residual amounts of pesticides into the soil or disposes tailings containing pesticides and wash water.

The most important sources of risks can be found with the manufacturers. This includes the large stores in which accidents and fires led to large-scale soil and groundwater contamination or such of water bodies. This happened with the fire at Sandoz AG in Basel in 1986 and with the fire in the operating unit Schönebeck of the former manufactor VEB Fahlberg-List, Magdeburg, in 1989. Additionally, in the East and West countries there are dumping sites where the by-products from lindane manufacturing are stored. The isomers alpha-, beta-,

Table 4. Overview of the deposition rates analyzed for atrazine, lindane and other compounds

Authors	Location Period	Deposition rates in mg ha^{-1} Atrazine	Lindane	Phenanthren
Braun [51]	Bavaria 1987–89	0.7	0.8	–
Gath et al. [48]	Hessen 1990–91	0.25	0.6	1.4
Herterich [44]	Bavaria 1989–90	0.13	–	–

Table 5. Concentrations of pesticides (1 mg kg^{-1}) in deeper soil layers on the premises of the factory Berlin Chemie [49]

	Depth (m)	HCH isomers alpha	beta	gamma	delta	DDT	DDD	DDE
1	30–33	738	1170	26	49	608	50	30
2	40–44	1080	3070	69	93	1063	54	132
3	70–74	43	60	2.5	3.7	73	8.3	2.2

1 – coarse and fine gravel
2 – lignite, poor clay
3 – medium sand and fine sand, poor clay

and delta-hexachlorcyclohexane (HCH), which have no insecticidal action, show partly higher water solubility (delta-HCH) or persistence than lindane and have higher K_{OC}- values. In the territory of the former GDR, Heinisch [49] describes the dumping sites and risks resulting from them and the contaminations determined in the environment. For instance, the production plants of the former manufactor VEB Berlin- Chemie were situated in a residential area and the dumping site pit "Regina" near Bad Freienwalde in the state of Brandenburg, Germany within a recreation area. Table 5 shows the concentrations found in the soil of the operating area in 1989.

This site also contaminated water bodies. Between 1985 and 1989, DDT, DDD and DDE were found in surface waters, sediments and fish along the rivers Spree and Havel and the Teltow-canal within the area and surroundings of Berlin. 24 samples were taken from the wells of the DDT and lindane manufacturer and from wells in its vicinity with depths of 2 m to 38.9 m. All samples contained alpha- and gamma-HCH up to 1590 µg l^{-1} and 6900 µg l^{-1}, respectively. Beta-HCH was present in two samples, delta-HCH in 13 samples (max. 2500 µg l^{-1}), hexachlorbenzene in 10 samples (max. 800 µg l^{-1}), DDT,

DDD and DDE in 5, 3 and 2 samples, respectively [50]. Monitoring has also been carried out at the DDT- and lindane-producing plants of the former, Chemiekombinat Bitterfeld, state of Sachsen-Anhalt, Germany, and its dumping site pits "Antonie" and "Freiheit" and at the former lindane-manufactor of the former VEB Fahlberg List Magdeburg and the dumping site Emden. In the groundwater collected around the pit "Antonie" HCH averaged 3.97 (0,56–8,74) $\mu g\,l^{-1}$ in 1986 and 1 $\mu g\,l^{-1}$ in 1991 [51].

Contamination due to agricultural practices in the former GDR, are supposed to occur as well because industrialization of plant production in the 1960s and 1970s was associated with the social division of labour in agricultural work and the separation of certain operations and transport from the production process itself. Fertilization and chemical plant protection measures, for instance, were to a large extent transferred from the plant- and fruit-producing enterprises to the agrochemical centres (ACC).

The agrochemical centres of the former GDR were service stations of agricultural enterprises and among others responsible for the storage and application of pesticides. Their proportion in the total area treated with pesticides is shown in Table 6. An agrochemical centre was responsible for nearly 20 000 ha arable land. Some agrochemical centres with subsidiaries had to cover up to 40 000 ha. According to the average service area and cultivation structure 50 to 150 tons of active ingredients had to be stored and applied. Because of the various amounts of inert ingredients this corresponds to a multiple quantity of products.

Table 7 illustrates the application volumes of selected active ingredients at the agrochemical centres Jänickendorf, Pritzwalk and Zossen, which are located in the state of Brandenburg, Germany [52]. Storing, pesticide application, the cleaning of plant protection equipment and disposal can have caused contaminations. In Brandenburg, 36 out of 66 former agrochemical centres were analyzed for the quantities of products handled and the procedures used. These criteria set the priorities for the evaluation of the problems arising from the past. Table 8 summarizes the most important criteria for the evaluation of those problems.

As a rule the search for contaminated sites started with the sampling of wells belonging to the agrochemical centres and of areas particularly exposed because

Table 6. The volume of chemical plant protection measures in the former GDR and the proportion of the agrochemical centres (ACC)

Year	Total (Mill. ha)	ACC (Mill. ha)	ACC (%)	Agrochemically (Mill. ha)	Agrochemically (%)
1980	9.3	7.38	79.4	2.051	22.0
1985	11.9	8.14	68.4	1.898	15.9
1989	14.5	9.05	64.6	1.059	7.6

Table 7. Comparison of selected herbicides and growth regulators supplied to the agrochemical centres Jaenickendorf, Pritzwalk and Zossen between 1986 and 1990 (active ingredient, kg $year^{-1}$)

Active ingredient (Herbicides from the GDR)	Jaenickendorf	Pritzwalk	Zossen
Chloral hydrate	4843	22206	9455
Chlorate	771	5646	1517
Dichlorprop	1256	1709	2916
2,4-D	1266	789	187
Dalapon	2225	1646	2330
Atrazine	761	1062	3427
Prometryn	1028	1635	1133
Simazine	415	2538	1368
Nitrofen	1609	1048	2613
Buminafos	46	2221	1465

Table 8. Important criteria for the evaluation of problems with respect to pesticides in agrochemical centres

A. Handled and applied quantities
   - total annual tonnage (1985 ... 1989)
   - applied quantity of product per crop and hectar (selected examples)

B. Local conditions and technologies
   - wells belonging to the agrochemical centres
   - storage of pesticides, storage of empty containers
   - mixing, filling, loading places and mixing, filling, loading technologies
   - removal of leftovers
   - washing places for application equipment
   - discharge and collection of wash and waste water
   - inactivation equipment
   - permanent and seasonal agricultural airfields
   - storage place for application equipment

C. Results from the sampling of suspected places
   - water samples from wells belonging to the agrochemical centre
   - soil samples: preferential leaching sites
     unpaved working places
     edges of paved working places
     joints and cracks in concrete surfaces

of their organizational structures and operation procedures. Attention is primarily directed to the degree of contamination of the wells, which were often only used for the preparation of spray liquids and the purification of equipment.

The findings of the water samples are given in Fig. 2. It shows that out of 27 well samples, 15% did not contain pesticides exceeding the EEC drinking water

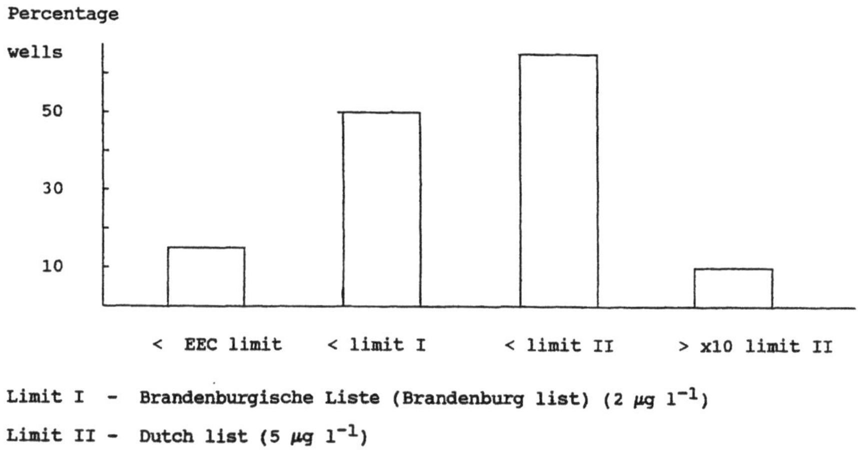

Fig. 2. Characterization of the pesticide content in groundwater samples from wells former agrochemical centres

limit of 0.1 $\mu g l^{-1}$ for single active ingredients and 0.5 $\mu g l^{-1}$ for the total amount of active ingredients. Another 50% of the well samples did not exceed the sanitation limit of 0.2 $\mu g l^{-1}$ set by the state of Brandenburg for the total amount of pesticides. Thus 65% of the groundwater samples came from wells which do not need sanitation. This percentage increases to 75% applying the limit of 5.0 $\mu g l^{-1}$ set by the so-called Dutch list for sanitation. This value was exceeded by more than 10 times in 4 samples (15%) from three sites the sanitation of which is necessary.

Soil samples from the premises of the agrochemical centres confirm that contaminated sites have caused groundwater contamination. The analysis of these groundwater contaminants is interesting. Chlorinated hydrocarbons, for instance DDT and HCH with their isomers, were hardly found. Whereas the active ingredients 2,4-D and dichlorprop of the substituted phenoxyalkanic acids were repeatedly determined. Among s-triazine herbicides, especially atrazine, prometryn and simazine were repeatedly found. Contrary to the result of a groundwater monitoring in the new Federal states of Germany (Sect. 3.2), atrazine does not rank first but follows prometryn.

According to the pesticide content the groundwater collected from wells varies from being of good drinking water quality to values exceeding the so-called sanitation limit (with the consequence of bioremediation) by almost a 100 times. Considerable differences of this range were also stated for soil samples. Table 9 shows the results of soil samples collected from places where the preparation of spray liquids and for the filling of plant protection machines took place in two agrochemical centres [53].

The loading places of the agricultural co-operative farms, which were used less often, show considerably lower residual levels. Only one soil sample from

Table 9. Content of pesticides in soil samples from mixing and filling places in agrochemical centres (ACC) and agricultural co-operative farms (ACF) [54]

Description of sample/ Depth of soil layer	Residues in mg kg^{-1} 0–30 cm	30–60 cm	60–90 cm	Active ingredients[a]
ACC – mixing place G	–[b]	3.3	0.2	1
ACC – mixing place F	48.2	15.8		8
ACF S – roadside	0.4			–
ACF K – leaching site	2.0	1.4		2
ACF G – near well	0.2	–	–	–
– discharge channel	<0.1	–	–	–
– leaching site	<0.1	–	–	–

[a] Exceeding the limit, requiring further investigation
[b] Crushed stone, no investigation

the leaching sites, which favour the penetration of soil, reached the sanitation limit, but had already fallen below it in the soil horizon of 30–60 cm.

With respect to all the results of the study from Beitz et al. [54] the most important active ingredients are also the persistent s-triazine herbicides simazine, atrazine and prometryn. Phenoxyalkanic acid herbicides could not be found in the soil in contrast to groundwater monitoring. Furthermore, chlorinated hydrocarbons, like nitrofen, and occasionally chlorinated hydrocarbon insecticides (derivatives and isomers of DDT and HCH) were detected.

The analyzed soil samples also show considerably varying pesticide contamination. Primarily, simazine, atrazine, prometryn and nitrofen were identified. The following findings caused by improper and careless handling of these substances illustrate this very clearly. Detailed descriptions can be found in the reports of Schmidt and Winkler [55] and Beitz and Krüger [56].

During the control measures against the nun moth (*Lymantria monacha*) in forests, larger quantities of the product "Aero Super" (active ingredients DDT and lindane) were stored at an agricultural airfield. On uncoupling the hose after each filling of the aircraft, type AN 2, 2.5 l of the product spilled on to the ground and seeped away. Tailings were run off into collecting pans, but not regularly. So they often overflowed and contaminated the soil.

After stopping operations on the airfield, the contamination of soil and groundwater was determined. Three holes were drilled for this purpose. The results (Table 10) show strong leaching of activc ingredients into the soil, groundwater also shows evident contamination [55].

On a routine check of surface waters, the water authority found a waste-dumping site near a water body. Since pesticide packings had been found, it became necessary to analyse soil and water for safety reasons. After clearing the site, samples (0–30 cm) were taken from the original soil (Table 11). Residues could not be found in the water sample from the above-mentioned surface water. The analyses clearly show the risks arising from improper disposal of pesticides.

Table 10. Contamination of soil and groundwater with DDT and DDE under the agricultural airfield of Neuendorf after control of the nun moth (*Lymantriamonacha*) [4]

Borehole 1 Barrel stand Depth (m)	Active ingredient ($mg\,kg^{-1}$) p,p′-DDE	o,p′-DDT	p,p′-DDT
0.0–0.2	0.33	0.25	9.9
0.2–0.4	0.03	n.d.	4.7
0.4–0.6	0.04	n.d.	6.5
0.6–0.8	0.9	18.0	63.0
0.8–1.0	0.3	3.0	49.0
1.0–1.2	0.8	4.5	27.0
1.2–1.4	0.7	0.3	3.5
1.4–1.6	0.03	0.2	1.0
1.6–1.8	0.01	0.05	0.5
1.8–2.0	0.03	0.08	0.6
2.0–2.5	n.d.	n.d.	0.08
Water ($\mu g\,l^{-1}$)	0.7	0.1	0.3
0.4–0.6	0.03	0.1	13.0
0.6–0.8	n.d.	0.05	1.0
0.8–1.0	0.01	0.2	1.7
1.0–1.2	n.d.	0.08	0.05
1.2–1.4	0.4	0.5	14.0
1.4–1.6	0.2	0.2	10.0
1.6–1.8	n.d.	0.2	11.0
1.8–2.0	0.5	0.3	6.5
2.0–2.5	n.d.	0.03	0.2
3.0–3.5	n.d.	0.02	0.5
4.0–4.5	0.01	n.d.	0.3
5.0–5.5	n.d.	0.03	0.5
6.0–6.5	n.d.	0.01	0.3
7.0–7.5	n.d.	0.01	0.2
8.0–8.5	n.d.	0.02	0.2
9,0–9.5	n.d.	0.03	0.4
Water ($\mu g\,l^{-1}$)	0.1	0.6	0.9
0.0–0.2	0.8	4.1	107.2
0.2–0.4	1.3	9.1	101.1
0.4–0.6	0.4	0.5	28.7
0.6–0.8	0.09	0.7	18.6
0.8–1.0	n.d.	0.01	0.2
1.0–1.2	n.d.	0.04	0.2
1.2–1.4	n.d.	0.01	0.2
1.4–1.6	0.02	0.06	0.5
1.6–1.8	0.02	0.06	0.5
1.8–2.0	0.02	0.07	0.5
2.0–2.5	0.01	n.d.	0.5
Water ($\mu g\,l^{-1}$)	0.1	0.1	0.7

n.d. – not detected

Table 11. Pesticide residues in soil samples collected from a landfill site near a water body

Active ingredient	Content ($mg\,kg^{-1}$) Soil sample 1	Soil sample 2
HCB	17.3	n.d.
α-HCH	29.3	15.5
β-HCH	26.6	16.1
γ-HCH	5.5	26.6
δ-HCH	25.8	22.5
p,p′-DDE	12.6	37.8
o,p′-DDT	79.9	617.2
p.p′-DDT	261.0	1434.0
Total HCH	87.2	80.7
Total DDT	353.5	2089.0

n.d. – not detected

Unless sanitation measures are taken immediately, pesticides may reach groundwater via the subsoil.

3 Occurence and Toxicological Evaluation of Plant Protection Products in Groundwater

3.1 Regulations and Guidelines

The greatest activities in regulating the maximum admissible concentrations (MACs) or pesticide limits in drinking water have taken place apart from by the special group of the WHO, in the USA and Europe. Several countries of the former COMECON also had lists with limits for pesticides in drinking water, e.g. the Soviet Union, Czechoslovakia, Hungary, Poland and GDR. The approaches were quite different and led to different MACs.

On comparing the differences between the limits of these countries and the EC Directive it is very important to refer to the kind of water (drinking water, groundwater or surface water). The EC Drinking Water Directive (No. 80/778/EEC) sets the standard for the quality of water for human consumption.

The directive covers a range of toxic substances, with pesticides and related products such as PCBs (polychlorinated biphenyls), considered as a single group. The maximum allowable concentrations are set as 0.1 $\mu g\,l^{-1}$ for each single substance in this group and 0.5 $\mu g\,l^{-1}$ for the sum of all pesticides and related products in a sample. The same directive sets MACs on an individual basis for other chemicals such as mercury (0.1 $\mu g\,l^{-1}$), cyanide (10 $\mu g\,l^{-1}$), zinc (10 $\mu g\,l^{-1}$), and nitrate (50 $mg\,l^{-1}$).

The very low MACs for pesticides, set, as they are, close to the technical limits of detection, presumably reflect an underlying wish to have what is effectively pesticide-free water in Europe [57]. However, because the levels take no account of the strongly varying toxicology of the about 300 individual pesticides concerned, they lack any scientific toxicological basis. The limits give no indication of the potential risk to human health. On the contrary they were set thus so that drinking water, in general should be free of pesticides. Therefore, the aim is dictated by the political motivation of environmental protection to create a clean environment for man.

Ruchay [58] thinks that the whole EC policy should be reconsidered in the light of the new Article 130 of the EEC treaty. According to the treaty, Community environment activities than will follow the principle of the prevention of environmental contamination and of the elimination of its source applying the perpetrator principle. Consequently an ecologically admissible use of pesticides in agriculture would be the guiding principle in the EC region. However, this casts doubts on its implementation.

The Directive 80/881/EEC prescribes for all "pesticides and similar substances" an admissible limit of 0.1 $\mu g\,l^{-1}$ per single substance and of 0.5 $\mu g\,l^{-1}$ for the sum of all substances. These standards have been incorporated into Annex 2 of the Drinking Water Regulations of the FRG, namely as "chemical substances for the treatment of plants and control of harmful organisms including toxic main degradation products of these substances". These limits have no sanitary relevance as they cannot be toxicologically substantiated [59]. For this reason they have a special quality among the "Limits for chemical substances" listed in Annex 2 of the Drinking Water Regulation in contrast with the limits of other substances mentioned there, for instance nitrate, lead, cadmium, mercury or carcinogenic arsenic, which were set only after thorough risk assessment with respect to human health. Nevertheless, in some cases they are far higher than those for pesticides.

The Drinking Water Regulations of Germany expressly sets the pesticide limit for single substances with an admissible error of the measured values of 0.05 $\mu g\,l^{-1}$. Accordingly only concentrations of more than 0.15 $\mu g\,l^{-1}$ can be regarded as exceeding the limit.

For the implementation of the Drinking Water Regulations with respect to potential pesticide contaminations, the Federal Health Office published a recommendation [60]. It contains-last amended on 21st June 1989 – a list of 36 active ingredients and the recommendation that these compounds can be found in drinking water (selected examples in Table 12). Looking at it critically, the levels are doubtful for a considerable number of the active ingredients mentioned because of their properties alone. Obviously, in some cases, reports on groundwater contamination are not verified or results of calculations with inappropriate simulation models have been used as selection criteria. Meanwhile, comprehensive monitoring and lysimeter tests have revealed that trans-

Table 12. Selected drinking water guideline levels for the Federal Republic of Germany [60] in comparison with the maximum allowable concentrations of the former GDR [61]

Active ingredient (a.i.)	FRG a.i. ($\mu g\,l^{-1}$)	Metabolites ($\mu g\,l^{-1}$)	GDR a.i. ($\mu g\,l^{-1}$)
Atrazine	3	3	2
Bentazone	10	–	20
Chlorfenvinphos	10	–	1
2.4-D	10	1	10
Dichlorprop	10	1	10
Dimethoat	10	–	10
Dinoseb	3	1	1
Endosulfan	3	1	5
Lindane	3	1	3
MCPA	1	1	5
Mecoprop	10	1	10
Metobromuron	3	0.1	10
Pirimiphos-methyl	10	–	5
Propazine	10	3	20
Simazine	10	3	20
TCA	10	–	10

gression of the limits can only be confirmed for a small number of these active ingredients. The recommendation contains, in addition to a few known metabolites, a large number of compounds which could theoretically formed from the 97 active ingredients. With the exception of the known desalkyltriazines these compounds have not been found so far in groundwater or drinking water in Germany.

Table 12 also compares selected active ingredients of the above-mentioned list of the Federal Health Office and drinking water limits set in the former GDR in 1988 [61]. Most values correspond with each other, which can be attributed to the employment of toxicological evaluation principles in both cases [23, 62, 63]. According to Lingk [63] the ADI concept provides sufficient security for human health from the toxicological point of view, particularly since in most cases the ADI limits, which provide high safety factors, are exceeded by contaminated drinking water by less than 1%. Table 13 demonstrates the non-compliance with the mentioned EC limits analyzed by Ohnesorge [59].

The WHO agreed to accept a rate of 10% of the ADI limits for pesticide contamination of drinking water within Europe which is based on the way of life. This rate, which seems to be reasonable from the toxicological point of view, gives the following admissible concentrations according to the formula:

$$\frac{10\% \text{ of ADI}\,(\text{mg}^{-1}\,\text{kg}^{-1}\,\text{d}) \times 70\,(\text{kg b.w.})}{2\,(\text{l drinking water}\,\text{d}^{-1})}$$

Table 13. Exseeding of ADI by pesticides in drinking water at a limit of 0.1 $\mu g l^{-1}$ [59]

Active ingredient	ADI ($mg^{-1} kg^{-1} d^{-1}$)	Exseeding (%) at 30 $ml\,kg^{-1}$ b.w. d^{-1}
Aldicarb	0.005	0.06
Atrazine	0.0007	0.42
Isoproturon	0.0025	0.12
Mecoprop	0.0025	0.12
Methabenzthiazuron	0.05	0.006
Simazine	0.005	0.06

Table 14. Selected drinking water guideline levels developed by the WHO [65] and changed values in 1993 [116]

Pesticide	Guideline level ($\mu g l^{-1}$) 1984	1993
Alachlor	0.3	20
Atrazine	2 (tentative)	2
Bentazone	25	30
HCB	0.01	–
MCPA	0.5	2
Methoxychlor	30	20
Metolachlor	5	10
Molinate	7	–
Pendimetalin	17	20
Propanil	179	20
Pyridate	60	100
Simazine	17	2
Trifluralin	170	20
2,4-D	100	30
DDT	–	2
Lindane	–	2

This leads to very different guideline levels for the individual pesticides which can easily be reached, especially since as a rule the other 90% of the ADI limits are by far not exceeded by other foodstuffs [59].

This approach can only be applied to pesticides without direct (genotoxic) or indirect carcinogenic mechanisms of action, i.e. for non-carcinogenic substances.

In contrast to the EC limits, the World Health Organisation has elaborated drinking water guideline levels on the basis of the toxicological properties of the substances [65]. The levels in Table 14 are used by many countries to evaluate the pesticide concentrations in drinking water and to fix their own limits. This was done for instance by the former COMECON states Czechoslovakia and

Hungary. Selected examples of the former GDR list which contained in total 43 active ingredients are presented in Table 12.

In the former Soviet Union, drinking water limits were set with regard to other principles. Toxicological evaluation of pesticides was carried out according to criteria deviating from the WHO and methods elaborated by Medved and his colleagues at the Allunion Institute for the hygiene and toxicology of pesticides, polymers and plastics of the USSR (VNIIGINTOKS) [64, 66]. On the other hand peculiarities of the drinking water obtained from surface waters in the southern republics which have intensive chemical plant protection measures were taken into account. According to Iskandarov [67] this practice leads to contamination of water bodies which have to be used for drinking water. Annex 8 of the current pesticide directory (1987) contains drinking water limits which are partly listed in Table 15. The new independent republics have entrusted new laboratories with the evaluation of pesticides and the revision of the old regulations, guidelines and MACs [68].

In the USA toxicological properties of a pesticide are decisive for the establishment of Maximum Contaminant Levels (MCL) or Lifetime Health Advisory Levels (HAL) in drinking water. Whereas non-carcinogenic substances are evaluated according to the principles of the WHO, carcinogenic pesticides are evaluated according to national procedures. For a number of important contaminants the EPA established the following values:

	MCL	HAL
Alachlor	proposed 2 $\mu g\,l^{-1}$	–
Atrazine	proposed 3 $\mu g\,l^{-1}$	3 $\mu g\,l^{-1}$
Bentazon	–	20 $\mu g\,l^{-1}$
DCPA	4000 $\mu g\,l^{-1}$	–

MCLs were used as a quality standard for the source drinking water. HALs are the maximum concentrations of a contaminant in water that may safely be consumed over an average human lifetime.

3.2 The Situation in Germany

An investigation to determine pesticides in drinking water has to be restricted to tapwater taken from the drinking water supply system. The Commission of the European Communities describes it in its Directive 80/887/EEC as "water for human consumption". Public discussion sometimes does not clearly distinguish here. In Germany by far the greater part of drinking water is taken from groundwater, findings of contaminants in groundwater are often equated with a contamination of drinking water. This is not correct because on the one hand the report seldom says whether the sample was collected in the catchment area and from the groundwater level used by a drinking water plant. Even if this is the case, it has to be taken into account that most of the organic substances are

Table 15. Maximum allowable concentrations (MAC) of pesticides in water in the former Soviet Union [68]

Pesticide group	Active ingredient	MAC ($mg\,l^{-1}$)	Base for limitation
Insecticides	Amitraz	0.05	1
Acaricides	Binapacryl	0.03	3
	Carbaryl	0.1	1
	Cyhexatin	0.001	2
	Diazinon	0.3	1
	Dichlorvos	1.0	1
	Dicofol	0.02	2
	Dimethoate	0.03	1
	Dinobuton	0.3	1
	Fenitrothion	0.25	1
	Phosalone	0.001	1
	Phosmet	0.2	1
	Trichlorfon	0.05	1
Fungicides	Captan	1.0	1
Soil	Carbendazime	0.1	1
Desinfectants	Folpet	0.4	1
	Iprodione	0.05	3
	Mancozeb	0.015	1
	Metam-sodium	0.01	1
	Metiram	2.0	3
	Zineb	0.03	1
Herbicides	Alachlor	0.1	1
	Atrazine	0.5	2
	Chloridazon	2.0	3
	2,4 D	0.1	1
	2,4-D amine salt	0.2	1
	2,4-D butyl-ester	0.5	1
	2,4-D octyl-ester	0.2	1
	2,4-DB	0.01	3
	Dalapon	2.0	1
	Diphenamide	1.2	3
	Diquat	0.02	1
	Diuron	1.0	1
	DNOC	0.5	3
	Glyphosate	0.1	1
	Lenacil	1.0	3
	MCPA	2.0	3
	Mecoprop	2.0	3
	Monolinuron	0.5	2
	Monuron	5.0	1
	Phenmedipham	0.5	1
	Prometryn	3.0	1
	Propazine	1.0	1
	Propanil	0.1	2
	Simazine	–	1
	TCA	5.0	2

1) organoleptic 2) sanitary requirements
3) toxicological base

more or less retained during their subsurface flow or remain at the water treatment plant or can be eliminated in another way.

Groundwater samples taken from pesticide-treated soils with a shallow water table [69] revealed a lot of positive findings – even when the transport of active ingredients from the surface soil due to drilling the hole can be absolutely excluded. However, they hardly give a reliable indication of the probable contamination or even the concentration of the contaminants in the drinking water of a water supply plant situated in this groundwater flow.

In the case of private wells, the risk of contamination of the immediate vicinity of a well must not be neglected especially in rural regions. Therefore, a positive finding requires a thorough examination before concluding that the groundwater has been contaminated as a consequence of the application of pesticides when following recommended procedures.

If samples are taken from a water supply plant, it can be raw water or clean water. The simplest method is to take samples at the tap of the consumer.

3.2.1 Validity of Results

The analytical determination of organic substances within the nanogram or picogram range involves sources of considerable errors. Because of a large number of ubiquitous interfering substances of anthropogenic, but also natural origin it is not possible to guarantee with a single method sufficient certainty of results even when operating in the most careful way and using high-quality separation and detection systems. Consequently tremendously high variations have to be taken into account on quantifying the measured signal. A quantitative analysis, however, requires a clear qualitative assignment of the signal. Only after a second analysis with another separation and detection systems can a positive result be considered a true finding [70]. Furthermore, as a rule, results have to be checked by efficient mass spectrometry [4]. UV spectrometry (HPLC or HPTLC) is of less significance in this respect.

An interesting contribution to this subject is the monitoring of groundwater contamination of 83 water samples from 21 water supply plants by Pestemer [71]. The 412 positive results found with the help of a modern, automatic analyzing program reduced to 94 findings after checking with another separation and detection procedure.

It can be supposed that the proportion of wrong positive findings of water analyses for pesticides was even much higher at the beginning of environmental analysis. Even recent analyses have produced a large number of wrong positive results. The NPS study (National Pesticide Survey) of the US-EPA reports that – depending on the analysis procedure – 22%–47% of the gas-chromatographically positive results could not be verified by subsequent mass spectrometry, whereas the percentage of samples with wrong negative values was below 1% [72]. Ring tests from 1991 to determine the efficiency of various laboratories in the analysis of water samples for the occurrence of triazine herbicides at concentrations relevant to the Drinking Water Regulation produced mainly wrong results [73].

3.2.2 Surveys and Monitoring Programmes

In Germany, groundwater has recently been the subject of numerous analyses for pesticide residues. Unfortunately, the extensive surveys were confined to well or raw water in water supply plants. Table 16 summarizes the comprehensive inquiries and programmes of the last few years.

The list of the Federal Association of Gasworks and Water Supply Plants (1988) on the "Occurrence of Active Ingredients of Pesticides in Wells, Bank-filtrate, Springs, Groundwater and Drinking Water" presents 170 positive findings for groundwater and raw water, but only one measured value for drinking water [74].

The publication Pesticides in Drinking Water by the Deutsche Forschungsgemeinschaft (German Research Society) [6] has presented 611 reports and publications on locations with positive findings in groundwater and raw water, but only two for drinking water. 1181 findings in raw water are in direct contrast to only 26 findings in drinking water. 4 drinking water values exceeded the limit of the Drinking Water Regulations.

The Institute for Water Research in Dortmund has elaborated a detailed study under the title "Occurrence and Detectability of Active Ingredients of Pesticides in Raw Water and Drinking Water of the Federal Republic of Germany" [75]. It contains the results of an inquiry with 1328 water supply plants. 196 positive results have been reported for the groundwater from various sources and 56 for drinking water, but only 5 exceeded the limit.

The three above-mentioned publications with data from literature or inquiries give, in general, only the positive findings of the sources so that the total number of analyzed samples are unknown. It is not possible to determine the percentage of positive findings or samples exceeding the drinking water limit. In the seven studies mentioned below, as carried out for instance by some German states, this is possible. Sometimes it is not obvious whether the analyses restrict themselves really to the final drinking water product or whether raw water had been included (this is especially understandable if treatment is not necessary). In the following, drinking water is understood to be water which has been treated – if necessary –, and supplied by water supply plants and available to the consumer through normal water supply systems.

The State of Baden-Württemberg has had a drinking water database since 1985 [76] (also for private well, raw water and industrial water). Until May 1988, 1232 analyses of public water supply plants found only the active ingredients atrazine and simazine, i.e. 162 positive atrazine findings with 56 samples exceeding the limit (4.5%) and 28 positive simazine findings with four samples exceeding the limit (0.3%).

Table 16. Results of monitoring on pesticides in drinking water (dw) and raw water (rw)

Source/ literature	Kind of data Results	Findings Number	Positive	> Limit	Samples Number	Positive	> Limit	Confirmation (MS, GC–MS)
[74]								
	rw	a	169	114	a	a	a	a
	dw	a	1	1	a	a	a	a
[6]	Literature							
	rw	a	611	275	a	1181	41	7 a
	dw	a	2	2	a	26	2	a
[75]	Investigation							
	rw	a	196	83	a	a	a	a
	dw	a	58	5	a	a	a	a
[76]	Monitoring							
	rw + dw atrazine 1232		162	56/4.5%	uk	uk	uk	yes
	rw + dw simazine 1218		28	4/0.3%	uk	uk	uk	yes
[77]	Monitoring							
	rw	315	216	101/33.0%	uk	uk	uk	uk
[78]	Monitoring							
	rw + bf[a]	190	41	22/11.5%	uk	uk	uk	uk
[79]	Monitoring							
	rw + bf[a]	206	uk	20	1534	143	44/2.9%	yes
[80]	Monitoring							
	dw	325	54 27[b]	4/1.2%	325	54 27[b]	4/1.2%	yes
[81]	Monitoring							
	dw	272	4	3/1.1%[a]	uk	uk	uk	uk
[70]	Monitoring							
	dw	545	96	11/2.1%[a]	545	96	11/2.1%[a]	p

[a] Bank filtrate
[b] Results of repeated monitoring
uk – unknown; p – partially

In January 1988, the "Drinking Water Programme" of the state of Bavaria covered 315 sampling sites in water supply plants and private wells [77]. 216 samples showed pesticide residues. In 101 cases a concentration of 0.1 μg was surpassed. Here, it is not clear how many samples had been taken from the drinking water supply system.

The Administrative Office for Water and Waste, North Rhine-Westphalia in its annual report [78], gave a series of findings for bank-filtrates and raw water, mainly atrazine and simazine, but also chlorotoluron. Unfortunately, results from analyses of drinking water are not known.

In the IPS study (1987) 1534 samples were taken from 206 sampling points. 12 674 water analyses were carried out. 35 active ingredients were included in the programme. This study was also restricted to raw water [79].

The results of an investigation of Ökolab were very interesting [80] because the 325 samples were exclusively taken from the taps of consumers. They were examined for triazine herbicides by ELISA test. 4 samples exceeded the limit of the Drinking Water Regulations. 54 samples were reported to be positive. A subsequent verification by mass spectrometry confirmed only 27 values.

In 1991, all the most important water supply companies of the new Federal states analysed the drinking water of the 30 biggest water supply plants in their territories. Out of 272 plants, only 3 plants, which take bank-filtrate from the River Elbe, exceeded the sum limit (0.5 $\mu g l^{-1}$) [81].

In 1990, in a joint operation of the water supply companies of the old and new states and the Institutes of Hygiene in the new states examined under the so-called Immediate Drinking Water Programme another 545 plants (water supply plants, public wells, elevated water tanks, mixed water zones, pumping stations, system samples in the new states for the presence of pesticides. In this case, the number of notifications was also remarkably low. Subsequent examinations confirmed the results. The limits are exceeded by plants on the river Elbe which use bank-filtrates [82]. Thus, Grohmann [70] arrives at the conclusion that the drinking water in the new states can be more favourably evaluated than in the old ones. Other authors [23] confirm this statement.

Wolter [83] evaluates 10 700 positive findings (10.3% of the analyzed samples) from 1991, 3300 (3.2%) of which exceed the EEC limit. It is very interesting to classify them, because 80.7% of them concern the triazines:

- 33.3% atrazine,
- 25.6% desethyl-atrazine
- 0.8% desisopropyl-atrazine,
- 12.0% simazine and
- 8.9% other triazines.

39.3% of the findings result from groundwater and 33.6% from drinking water.

3.2.3 East Germany (Former GDR)

The former GDR did not carry out systematic monitoring of groundwater and drinking water for pesticide contamination. The present investigations result from research and monitoring tasks carried out by institutions of health, water resources and the agricultural sector over a limited period.

Systematic studies by the water resources sector on the pesticide load of groundwater used for drinking water only started in 1989. These took place more or less by order of Minister of Environment Protection and Water Resources of the GDR with the aim of testing the new drinking water standard which contained i.e. the limits of the Residue Limits Regulation of the GDR of 30th June 1988 [61] for pesticides (Table 12). As residue analysis was not appropriate for large scale monitoring, there had to be restrictions made with respect to sampling sites, active ingredients and laboratories.

The studies concentrated on some main active ingredients (lindane, triazines, nitrofen, phenoxyalkanic acids) and at first only on triazines. In 1989/90, the Water Resources Administration Oder-Havel Potsdam (the State of Bradenburg) performed a survey on simazine, atrazine, propazine and prometryn in 70 water supply plants which did not reveal any contamination.

The results showed that even without reaching the EC limit, there is no acute need for action on health grounds. In 1991, the results were confirmed, at least for Bradenburg, by investigations under the research Programme "Immediate Research Programme for the Determination of the Drinking water Contamination" of the Federal Minister for Environment. 141 samples were taken from 141 water supply plants. The limits were not exceeded by more than 0.1 $\mu g l^{-1}$.

In 1990, the Biological Research Centre, Kleinmachnow examined shallow water tables within the catchment area of various water wells in Saxony. 9 drinking water wells were selected which are situated in areas of intensive agriculture of horticulture and which could be precisely assigned to a drinking water protection zone or a catchment area of a drinking water well. As far as possible, data on cultivation and pesticide use from various places and at different times were gathered with respect to land use in the flow area of the measuring points. This was the first time on the territory of the former GDR that an attempt had been made to show the impact of pesticides on groundwater in dependence on both the corresponding hydrogeological conditions and cultivation and the use of pesticides. The results are summarized in Tables 17–21.

The studies were finished in October 1990 by a third sampling procedure the results of which are presented in Table 22.

The results clearly show the connection between above-ground use of a water catchment area and the contamination of the ground water below it.

Since 1981, hygienic problems of triazine application had been dealt with under the research Programme "Medicinal Aspects of Environment Protection" [82]. Special attention had been given to triazine residues in drinking water. Because of random groundwater sampling from geologically different regions of the GDR it had been recommended in 1981 to perform regular control checks

Table 17. Occurrence of pesticides in groundwater with regard to hydrogeological conditions and agricultural use

Sampling sites 1 and 2:	Coswig	Coswig
Protection zone:	IV	II
Depth to water table:	5.4 m	4.7 m
Thickness of the covering layer: (A-horizon)	0.5 m	1.0 m
Agricultural use:	decorative plants, bog-soil crops	strawberries
Period:	since 1980	1985–1989

Active ingredient	Date of sampling/concentration ($\mu g l^{-1}$)			
	Sample 1		Sample 2	
	02–04 April	16–10 July	02–04 April	16–10 July
Simazine	n.d.	1.4	<0.1	<0.1
Atrazine	<0.1	n.d.	<0.1	n.d.
Propazine	<0.1	n.d.	< 0.1	n.d.
Prometryn	n.d.	<0.1	n.d.	n.d.

n.d. - not detected

Table 18. Occurrence of pesticides in groundwater with regard to hydrogeological conditions and agricultural use

Sampling sites 3 and 4:	Meißen	Weinböhla
Protection zone:	III	III
Depth to water table:	4.7 m	4.8 m
Thickness of the covering layer: (A-horizon)	2.0 m	1.5 m
Agricultural use:	apple	field forage
Period:	1981–1989	Until 1989

Active ingredient	Date of sampling/concentration ($\mu g l^{-1}$)			
	Sample 3		Sample 4	
	02–04 April	16–10 July	02–04 April	16–10 July
Simazine	0.31	n.d	0.25	0.8
Atrazine	0.19	n.d.	0.17	< 0.1
Propazine	<0.1	n.d.	n.d.	n.d.
Prometryn	n.d.	n.d.	n.d.	n.d.

n.d. – not detected

Table 19. Occurrence of pesticides in groundwater with regard to hydrogeological conditions and agricultural use

Sampling sites 5 and 6:	Coswig	Coswig
Protection zone:	III	III
Depth to water table:	6.0 m	6.2 m
Thickness of the covering layer: (A-horizon)	1.0 m	0.5 m
Agricultural use:	strawberries, sour cherries	cauliflower, strawberries kohlrabi
Period:	1980–1989	1980

Active ingredient	Date of sampling/concentration ($\mu g\, l^{-1}$) Sample 5		Sample 6	
	02–04 April	16–10 July	02–04 April	16–10 July
Simazine	< 0.1	< 0.1	0.18	0.6
Atrazine	<0.1	n.d.	<0.1	n.d.
Propazine	n.d.	n.d.	n.d.	n.d.
Prometryn	n.d.	n.d.	n.d.	n.d.

n.d. – not detected

Table 20. Occurrence of pesticides in groundwater with regard to hydrogeological conditions and agricultural use

7th Sampling site:	Brockwitz	Weinböhla
Protection zone:	III	III
Depth to water table:	5.8 m	5.1
Thickness of the covering layer: (A-horizon)	1.5 m	1.0 m
Agricultural use:	pasture, strawberries	strawberries, forage
Period:	1980–1989	1985–1989

Active ingredient	Date of sampling/concentration ($\mu g\, l^{-1}$) Sample 7		Sample 8	
	02–04 April	16–10 July	02–04 April	16–10 July
Simazine	n.d.	0.1	0.12	2.9
Atrazine	< 0.1	n.d.	< 0.1	n.d.
Propazine	<0.1	n.d.	<0.1	n.d.
Prometryn	n.d.	n.d.	n.d.	n.d.

n.d. – not detected

Table 21. Occurrence of pesticides in groundwater with regard to hydrogeological conditions and agricultural use

Sampling site 9:	Sörnewitz
Protection zone:	III
Depth to water table:	1.1 m
Thickness of the covering layer: (A-horizon)	1.5 m
Agricultural use:	cereals
Period:	not known

Active ingredient	Date of sampling/concentration ($\mu g l^{-1}$) Sample 9	
	02–04 April	16–10 July
Simazine	< 0.1	1.9
Atrazine	< 0.1	n.d.
Propazine	n.d.	n.d.
Prometryn	n.d.	0.5

n.d. – not detected

Table 22. Occurrence of pesticides in groundwater samples taken from sampling sites 1–9, third sampling October 1990

Sampling site	Simazine ($\mu g l^{-1}$)	Atrazine ($\mu g l^{-1}$)	Propazine ($\mu g l^{-1}$)
1	0.11	n.d.	n.d.
2	n.d.	0.01	0.01
3	0.79	0.04	0.01
4	n.d.	n.d.	n.d.
5	0.50	0.01	n.d.
6	n.d.	0.02	0.01
7	5.0	n.d.	n.d.
8	0.46	0.02	0.01
9	0.05	n.d.	0.01

n.d. – not detected

over a longer period. Concentrations in drinking water were between 0.01 $\mu g l^{-1}$ and 0.1 $\mu g l^{-1}$.

Thereupon, the Ministry of Health initiated the 4-year research Programme "Pesticides in the Environment". It was aimed to assess the exposure of the population to triazines in foodstuffs and drinking water. Samples were taken from public water supplies and private wells in six districts in the Magdeburg area. They clearly showed varying concentrations of triazines depending on the agricultural use of triazines, the geological class of protection and the distance from agrochemical centres. The results for simazine, atrazine, propazine and prometryn are summarized in Table 23. That 19% of private wells have concentrations of more than 1 $\mu g\,l^{-1}$ is remarkable. However, all concentrations are below the triazine MACs of the former GDR.

To determine the influence of the kind of water (groundwater, bank-filtrate, surface water), seasons and other factors on the triazine content of water, seven water supply plants were examined (Table 24).

The water supply plant Colbitz is one of the biggest in the former administrative area of Magdeburg and the supplied drinking water to the city and the surrounding districts from an artificial catchment fed by leaching surface water. The sampled water wells had a depth of 70 m and were assigned to the geological protection class I (not protected, at high risk). The water supply plant Burg raises groundwater from a depth of 20 m (geological protection class I). Although this water supply plant is situated in an agricultural area, contamination in the protection zones is low because of the depth of the well and restrictions as regards the use of pesticides.

Table 23. Triazines findings in public water supplies and private wells in the district Magdeburg (state of Sachsen- Anhalt, Germany) [82]

Water supply	Sum of triazines ($\mu g\,l^{-1}$)				
	<0.02	0.02–0.1	0.1–1.0	1.0–10.0	>10.0
PWS	11	1	4	0	0
PW	43	12	21	8	0

PWS – public water supplies; PW – private wells

Table 24. Influence of the kind of water, season and other factors on the triazine contents in water supply plants in the Magdeburg (State of Sachsen-Anhalt, Germany) [82]

Place	Kind of water	Triazine contents ($\mu g\,l^{-1}$)				
		February	April	June	July	September
Colbitz	Groundwater (PWS)	0.03	0.03	n.d.	n.d.	n.d.
Burg	Groundwater (PWS)	n.d.	n.d.	n.d.	0.05	n.d.
Gnadau	Shallow	0.13	0.02	0.04	0.03	0.14
Pömmelte	Groundwater	0.18	0.21	0.11	0.17	0.39
Wespen	(PW)	3.64	3.75	1.00	2.90	3.57
Barby	Bank-filtrate (PWS)	0.95	0.97	0.71	0.67	0.21
Magdeburg-Buckau	Surface water	2.48	0.95	4.12	1.86	5.23

PW – private wells; PWS – public water supply; n.d. – not detected

In contrast to this, the private wells in Wespen, Gnadau and Pömmelte take water from shallow groundwater levels of < 10 m. Hydrogeologically-inappropriate sites involve an acute risk of contamination outside water protection areas. Significantly higher quantities of triazines were also found in the water of the central water supply plant Barby which treats bank-filtrates from the River Elbe. Even higher concentrations were found in the industrial water supply plant Buckau, which takes water from the River Elbe. All analyzed triazine concentrations were also below the MACs of the former GDR.

As a consequence of the research report the work was continued. In 1986, 100 water samples from private wells were examined. 54% of the samples did not contain triazines, but 8% more than 2 $\mu g l^{-1}$.

Studies in 1987 also have showed a decline in triazine residues in the water samples. This tendency was confirmed by sampling four times (May, June, September, and November). Concentrations were between 0.02 $\mu g l^{-1}$ and 0.1 $\mu g l^{-1}$, in private wells. Additionally, in the Burg district, two agricultural areas which had been treated with triazines over several years were chosen as a result of their cultivation and pesticide application schemes. Taking into account the groundwater flow direction samples were taken from nearby water supply plants. In general, triazines could not be found.

The last investigations under the mentioned research Programme "Pesticides in the Environment" took place in 1988. Triazine concentrations were below 0.02 $\mu g l^{-1}$.

These investigations stated unanimously that unprotected groundwater layers and private wells can be evaluated as potentially at risk. This is mainly due to the lack of water protection areas or their being to small.

Similar results (Table 25) were reported by Gorf [84]. From 1975 until 1991, he systematically examined 2137 water samples and found pesticides in 245 cases. It is very interesting that the spectrum of active ingredients changed remarkably during this period. Whereas until 1980 the organochlorine insecticides DDT, lindane and methoxychlor were mainly found, in the following years triazines and phenoxyalkanic acids were found which can be attributed to a change in cultivation organisations and the prohibition of DDT and restrictions for other organochlorine compounds. Furthermore, a decline in the absolute contents of active ingredients in samples and in the number of contaminated samples in relation to the total number was noticed. This is attributed to increased competence and more responsible handling of pesticides because of intensive training of farmers.

3.2.4 Sources of Contamination

On trying to explain the lower degree of contamination of drinking water in the new Federal states, one should remember that agriculture in the former GDR was centralized. Pesticides were applied by comparatively few agrochemical centres and they had partially good facilities for the disposal of tailings and

Table 25. Contamination of water samples in the former district of Gera (state of Thuningen, Germany) including accidents [84]

Period	Samples	Standing water body	Outfall ditch	Groundwater
1975–1980	Number	129	8	246
	Positive[1)]	12	7	56
1981–1990	Number	173	117	859
	Positive	37	26	84
1991	Number	12	13	643
	Positive	5	1	35
1975–1991	Number	314	138	1748
	Positive	54	34	175
		(17.2%)	(24.6%)	(10%)

[1)] higher than detection limit of TLC (to 1980) or gLC

Degree of contamination

Year	1985	1986	1987	1988	1989	1990	1991
(%)	1.4	5.0	6.5	19.1	2.7	3.2	4.1

rinsings. In the old Federal states, the number of farms applying pesticides are many times greater and consequently so is the number of point sources which, for example, cannot be avoided when cleaning the spray equipment. This leads us to the conclusion that normal agricultural practice of pesticide application is not responsible for groundwater contamination, but mainly point sources; because with respect to other factors which have to be taken into account (such as application rate, treated region, subsoil) the former GDR and the old Federal Republic of Germany did not differ significantly.

3.3 The Situation in Western Europe

The greatest efforts in regulating the levels of pesticides in water have been made in Europe and the USA, although the approaches have been quite different. In the USA, groundwater contamination has received the most attention since most of the drinking water in the USA is drawn from wells. In Europe, however, equal emphasis is given to water drawn from surface or underground supplies. Table 26 demonstrates the wide variation between countries in the amount of water taken from each source.

In all the EC member states, current practices lead to an intolerable use of groundwater sources. The most important problems in connection with groundwater contamination occur in agricultural and industrial centres. The range

Table 26. Percentage drinking water taken from groundwater and surface water [57]

Country	Groundwater (%)	Surface water (%)
Austria	99	1
Belgium	67	33
Denmark	99	1
Finland	49	51
France	65	35
Germany	72	28
Great Britain	28	72
Italy	87	13
Luxembourg	73	27
The Netherlands	70	30
Norway	5	95
Spain	22	78
Sweden	22	78
Switzerland	83	17

and complexity of the problems vary from region to region. The RIVM study [46] found out that the most important problems result from the

- pollution from pesticides and nitrates;
- pollution from industrial and urban areas;
- overexploitation and intensified drainage leading to lowering of groundwater levels
- point pollution from illegal or improper dumping of municipal, industrial mining and hazardous waste.

In industrial and urban areas there are also other sources of contamination, such as waste disposal sites, this means landfills, surface impoundments, hazardous waste sites, dredged harbour sediments and sewage sludge disposal, and mining spoils.

Advanced agriculture has turned into a source of groundwater pollution to a great extent. In some EC member states, fertilizers and pesticides are applied in such quantities that the nitrates and pesticides leach towards the groundwater. The leaching of pesticides and their metabolites depends on the chemical structure and physical properties of the compound and the properties of the soil. Simulation models describe the interaction between the different processes. In RIVM study [46], the simulation model PESTLA [85] was used to calculate the amount of pesticides leaching into the groundwater using a standard scenario consisting of

- application rate of atrazine 1 kg ha^{-1},
- application time in May,
- orthic podsol soil, and
- weather data for 1980 (source CORINE).

For the extrapolation for different soils, sorption constants adjusted according to the different organic matter contents were used. Extrapolation of different regions was done by adjusting the degradation constant to the prevailing average temperature. Expected concentrations in the uppermost layer of groundwater were calculated by dividing the amount leached by the rainfall surplus.

The leaching amounts represent values at a depth of about 1 m below the ground surface. In the case of a thick unsaturated zone and a greater depth, the calculated and observed values may differ.

The model calculation in the study showed that 75% of the European soils are sensitive to leaching of pesticides because in all these soils the value of 0.1 $\mu g\,l^{-1}$ is exceeded.

In principle, it is possible to calculate the sensitivity of a soil to each pesticide if the basic data are available. It is also possible to classify the pesticides in relation to atrazine. The results of such a comparison with some important pesticides are given in Table 27. The pesticides in column 1 show practically no leaching and those in column III a five times higher leachability compared to atrazine.

The groups following the pyrethroids contain only selected active ingredients which are representative of other pesticides of the same chemical structure, e.g. 2,4-D for the phenoxyalkanic acids or atrazine for the other s-triazines. The authors point out that the model is not able to simulate the behaviour of ionizable compounds and inorganic chemicals. Heavy metals presumably pose more a threat of accumulation than of leaching.

The study combined the apparent load of the pesticides of the groups I to III with the calculated leaching of atrazine on the one hand and the geological conditions (top soil 0–1 m) in the EC member states on the other hand. Starting from this assumption, the EC pesticides limits for drinking water are exceeded in 65% of arable land area in the EC region and even by 10 times in 25% of the area.

Table 27. Leachability of selected pesticides in relation to atrazine [46]

Low leaching	in the order of magnitude	Leachability greater
Metam-sodium	2,4-D	Aldicarb
Copper compounds	Metribuzin	Bentazon
Pyrethroids	Thiameturon	Glyphosinate
Glyphosate	Triclopyr	TCA
Paraquat	Linuron	Zineb
Mineral oils	Isoproturon	Metalaxyl

In deeper groundwater layers of various regions within the EC the average pesticide concentrations are expected to be lower because of

- dilution with water from beneath forests and nature reserves,
- possible transformation of pesticides in the saturated zone.

These concentrations can be verified by groundwater monitoring in the EC member states. Both the model calculation and the analyzed pesticide concentrations give a more realistic picture of the situation caused by pesticide use in the EC. An overview of all data reported in the RIVM study [46] is given in Table 28. The results for Germany are described in Sect. 3.2.

For a number of pesticides registered in EC member states, analytical methods appropriate for the determination at the 0.1 $\mu g l^{-1}$ level are not available. Because of this and the expensive analyses, a representative data on the occurrence in groundwater is available only in three countries and a smaller number in another three countries (Table 29). The analysis of the data revealed that most of the contaminated samples contained triazines and phenoxyalkanic acids. Atrazine is found in groundwater throughout Europe. The EC limit for drinking water is often exceeded by a factor of 10 to 100. A direct comparison between the computed leaching and monitoring data is not possible. The paucity of monitoring data means that we cannot answer the question over the whole period of application.

The UK Ministry of Agriculture, Fisheries and Food [86, 87] has published reports on the contamination of ground- and surface waters with atrazine and simazine. Several surveys carried out in the UK show that atrazine and simazine are the most frequently found pesticides in groundwater. For the period up to 1988, the measured concentrations of atrazine and simazine had a typical range of 0–0.5 $\mu g l^{-1}$ and 0–42 $\mu g l^{-1}$, respectively. The result of 9 boreholes in N-England where atrazine was found at concentrations between 10 $\mu g l^{-1}$ and 110 $\mu g l^{-1}$ (mean 31 $\mu g l^{-1}$) were also given. Long-term studies by the authorities on pesticides have shown a widespread occurrence of atrazine and simazine in groundwater. In Cambridgeshire, for instance, out of 42 samples form boreholes, 17 contained atrazine and 20 simazine.

The frequent detection of atrazine and simazine did not correlate to the quantity applied in the catchment area for agricultural purposes (estimated to be 0.3 tons per annum for atrazine and 0.9 tons per annum for simazine). The contamination is believed to originate from use for non-agricultural, total weed control on industrial sites, railways, and roads.

The results of groundwater monitoring in 1990 are summarized in Table 30 and demonstrate a large number of contaminated samples of groundwater with values higher than the Maximum Admissable Concentration set by the EC. The inexact mean of the positive samples of water supply companies amounts to 27% and 16% of atrazine and simazine. The highest atrazine detection frequencies in drinking water supplies were reported in the Thames (50.3%) and Anglian (34.0%) areas. In case of simazine, 34.1% of the samples from the Thames were contaminated and 19.9% of the samples from Anglian.

Table 28. Monitoring data of various compounds from Europe [46]

Country	Italy		England		Denmark		France		The Netherlands	
Compound	Number of positive data	Max conc. ($\mu g\,l^{-1}$)	Number of positive data	Max conc. ($\mu g\,l^{-1}$)	Number of positive data	Max conc. ($\mu g\,l^{-1}$)	Number of positive data	Max conc. ($\mu g\,l^{-1}$)	Number of positive data	Max conc. ($\mu g\,l^{-1}$)
Alachlor	3	1.68							1 +	0.3
Aldicarb									3	0.45
Aldicarb-sulfone									53	61
Aldicarb-sulfoxid									25	26
Amitrole									11 +	6.1
Atrazine	2175	3.8	32	0.48	53	7.8		0.225	85	7.4
Atrazine-DE									84	8.9
Atrazine-DEDIP									10 +	0.37
Atrazine-DIP									52 +	0.98
Bentazone	>25	40							89 +	1.1
Bentazone-AIPA										0.2
Bromacil									2 +	14
Bromoxynil					3	0.01				
Captan									1	0.05
Carbendazim									1	0.1
Cyanazine									1 +	0.6
Cypermetryn							1 +	0.075		
D(2,4-D)			3	0.56	9	1				
DCPa-1,2									73 +	200
DCPa-1,2,3									39 +	13
DCPa-1,3									39 +	10
DCPe-1,3									9 +	80
DCPe-CAA									1	2.3
Dichlobenil									14	0.83
Dichlobenil-BAM									16	180
Dichlorprop			1	0.2	7	1.36				

Dimethoate								0.07	1	
Dinoseb									26 +	9.2
Diuron									16	2
DNOC									1	0.05
Ethofumesate								0.05		
Ethoprophos									8 +	1.5
ETU									50 +	42
γ-HCH (lindane)								1	0.16	
Hexazinone					39	42.7				
Ioxynil					1	0.09				
Isoproturon					6	0.15				
MCPA		>0.1	1	0.42	13	0.17			1 +	0.02
MCPB			7	0.27						
Mecoprop			1	0.36	2	0.06			4 +	2.2
Metalaxyl									1	0.25
Metamitron									2 +	4.4
Metoachlor	1	<0.1							31	0.42
MITC									24 +	1.5
Molinate	272	7								
Oxamyl									1	0.2
Parathion-ethyl									1 +	0.03
Pendimethalin	1	<0.1								
Propazine	1	0.4	4	0.47						
Simazine	31 +	10	14	0.42	8	0.09	4 +	170	6 +	1.1
T(2,4,5-T)			5	0.5						
Terburtryn									1 +	2.4
Vinclozolin							1 +	0.06		

† Further findings known, but exact quantity unknown

Atrazine-DE: desethyl atrazine
Atrazine DEDIP: desethyl deisopropyl atrazine
Atrazine DIP:desisopropyl atrazine
Bentazone-AIPA: anthranilacid-isopropylamid
DCPa 1,2: 1,2-dichloropropane
DCPa-1,2,3: 1,2,3-dichloropropane
DCPa-1,3: 1,3-dichloropropane
DCPe-1,3: 1,3-dichloropropene
DCpe-CAA: chlorallyl alcohol
dichlorbenil-BAM: 2,6-dichlorobenzamide
ETU: ethylene thiourea
MITC: methylisothiocyanate

Table 29. Overview of the contaminated groundwater samples reported in EC member states [46]

Country	Contaminated samples	Percentage (%) Phenoxycarbonic acids	Triazines	Atrazine
Italy	2509	<1	88.1	86.7
Germany	1329	7.9	38.8	23.3
Netherland	784	<1	32.9	30.9
Denmark	141	22.0	43.3	32.3
UK	67	26.9	74.6	47.8
France	9	7.9	38.8	23.3

Table 30. Percentage of water samples analyzed for atrazine and simazine exceeding the EC limit in England and Wales in 1990 [86, 87]

Herbicide	Water service companies	Water supply companies
Atrazine		
Number of samples	19297	3923
Range of excess in % (mean)	0 ... 50.3[a]	5.1 ... 74.4 (about 27)
Simazine		
Number of samples	16796	3925
Range of excess in % (mean)	0 ... 34.1[b]	0.9 ... 60.3 (about 16)

[a] 0%, one of 10 companies
[b] 0%, three of 10 companies

3.4 The Situation in the United States of America

Approximately 545 million kg of pesticides are sold annually with about 70% used in agriculture, 23% in forestry, industry and under government programmes, and 7% in households and gardens [18]. In the USA, maize represents one of the most important crops from the viewpoint of pesticide use. Consequently may studies deal with the behaviour of widely used atrazine, carbofuran and other pesticides. Ritter et al. [88] analyzed soil and groundwater samples from 1984 to 1986 in Pennsylvania, Delaware and Maryland, when atrazine, simazine, metolachlor, cyanazine, dicamba and carbofuran were applied to no-till and conventionally-irrigated maize. Delaware and Maryland have a larger percentage of their cropland under conservation tillage than other states. In southern Delaware and also in Georgetown many of the soils are very sandy with low organic matter, and crop production therefore depends upon frequent rainfall or irrigation.

For the field studies, three monitoring wells with a depth of 3.1 m were sampled monthly in each plot. Five monitoring wells were installed outside the plots of 0.5 ha. A summary of atrazine and simazine concentrations is presented in Table 31. The data demonstrate the possible leaching of both triazines into groundwater under different weather conditions and sandy soils (organic matter 1–1.5%) with a water table at a depth of 1.5 m to 3 m.

The maximum atrazine concentration of 54 $\mu g\,l^{-1}$ was found in 1984 at an application rate of 4.48 $kg\,ha^{-1}$. Simazine concentrations of 50 $\mu g\,l^{-1}$ and 67 $\mu g\,l^{-1}$ were analyzed with a rate of 1.68 $kg\,ha^{-1}$ in the same year. More rainfall occurred in 1984 and 1986 shortly after the application which is the reason for the rapid movement through the macropores to the groundwater.

In 1985 and 1986, the average and the maximum concentrations of metolachlor in groundwater with 4 $\mu g\,l^{-1}$ and 12 $\mu g\,l^{-1}$, respectively, were not so high as atrazine simazine. Carbofuran (2.24 $kg\,ha^{-1}$) were detected in one of nine monitoring wells in 1984 and cyanazine was found in three years only once (1986).

In comparison with these pesticides, dicamba and alachlor leached into groundwater if sufficient rainfall occurred shortly after their application. The average concentrations of dicamba amounted to 6.5–23.5 $\mu g\,l^{-1}$ 12 days after application in 1985 at a post-emergence rate of 0.28 $kg\,ha^{-1}$. Average concentrations of alachlor between 0.2–5.5 $\mu g\,l^{-1}$ were found 24 to 116 days after applications with 2.24 $kg\,ha^{-1}$.

The results of this field monitoring demonstrate the possibility of the contamination of groundwater by several pesticides under unfavourable practical conditions. Atrazine, simazine and dicamba move more rapidly than carbofuran, cyanazine and metolachlor.

This knowledge of the contamination of groundwater by pesticides is important for the evaluation of the risk to human health in major regions of maize production in the United States. Over 50 million people live in regions with potential pesticide contamination of groundwater (e.g. Eastern Seaboard, Gulf coast, Upper Midwest, Corn Belt, lake states, the Northeast). Of these people, 19 million obtain their water from private wells which are vulnerable to contamination because they are not as deep as regulated public system wells and are often not very well-built [18].

In general, the United States relies strongly on groundwater. More than 97% of rural drinking water taps are supplied from underground sources, and in 1980, 40% of the population was served by public water supplies using groundwater (nearly 74 million people).

To estimate the areas in the United States potentially at risk to groundwater contamination from pesticides and fertilizers, areas of potential contamination were defined using data on actual levels of contaminants in groundwater. For the simulation of potential contamination two data sources were combined, the U.S. county-level "Pesticide DRASTIC" assessment (D = depth to the water table; R = net recharge; A = aquifer media; S = soil media; T = topography;

Table 31. Summary of medium concentrations of atrazine and simazine in monitoring wells after pre-emergence use to non-till (A-C) and conventional irrigated maize [88]

Days from use	Atrazine concentrations in $\mu g\,l^{-1}$ plot A 1.12	plot B 2.24	plot C 4.48	plot D 2.28 $kg\,ha^{-1}$	Simazine concentrations in $\mu g\,l^{-1}$ plot A 2.24	plot B 2.24	plot C 2.24	plot D 2.24 $kg\,ha^{-1}$
1984								
24	1.6	0	0.9	–	9.3	1.0	21	–
59	1.6	0	19	–	19	0	1.0	–
88	0.7	0.3	5.0	–	28	1.5	2.3	–
116	0.3	0.7	0.3	–	2.7	0.7	0	–
159	0	0	0.3	–	1.0	0	0	–
180	0.3	0	0	–	0.7	0	0	–
1985								
39	0	0	0	0	0.3	0.7	0	1.0
56	0	0	0	0	1.7	0	0	0
147	0	0	0	0	0.7	0	0	1.0
183	0	0	0.3	0	1.0	0	0.7	0.3
205	0	0	1.5	0	1.0	0	0	0.7
240	0	0	3.0	0	0.3	0	0.7	0
283	0	0	1.3	1.1	4.3	2.0	0.7	0.3
385	1.7	1.0	1.3	2.0	2.7	1.3	0	1.0
1986								
24	3.0	1.0	1.7	10.7	0.7	1.0	0	1.0
47	0	2.1	1.0	7	3.3	0–10	0	0–5

I = impact of the vadose zone; C = hydraulic conductivity) and the county-level pesticide use. The analysis was limited to those 38 pesticides which are listed in Table 32 and included all chemicals with the EPA priority A and B categories (potential leachers). 361 counties come under both the high DRASTIC score and high pesticide use categories. 17 pesticides were found in the groundwater of 23 states in mid 1986, with most of them located along the Eastern Seaboard, in the Midwest, and in some agricultural areas of the West [18]. The results demonstrate the connection between the volumes of pesticides applied and the number of findings with approximately similar chemical and physical properties of the pesticides, e.g. ametryne, atrazine and propazine.

The data indicate that groundwater contamination from agricultural chemicals is not national in scope. Areas of potential contamination appear to be regional. In three-fourths of the 1437 potentially contaminated counties, pesticide and nitrate problems are not predicted simultaneously [18].

In 1990, the EPA completed its five-year National Survey of Pesticides in Drinking Water Wells (NPS). It is the most extensive survey ever undertaken to evaluate the presence of pesticides, their metabolites, and nitrates in drinking water wells in the United States [89].

Based on the results of NPS, the EPA assumes that about 52.1% of the 94 600 community water system wells contain nitrate; about 10.4% contain one or more pesticides, and about 7.1% both. With respect to the approximately 10.5 millions rural domestic wells, the EPA estimates that about 57% contain nitrate, 4.2% one or more pesticides, and about 3.2% both.

The toxicological evaluation of these data by the EPA led to the result that only 0.6% of rural domestic wells (approximately 60 900) contain at least one pesticide above the Maximum Contaminant Level (MCL) or Lifetime Health Advisory Level (HAL). In the drinking water of community water system wells not one sample was analyzed with pesticides above the MCL or HAL [90].

In the rural domestic wells, five pesticides were detected at levels above their respective MCLs or HALs. These were alachlor, atrazine, dibromochloropropane (DBCP), ethylene dibromide (EDB) and lindane.

An overview of the results of NPS is given in Table 33. The most frequently detected pesticides were the DCPA acid metabolites (average in rural domestic wells approximately 0.43 $\mu g l^{-1}$) and atrazine. The selective herbicide DCPA was reported by the EPA in 1958 and the primary use (> 80%) is on turf such as golf courses and home lawns. DCPA acid metabolites have a high potential for migrating through the soil due to their water solubility.

The Phase I Report of the NPS refers to results of the EPA laboratories which have analyzed alpha-chlordane, gamma-chlordane and beta-HCH at concentrations below the minimum reporting limits. In addition 4-nitrophenol was also detected, but national estimates could not be made due to limitations of the laboratory analytical method for this chemical.

The Phase II Report discusses the EPA assessment of the causes and factors associated with the occurrence of pesticides and nitrate in drinking water wells [91].

Table 32. Pesticides with high-priority categories for the EPA's national survey of pesticides in well water [18]

Pesticide	Estimated use[a]	Number states[b]	EPA description
Acifluorfen	1.399	0	Leacher
Alachlor	85.015	4	Leacher
Aldicarb	2.271	15	Mobile, marginal persistence
Ametryn	96	0	Leacher
Atrazine	77.316	5	Leacher
Bentazon	8.410	0	Leacher, toxicological concern
Bromacil	1.234	1	Leacher
Butylate	55.095	0	Mobile, uncertain persistence, toxicological concern
Carbofuran	7.695	3	Leacher
Chloramben	6.069	0	Leacher
Chlordane	11	0	Persistent
Cyanazine	21.626	2	Leacher
Cycloate	52	0	Mobile, uncertain persistence, toxicological concern
Dalapon	261	0	Leacher
DCPA	196	1	Leacher
Dicamba	4.158	0	Leacher
2,4-D	37.217	0	Marginal leacher, heavy use
Dinoseb	8.835	1	Leacher
Diphenamid	698	0	Marginal leacher, toxicol. data gaps
Disulfoton	2.105	0	Leacher
Diuron	1.861	0	Leacher
Fenamiphos	348	0	Moderate leacher, toxicol concern
Fluometuron	2.943	0	Leacher
Hexazinone	11	0	Leacher
Maleic hydrazide	287	0	Leacher, toxicological data gaps
MCPA	9.861		Marginal leacher, possible occurrence in groundwater
Methomyl	425	0	Leacher
Metolachlor	37.940	2	Leacher
Metribuzin	10.603	1	Leacher
Oxamyl	51	2	Leacher
Picloram	549	0	Leacher
Pronamid	83	0	Leacher
Propazine	1.287	0	Leacher
Propham	445	0	Leacher
Simazine	3.975	3	Leacher
2,4,5-T	204	0	Marginal leacher
2,4,5-TP	7	0	Marginal leacher
Terbacil	833	0	Leacher

[a] Thousand tons
[b] Number of states in which pesticides is found in groundwater

Table 33. Estimated number and % of contaminated water wells in the USA [89]

Contaminant	Community water systems wells (Number)	(%)	Rural domestic wells (Number)	(%)
Nitrate	49300	52.1	5990000	57
DPCA[a]	3010	6.4	264000	2.5
Atrazine	1570	1.7	70800	0.7
Prometron	520	0.5	25600	0.2
Simazine	1080	1.1	25100	0.2
Dibromochloropropane[b]	370	0.4	38400	0.4
Ethylene Dibromid[b]	–	–	19200	0.2
γ-HCH	–	–	13100	0.1
Hexachlorobenzene[b]	470	0.5	–	–
Alachlor	–	–	3140	<0.1
Betazon	–	–	7160	0.1
Dinoseb[b]	25	<0.1	–	–
Ethylenethiourea	–	–	8470	0.1

[a] Dimethyl tetrachloroterephthalate and its acid metabolites
[b] Registration cancelled by EPA

4 Occurrence and Ecotoxicological Evaluation of the Contamination of Water Bodies

4.1 Ecotoxological Evaluation of the Occurrence of Pesticides in Water Bodies

In cases of possible contaminations of water with pesticides groundwater and surface water are of equal significance. Especially in conurbations, the population is partially served with drinking water from surface waters, i.e. from rivers, lakes and dams. Detractions from water quality can be locally defined due to dumping sites, contaminated sites or industry as well as due to non-point pollution by agriculture.

As early as the 1960s and 1970s the vulnerability of surface water to pesticides was recognized and investigated. The investigations were directed at the possible side effects of pesticide application on water bodies – namely bioaccumulation in fish, zooplankton, aquatic plants etc. [92–98]. The consequences of drinking water withdrawal and questions of exceeding limits and an improved management in drinking water protection areas were discussed [99–103].

The interaction between groundwater and surface water is enhanced where rivers and lakes are located in permeable deposits. In most places they drain the surroundings. However, in other areas the groundwater level is lower than the surface water level and water intrudes from the river into the soil. Due to changes in the river water level huge amounts of water can infiltrate and seep into the river banks.

4.2 Selected Data on the Situation in the EEC Member States

The data on the contamination of groundwater with triazines in United Kingdom are very interesting. For this reason, data on the atrazine and simazine contents are presented here [86, 87] to draw comparisons with these very typical contaminants.

Table 34 summarises the data on levels of atrazine and simazine in surface waters in the UK from 1989 to 1990. The nature of sampling sites, sampling frequency, month of sampling and detection limits vary considerably between the different regions. However, some general regional and temporal trends in the level of atrazine and simazine can be elucidated:

- For both active ingredients the levels are generally lower in Scotland and Wales than in England.
- The maximum levels commonly occur in June or July but peaks were detected between April and August for atrazine and March to November for simazine.
- The maximum levels of simazine found in surface waters were higher than those of atrazine in England (20.2 $\mu g\,l^{-1}$ respectively) and in Scotland the levels were 3.86 $\mu g\,l^{-1}$ and 4.23 $\mu g\,l^{-1}$.

Groundwater contamination with triazines has various origins and can be explained with the non-agricultural use of these herbicides.

Table 34. Reported concentrations of atrazine and simazine in UK surface waters (1989–1990) [86, 87]

	England		Wales		Scotland	
	Atrazine	Simazine	Atrazine	Simazine	Atrazine	Simazine
Sites	246	177	24		>176	
Samples	988	550	163		>1606	
Detection Rate[a] (% of total)	74	57	8	5	<7	<4
Samples >0.1 $\mu g\,l^{-1}$ (% of total)	50	29	<1	0	<3	<3
Samples >2 $\mu g\,l^{-1}$	2.9	0	0	0	0.06	0.06
Maximum ($\mu g\,l^{-1}$)	10.2	4.29	0.17	0.05	4.23	3.86
Month	June/July	June	September	June	–	–

[a] England 0.01 to 0.05 $\mu g\,l^{-1}$
Wales 0.01 $\mu g\,l^{-1}$
Scotland 0.025 to 0.05 $\mu g\,l^{-1}$

Besides the non-agricultural use of herbicides, waste water from the manufacturers have repeatedly contributed to single or continuous pollutions of rivers. Typical examples are the accidents with endosulfane at plants at the River Rhine in Germany in 1960s or the double accident on 31 October 1986 in Switzerland. Nearly 4 h before the fire at Sandoz AG in Schweizerhalle, about 400 kg of atrazine flowed into the river from Ciba-Geigy. The water for extinguishing the fire contained pesticides and rinsed into the Rhine [104].

The impact of agricultural production on the drinking water of one of the biggest water plants of the Federal Republic of Germany in Haltern is described by a report published in 1991 [105]. 75% of the drinking water originates from surface water, which comes for the most part from reservoir lake. Routine analytical checks of the raw and drinking water cover more than 80 active ingredients. Principle contaminants are atrazine (750–2500 $ng\,l^{-1}$), chlorotoluron (600–2500 $ng\,l^{-1}$) and isoproturon (500–2500 $ng\,l^{-1}$). Considerable deviations of concentrations within a monitoring year are evident. Particularly during the main application periods and after heavy rain, concentrations increased to a maximum of more than 2500 $ng\,l^{-1}$. The tendency in the long run (1987–1991), however, was for the concentrations to decline. This results from, among other things, changes in the production management of agricultural enterprises in the catchment area of water supply systems (prohibition of pesticide use in drinking water protection areas, changes in the product range by replacing with easily leached active ingredients, and improving plant protection consultation).

Furthermore, it has been found out that not only agriculture and forestry can cause water pollution. The presence of diuron (max. 1000 $ng\,l^{-1}$) can be attributed to its use as a total herbicide on railway tracks by the German railways [105].

4.3 Selected Data on the Situation in Eastern Europe

In Central Europe, the Elbe belongs to the rivers with the highest pollution. This is evident from the reports of ARGE Elbe (Working group for the purity of the Elbe) and studies in the former GDR. Table 35 summarizes selected data on DDT and HCH in the Elbe and the spectacular data from waste water samples taken by Greenpeace. The mentioned companies may not be the manufacturers of these active ingredients. Table 24 contains, for comparison, the triazine concentrations found in the Elbe which were below the MAC of the former GDR.

The impact of the manufacturing and processing of DDT on adjacent water bodies is reported by Heinisch et al. [106]. The route from the former VEB Berlin-Chemie to the Berlin water bodies was followed. In the Teltow-canal (ca. 1 km from the factory) annual measurements between 1984 and 1989 showed on average total concentrations between 1.0 $\mu g\,l^{-1}$ (1985) and 0.2 $\mu g\,l^{-1}$ (1988, 1989) with a maximum of 3 $\mu g\,l^{-1}$ in 1985 [47]. DDT and DDD concentrations

Table 35. Concentrations of DDT, HCH, HCB and PCBs (in $\mu g l^{-1}$) in surface water and waste waters (incl. sediments) of chemical companies along the River Elbe [46]

Place of Sampling	surface water				waste waters[a]	
Zellstoffwerke Pirna	1988	05	HCH	0.018	γ-HCH	0.32
			HCB	0.040	HCB	0.27
					DDT	1.71
Pharmazeutische Werke Germed Radebeul					HCB	17.0
					PCB 52	4.2
					DDT	92.0
Chemiewerk Nünchritz bei Riesa	1988	03	HCH	0.063	HCB	0.52
Fahlberg-List Magdeburg	1988	06	HCH	41.5	HCH	90.6
			HCB	7.04	HCB	0.68
					DDT	7.05
					PCB 101	7.55

[a] results of the Greenpeace expedition in March 1990

were still found in Potsdam (ca. 25 km from the source) and still amounted to ca. 5 $ng l^{-1}$ in the lake Jungfern see.

Lindane contamination of inland waters in the former GDR was analyzed bei Heinisch [49] on the basis of surveys carried out by Stechert for the Schwerin lakes (State of Mecklenburg-Vorpommern). From 1977 to 1990, 7.4 (5–67) samples per year were studied on average. An annual mean of 5 $ng l^{-1}$ in 1979 and 50 $ng l^{-1}$ lindane in 1987 was found. According to Mohnke the draining off of HCH isomers from these lakes and other rivers of the bordering states (incl. the former GDR) into the Baltic Sea [47] leads in the period between 1980 and 1984 near the coast of the GDR to a total average content between 5.8 $ng l^{-1}$ and 10.1 $ng l^{-1}$ with maxima of 8.81 $ng l^{-1}$, 3.62 $ng l^{-1}$, 8.02 $ng l^{-1}$ and 0.45 $ng l^{-1}$ for alpha-, beta-, gamma- and delta-HCH, respectively. The occurrence of herbicides in Lake Miedwie (Poland) was reported by Dabrowski et al. [107]. Their study was initiated by findings of DDT in the River Weichsel and its tributaries with concentrations of up to 1 $mg l^{-1}$ [108]. The aim of this work was an analytical monitoring of pesticide levels in the water and sediment of the lake and its most important inlets from 1984 to 1985 to obtain information on the entry of active ingredients. Among the active ingredients surveyed – 2,4-D, MCPA, linuron, chlorotoluron, metabenthiazuron, chloridazon, lenacil, phenmedipham and atrazine – only the first four were important (Table 36).

The occurrence of pesticides in the largest lake of Hungary, Balaton, is described by Karoly et al. [109]. The monitoring carried out since 1975 has analyzed 260–250 water samples annually 10–28% of which showed residues. The active ingredients found were the triazines atrazine, prometryn and sima-

Table 36. Herbicidal residues in the water and sediment of Lake Miedwie [107]

	N	n	1984 Concentration ($\mu g\,kg^{-1}$)	N	n	1985 Concentration ($\mu g\,kg^{-1}$)
Water						
MCPA	88	86	0.05 ... 1.32	88	86	0.03 ... 0.61
2,4-D	88	47	0.01 ... 1.38	88	80	0.05 ... 1.08
3,4-dichloraniline (linuron)	88	9	0.2 ... 0.9	88	5	0.1 ... 1.0
Sediment						
MCPA	60	45	0.3 ... 24.0	62	53	2.0 ... 15.5
2,4-D	60	8	3.0 ... 7.0	62	35	2.0 ... 31.0
3,4-dichloraniline (linuron)	54	36	1.0 ... 20.0	62	26	6.0 ... 43.0

N– number of samples; n – number of samples with residues

zine, the phenoxialkanic acids 2,4-D, 2,4,5-T and MCPA, but also organophosphorous and organochlorine insecticides, DNOC and a number of herbicides from various groups of active ingredients.

Under the Global Environmental Monitoring System (GEMS), so-called background stations were established and environmental contamination with important pollutants like DDT and HCH were studied in Eastern Europe. In the former Soviet Union, for instance, Lake Baikal, which is considered clean, was also included. Lindane and DDT were indeed found. Average concentrations of DDT and lindane were 7.6 $ng\,l^{-1}$ and 12 $ng\,l^{-1}$ between 1980 and 1986. An overview of the results of the other background stations is shown in Table 37. It was observed that self-purification processes, mainly due to sedimentation of suspended particles, cut down anthropogenic contamination of water in lakes and rivers. 20–30% of the mentioned pesticides have been found to migrate in the suspended material in rivers with moderate water turbidity [110].

4.4 Toxicological and Ecotoxicological Evaluation

The contamination of the water in rivers, lakes, and other water bodies with pesticides is measured when it is used as raw water for drinking water. In this case the maximum allowable concentration of 0.1 $\mu g\,l^{-1}$ for a single active ingredient and 0.5 $\mu g\,l^{-1}$ for all pesticides in total has to be kept. For the ecotoxicological evaluation of pesticides in water bodies, the EC limits are outdated.

Several proposals for ecotoxicological evaluation are described in the literature, among others also limits for the concentration of pesticides in water bodies, e.g. the Federal Environmental Office demands the same concentration as for drinking water. Such a concept is not realistic because the behaviour of

Table 37. Mean concentrations over many years (1980–1986) of DDT and lindane in water of lakes and rivers in background areas of the former COMECON countries [110]

Station	Country	Concentration ($ng\,l^{-1}$) DDT	Lindane
Berezina (river)	Soviet Union	23	4.5
Oka-Terrace (river)	Soviet Union	105	19
Caucasus (river)	Soviet Union	57	16
Borovoye (lake)	Soviet Union	20	17
Chatkalsk (river)	Soviet Union	2	2
Sary Chelek (river)	Soviet Union	1.6	6
Lake Sary Chelek	Soviet Union	7.5	2
Abramov Glacier	Soviet Union	2	2
Barguzin (river)	Soviet Union	35	19
Astrakhan (river)	Soviet Union	10	2.4
Rapotamo (river)	Bulgaria	30	2.0
Neuglobsow (lake)	GDR	75	11
Boretska Forests (lake)	Poland	1.0	0.3

pesticides in water bodies is influenced by many fctors (e.g. partition between water and sediment, water and organisms) and the sensitivity of aquatic organisms varies strongly. The toxicity of pesticides to aquatic organisms is important for the evaluation, but also such properties as the solubility in water, the $P_{O/W}$, the bioconcentration factor (BCF), and the degradability.

The Joint EPPO/Council of Europe Panel on environmental risk assessment of plant protection products [111] has worked out a general approach.

The proposals are based on the requirements of the International Workshop on Terrestrial Field Testing of Pesticides, held in Cambridge in 1988. The Joint Panel affirmed the necessity of establishing an internationally harmonized decision-making scheme to evaluate the side-effects of pesticides on the environment. In this connection the EPPO proposed twelve so-called end-points of risk assesment, for instance

- groundwater (presence of certain concentrations of active ingredients, accumulation),
- surface water (presence of certain concentrations, persistence),
- aquatic sediments (presence of certain concentrations, persistence including bound residues),
- aquatic organisms (individual mortality in fish, population effects, reproduction effects, bioaccumulation, indirect effects).

The programme provides the basis for working out the criteria for assessing the water quality and general guidelines on data interpretation and decision-making, terminology, and classification of ecotoxicity. It included the definition of an ecotoxicological classification related to this risk and the connection between hazard and risk. These problems are also discussed among expert groups of other international organizations (e.g. OECD, SECOTOX and SETAC).

The ecotoxicological evaluation of pesticides must always be based on data containing

- laboratory data obtained from ecotoxicological short-term and long-term studies on toxicity and
- field data obtained from ecosystem monitoring or post-registration monitoring.

Säle and Knauf [112] define the basis and terms important for hazard and risk assessment which in case of a general evaluation necessarily lead to a risk-benefit assessment. Whereas laboratory studies are primarily based on the known model organisms, field studies should above all concentrate on the evaluation of ecosystem compartments. They have to include the transition areas belonging to them [113]. Furthermore, the biotic and abiotic processes in a compartment have to be taken into consideration to analyze side effects. This complexity makes it difficult to elaborate harmonized assessment schemes and involves time and again the introduction of safety factors. This is exacerbeted by the fact that numerous authors question the application of laboratory data to field conditions. Such a safety evaluation is a worst case which represents the unfavourable conditions for the model organisms (laboratory) of the compartment (field study). A worst case can, for instance, be simulated by an excessive concentration. Fischer and Heimbach [114] present, for aquatic systems, examples with good conformity between laboratory and field conditions and examples with contrary results.

To assess a hazard potential and to evaluate a possible risk toxicity, data obtained under laboratory conditions have to be compared with environmental concentrations. The assumed worst case has to confrom to real condiions if the data are to be used for the evaluation of concentrations determined by chemical monitoring [115]. On evaluating environmental concentrations, quantities of pesticides acting on organisms or ecosystems are as a rule overestimated, their degradation or biological availability underestimated. Despite the theoretical difficulty of applying laboratory data on a few species to the large number of species of an ecosystem, in practice the laboratory data prove to be sufficient for an evaluation of the possible effects of a pesticide [114]. This requires international agreement. Practicable guidelines for the evaluation could then be derived.

5 Summary

The contamination of groundwater and surface water with pesticides can be attributed to improper handling of these environment chemicals and accidents in agriculture, industry and trade organizations but also to their use in accepted agricultural practice. Their significance for human health and the environment largely depends on the quantity applied annually in a country or within an area, the toxicological and ecotoxicological properties of the pesticides and the persistence in the environment, especially in soil and water. The current situation of the world-wide sales of pesticides in the different regions of the world[1].

The contamination of groundwater as a consequence of the use of pesticides has been illustrated by examples, and the individual factors influencing leaching behaviour are classified. The most important mathematical models and models for computer simulation are described[2].

The risk of the atmospheric deposition of pesticides is lower than assumed because the maximum deposition rates for lindane and atrazine have been determined as 1.9 g ha^{-1} yr^{-1} and 8.1 g ha^{-1} yr^{-1}, respectively.

The pollution of groundwater from contaminated sites is of local importance. The most important sources are the manufacturers which has been demonstrated with DDT and HCH producer[3] In the agrochemical centres, long intensive use of pesticides in the agriculture of the GDR has also left behind contaminated sites.

The regulations and guidelines for maximum allowable concentrations in drinking water are described for the EC, WHO, Germany, the former Soviet Union, and the USA. The special problems of the toxicological relevance of the EC limit afr of 0.1 $\mu g\,l^{-1}$ per single active ingredient and of 0.5 $\mu g\,l^{-1}$ for the of all pesticides discussed and compared with the limits of WHO and of the USA.

An overview of the contamination of groundwater and drinking water in West and East Germany, Western Europe and the USA is given. With regard to the evaluation of the quantity of contaminated samples or counts the results of various monitoring programmes in the mentioned countries and selected results of special research tasks after the use of pesticides are analyzed. The most important result in West Germany is the classification of the contaminants because 80.7% of them belong to the triazines, especially atrazine (59.7%). In East Germany (former GDR) too, the triazines play an important role. But the number of contaminated samples is smaller. It is difficult to compare the results of the EC member states because of the very different number of analysed samples.[4] The most important pesticides are the triazines and the phenoxy-alkanic acids. In the United States in 1990. Pesticides were found in 4.2% of rural domestic wells and in 10.4% of community water system wells. The main pollutants are the DCPA acid metabolites and atrazine.

The contamination of water bodies is described by selected data from Western and Eastern Europe. The results of the contamination of surface waters in UK with atrazine and simazine are very interesting in comparison with the data on groundwater contamination with both triazines in the same country. Furthermore, the situation at the contamination by chloroorganic pesticides in the River Elbe, Teltow-canal, Schwerin-lakes and the Baltic Sea (all former GDR), in Poland, Hungary and the former Soviet Union is analyzed. In the latter case results of the Global Environmental Monitoring System (GEMS) are integrated.

The ecotoxicological evaluation of the contamination of water bodies with pesticides is difficult because international harmonized guidelines are lacking. Therefore, current projects in Europe or Germany are referred to.

6 References

1. FAO (1976) Report of Ad Hoc Government Consultation on International Standardization of Pesticides Registration Requirements, FAO Rome
2. FAO (1979) Report of Expert Consultation on Environmental Criteria for Registration of Pesticides, FAO Rome
3. FAO (1986) International Code of Conduct on the Distribution and Use of Pesticides, FAO Rome
4. Beitz H, Schmidt HH, Hoernicke E, Schmidt H (1991) Mitt aus der Biologischen Bundesanstalt für Land- und Forstwirtschaft Berlin-Dahlem H. 274
5. IVA (1992) Jahresbericht 1991/1992. Industrieverband Agrar e. V., Frankfurt/Main, 5
6. Deutsche Forschungsgemeinschaft (1990) Pflanzenschutzmittel im Trinkwasser Mitteilung XVI der Kommission für Pflanzenschutz-, Pflanzenbehandlungs- und Vorratsschutzmittel. Verlag Chemie Weinheim.
7. Beran F, Guth JA (1965) Pflanzenschutzberichte Wien 23: 65
8. Leh HO (1968) Nachr-B1 Dt Pflanzenschutzdienst, 20: 99
9. Heinisch E (1969) Fortschritte der Wasserchemie 11: 9
10. Lingelbach H, Borris W (1968) Z ges Hyg und Grenzgebiete 14: 889
11. Synek H, Tibensla M, Verssidova M (1973) Bratisl Lek Listy 59: 203
12. Meyer L, Freimuth M, Engst R (1974) Zur Oxydation insektizider Thiophosphorsäureester. Vortrag, Hauptjahrestagung der Chem. Ges. DDR, Leipzig
13. Engst R, Knoll R (1973) Die Nahrung 17: 837
14. Vrocinskiy KK, Zemkov GV (1977) Gig i San, Moskau 1: 99
15. Althaus H (1968) gwf-Wasser/Abwasser 109: 988
16. Hellmann (1973) Dt Gewässerkundl Mitt 17: 47
17. Anonymous (1987) Agricultural chemicals in ground water: Proposed pesticides strategy. U.S. Environmental Protection Agency, Office of pesticides and toxic substances (December 1987)
18. Nielsen EG, Lee LK (1987) US Department of Agriculture Staff Report 18.
19. Pivetz B, Steenhuis TS (1989) Pesticide nitrate, and tracer loss in agricultural tile lines and to ground water under conventional and conservation tillage. Department of Agricultural & Biological Engineering Cornell University, Ithaca, NY 14835 Paper No. 89-2579, Presentation at the 1989 International Winter Meeting of the American Society of Agricultural Engineers
20. Kimbal CG, Goodman JPE (1989) Non-point source pesticide contamination of shallow ground water. Paper No. 892529, Presentation at the 1989 International Winter Meeting of the American Society of Agricultural Engineers
21. Leonard RA, Shirmohammadi A, Johnson AW (1989) Pesticide transport in shallow ground water. Paper No. 86-2508, Presentation at the 1989 International Winter Meeting of the American Society of Agricultural Engineers
22. Helling CS, Gish JT (1986) ACS symposium Series No. 315, American Chemical Society, Washington, DC
23. Beitz H, Schmidt H (1989) Nachr-Bl Pflanzenschutz DDR, Berlin 43, 89

24. Group of Authors (1991) Rahmenempfehlung für die landwirtschaftliche Bodennutzung in wasserschutzgebieten der Bundesländer Mecklenburg-Vorpommern, Brandenburg, Sachsen-Anhalt, Sachsen und Thüringen. Forschungszentrum für Bodenfruchtbarkeit Müncheberg, gefördert mit Mitteln des Bundesministers für Forschung und Technologie
25. Aller L, Bennet T, Lehr J, Petty R (1985) DRASTIC: A standardized system for evaluating groundwater pollution potential using hydrogeological settings. USEPA Off. Res. & Devel., Washington, DC, EPA/600/2-85/018, 384 p
26. Rao PSC, Hornsby A, Jessup RE (1985) Soil and Crop Sci Soc Florida Proc 44: 127
27. Blume HP, Brümmer G (1987) Landwirtsch. Forsch. 40: 41
28. Jury WA, Focht DD, Farmer WJ (1987) J Environ Qual 16: 119
29. Binner R, Wolf N, Schmidt H, Beitz H (1990) Nachr-Bl Pflanzenschutz DDR 44: 97
30. Boehncke A, Siebers J, Nolting H-G (1989) Verbleib von Pflanzenschutzmitteln in der Umwelt – Exposition, Bioakkumulation, Abbau-Teil B. Forschungsbericht 89-126 05 008/02
31. Oberwalder Ch, Gießl H, Irion L, Kirchhoff J, Hurle K (1991) Nachrichtenbl deut Pflanzenschutzd (Braunschweig) 43: 185
32. Robinson E, Fox LL (1978) J Air Poll Control Assoc 28: 1015
33. Eisenreich SJ, Looney BB, Thorton JD (1981) Environ Sci Technol 15: 30
34. Mücke W, Steinmetzer H-C, Stumpp J, Baumeister W, Boneberg R, Vierle O (1991) UWSZ-Z Umweltchem Ökotox 3: 176
35. Herrmann (1984) Erdkunde 38: 55
36. Dulnker JC, Bouchertall F (1989) Environ Sci Technol 23: 57
37. Dommröse AM, Figge K (1988) Fresenius Z Anal Chem 332: 606
38. Pankow JF (1987) Atmospheric Environment 21: 2275
39. Sehmel GA (1980) Atmospheric Environment 14: 983
40. Cautreels W, Cauwenberghe van K (1978) Atmospheric Environment 12: 1133
41. Brun GL, Howell GD, O'Neill H (1991) Environ Sci Technol 25: 1249
42. Siebers J, Gottschild D, Nolting H-J (1991) Nachrichtenbl deut Pflanzenschutzd (Braunschweig) 43: 191
43. Buser H-R (1990) Environ Sci Technol 24: 1049
44. Herterich R (1991) Z Umweltche Ökoto 3: 196
45. Schleyer R, Renner I, Mühlhausen D (1991) WaBoLu-Heft 5, Berlin (1991)
46. RIVM (1991) Sustainable Use of Groundwater Report No 6000250001, pp 35
47. Baart AC, van Diederen HSMA (1991) TNO Environmental and Energy Research, Report No R 91/219, Delft
48. Gath B, Jaeschke W Rickert I, Zietz E (1992) Nachr- Bl Dt Pflanzenschutzdienst 44: 57
49. Heinisch E (1992) Umweltbelastung in Ostdeutschland, Wissenschaftliche Buchgesellschaft Darmstadt
50. Heinisch E, Klein S, Giese F, Terytze K, Priesemuth B, (1989) Beiträge bisher nicht oder nur wenig untersuchter Schadstoffe in Berlin – Potsdamer Gewässern. Forschungsbericht, Inst für Geographie und Geoökologie der AdW Berlin
51. Braun L, Krapp L (1992) WHL, Wasser, Luft und Boden 36, 90
52. Beitz H, Riebel A, Buhr L (1992) Abstracts of Joint Meeting of SETAC – Europe Ecology/Environmental Fate/Toxicology Bio- remeditation, Potsdam June 21 to 24
53. Beitz H, Buhr L, Riebel A (1992) Mitt aus der Biologischen Bundesanstalt für Land- und Forstwirtschaft Berlin-Dahlem 283: 124
54. Beitz H, Blumrich H, Buhr L, Riebel A (1992) Erfassung der durch Planzenschutzmittel verursachten Altlasten und Erarbeitung von Sanierungsmaßnahmen. Forschungsbericht Ministerium für Umwelt, Naturschutz und Raumordnung des Landes Brandenburg FM/LH/92-338.15/38–20
55. Schmidt H, Winkler R (1991) Bewertung der Belastung des Bodens und der Ökosysteme der ehemaligen DDR mit Pflanzenschutzmitteln. Umweltbundesamt Forschungsbericht 91- R IX 12

56. Beitz H, Krüger B (1991) Einsatzbegrenzung von neuen Pflanzenschutzmitteln und Auswirkungen auf Ökosysteme. Umweltbundesamt, Forschungsbericht R VIII 1
57. Roberts T (1989) GIFAP Bull 15: 1
58. Ruchay D (1991) Gesunde Pflanzen 43: 16
59. Ohnesorge FK (1991) Gesunde Pflanzen 43: 26
60. BGA-Empfehlungen (1989) Empfehlungen des Bundesgesundheitsamtes zum Vollzug der Trinkwasserverordnung (TrinkwV) vom 22. Mai 1986 Bundesgesundhbl. 32, 290
61. Anordnung über Rückstände der Wirkstoffe von Pflanzenschutzmitteln in Lebensmitteln–Rückstandsmengen-Anordnung vom 30. Juni 1988, GBL. DDR 1988, Sdr Nr 1311
62. Beitz H, Hamann W (1990) Nachr-Bl Pflanzenschutz 44: 16
63. Lingk W (1991) Gesunde Pflanzen 43: 21
64. Kagan JS (1981) Tagungsbericht Akad. Landwirtschaftswiss. der DDR, Berlin 187: 13
65. WHO (1984) Guidelines for Drinking-Water-Quality. Vol 1: Recommendations WHO Geneva
66. Medved LI (1974) Spravochnik po Pesticidam. Izdatelstvo Uroshaiy Kiev
67. Iskandarov (1992) private information
68. Minzdravom (1983) PDK v vode vodoemov sanitarno- bytovovo vodopolzovanija, Priloshenije 8 in: Gos agropom komitet SSSR (1988) List of chemical and biological plant protection products for the use in the agriculture in 1986/90
69. Stock R, Friesel B, Milde G (1978) Schr Reihe Verein WaBoLu 68: 209
70. Grohmann A (1992) Bundesgesundhbl. 35: 89
71. Pestemer W (1991) Umweltbundesamt Berlin, FuE-Bericht 10202213/01
72. Worthy W (1991) Chemical and Engineering News, May 6, 46
73. RCC (1991) Project Nr. 29 32 42, RCC Umweltchemie AG, CH-4452 Itingen/BL
74. Bundesverband der Deutschen Gas- und Wasserwirtschaft (1988) Bericht zum Stand April 06 1988
75. Zullei-Seibert N (1990) Veröffentlichungen des Inst für Wasserforschung GmbH Dortmund und der Dortmunder Stadtwerke AG Nr 39, 1990
76. Roth M (1989) Schr Reihe Verein WaBoLu 79: 365
77. Amann W, Schuster M, Gilsbach W, Kees H, Rappel A (1989) Schr Reihe Verein WaBoLu 79: 151
78. Friege H, Kalthoff H (1989) Landesamt für Wasser und Abfall Nordrhein-Westfalen, Jahresbericht '88, 7, Düsseldorf
79. Industrieverband Pflanzenschutz (1987) Pflanzenschutzmittel-Wirkstoffe im Trinkwasser. Ergebnisse einer Untersuchungsreihe der Pflanzenschutzindustrie IPS, Frankfurt/Main
80. Greiner-Schuster E (1989) Öko-Test-Magazin 12/89: 26
81. Bundesministerium für Gesundheit (1992) Bericht der Regierung der Bundesrepublik Deutschland an die Kommission der EG über Maßnahmen zur Erreichung der Qualitätsnormen der Richtlinie 80/778/EWG bis 31. Dezember 1995 im Trinkwasser der Bundesländer Brandenburg, Mecklenburg-Vorpommern, Sachsen, Sachden-Anhalt und Thüringen. Bonn, 1992
82. Hellwig A, Kühne V, Schumann G et al. (1989) Pestizide in der Umwelt. FuE-Bericht, Bezirkshygieneinstitut Magdeburg
83. Wolter R (1992) 18. Seminar Pesticides in water UTECH, Umwelttechnologieforum, Berlin, 19th February 1992
84. Gorf A (1991) International East-West Symposium on contaminated areas in Eastern Europe. Abstracts, papers, posters: 53
85. Boesten JJTI, van der Linden AA (1991) J Environ Qual 20: 425
86. MAFF (1992a) Food and Environment Protection Act 1985, Part III, Control of Pesticides Regulations 1986, Evaluation on Atrazine, May 1992 Ministry of Agriculture Fisheries and Food, London No 51

87. MAFF (1992b) Food and Environment Protection Act 1985, Part III, Control of Pesticides Regulations 1986, Evaluation on Simazine, May 1992 Ministry of Agriculture Fisheries and Food, London No 51
88. Ritter WF, Chirnside AEM, Scarborough RW (1987) International Winter Meeting of the American of Agricultural Engineers, Illinois, December 15–18, Paper No. 87-2630
89. EPA (1990) National Pesticide Survey Phase I Report EPA 570/9-90-015
90. EPA (1989) Drinking water Health Advisory: Pesticides Michigan Lewis Publishers
91. EPA (1992) National Pesticide Survey Phase II Report EPA 570/9-91-020
92. Rogowskaja AP, Dubinskaja GO (1969) Gig. i. san. 34: 76–77
93. Mestres R, Heuaux P, Vaquer A (1972) Ann Hyg Langue Franc Med et Nutrit 8: 71–90
94. Miles JRW, Harris CR (1973) Pestic Monit J 6: 363–368
95. Hurlbert SH (1975) Secondary effects of pesticides on aquatic ecosystems Res Rev Heidelberg 57: 81–184
96. Laska AL, Bartello CK, Laseter JL (1976) Bull Environ Contam Toxicol New York 15: 535–542
97. Börner K (1977) Z Pflanzenkrankh u Pflanzenschutz Stuttgart 84: 705–718
98. Weser SD, Day EW, Burger RO (1979) J Agr Food Chem Washington 27: 1067–1072
99. Bindra OS (1972) Pesticides 12: 77–82
100. Heidler G (1973) Nachrbl Deut Pflanzenschutzdienst 25: 91–93
101. Bitter B (1974) Einsatz von Herbiziden, Fungiziden und Insektiziden in der Forstwirtschaft unter Berücksichtigung der Sonderbelange der Wassereinzugsgebiete von Trinkwassertalsperren DVGW-Schriftenreihe Wasser Nr 1
102. Paluskova O, Sackmauerova M (1975) Water pollution with organochlorine insecticide residues in Slovakia VIII Internat Pflanzenschutzkongreß Moskau 21–27th August Sect IV 163–169
103. Bowner KH (1982) Pestic Sci Oxford 13: 623–638
104. Anonymus (1987) Atrazin, eine Dokumentation CIBA-GEIGY Division Agro
105. Anonymus (1991) Kooperation Landwirtschaft und Wasserwirtschaft im Einzugsgebiet der Stevertalsperre Bericht
106. Heinisch E, Koch R, Klein S, Terytze K, Giese F, Krüger A, Ladewig B, Hoernicke E, Volland M, Huschek G, Schmidt H (1990) Wiss Mitt, Inst für Geogr und Geoökologie AdW DDR 33
107. Dabrowski J, Krause A, Waliszewski S, Stasieczek Z, Filary Z (1987) NachrBl.Pflanzenschutz DDR 41: 60–63
108. Twardowska J, Kozlowska K (1990) Pol Arch Hydrobiol 27: 187–213
109. Karoly G, Fernczi J, Orosz F, Visi E (1991) Model and field experiments for assessing the behaviour of pesticides in the environment, East-West Symposium on contaminated areas in Eastern Europe Gosen November 25–27: 27
110. Anonymus (1988) Analytical review of background environmental pollution in the region of Eastern European CMEA member-countries (1980–1986) Moskau
111. Report of the Joint EPPO/Council of Europe Panel on Environmental Risk Assessement of Plant Protection Products (1990), CETIOM, Paris 1990 06-06/07
112. Säle M, Knauf W (1991) Gesunde Pflanzen 43: 411
113. Knauf W, Säle M (1991) Gesunde Pflanzen 43: 159
114. Fischer R, Heimbach F (1991) Gesunde Pflanzen 43: 270
115. Beitz H (1992) Mitt Biologische Bundesanstalt für Land- und Forstwirtschaft, Berlin- Dahlem, 279: 64
116. Anonymûs (1993) Gesunde Pflanzen 45: 202

The Mobility of Pesticides in Soil – Studies to Prevent Groundwater Contamination

David Bewick

Zeneca Agrochemicals, Jealott's Hill Research Station, Bracknell, Berkshire, RG12 6EY, UK

Extensive studies on the fate of pesticides in surface soils have been conducted for many years but, until recently, little has been done to understand the behaviour of these compounds beyond the root zone. However, concerns about the risks posed by residues of pesticides in ground and surface waters have stimulated research and there are recent studies which clearly indicate the potential for microbial degradation at considerable depth. Small, but significant, populations of microorganisms have been detected at all points of the soil profile down to, and beyond the water table and there is increasing evidence that these are able to degrade a range of compounds, including pesticides. Degradation rates are usually low compared to surface soils, but they can be sufficient, in view of the long residence times involved, to reduce significantly the concentrations of pesticide residues in ground waters. However, further work is needed if we are to develop a full understanding of the microbial ecology of the subsurface environment and the mechanisms of associated xenobiotic degradation. Only if we are successful in this, we will be able to predict the mobility of pesticide residues with confidence and retain the flexible use of pesticides without jeopardising our water reserves.

Chemistry of Plant Protection, Vol. 9

1 Introduction

In recent years, the realisation that the approved use of pesticides could result in the contamination of valuable supplies of both surface and groundwater has aroused increasing public concern and scientific curiosity. There have been many well authenticated reports of the detection of residues of pesticides in groundwater, mainly in the USA [e.g. Ritter, 1] and Europe [e.g. Liestra and Boesten, 2; Fielding, 3], and frameworks for risk assessment and pesticide regulation have been developed to ensure the safety of humans and the environment. The approach which has been adopted in the USA and by the World Health Organisation [4] has been to ensure that any residues of pesticides are below the concentration likely to present any safety concerns whilst, in the European Community [5], an arbitrary threshold level of $0.1\ \mathrm{ug\,l^{-1}}$ has been set for residues of all pesticides in drinking water, irrespective of their toxicity. Whichever approach is taken, the prediction of the likely mobility of pesticides, in use, has become a key step in the process.

Fundamentally, the extent of leaching or runoff of a pesticide from soil is determined by the relative rates of degradation and movement of the compound in question. However, these processes are influenced by many different factors associated with both the properties of the pesticide and its environment, all of which must be understood in their own right and integrated in any mobility assessment. The purpose of this chapter is to review those factors which influence the mobility of pesticides in soil, with particular emphasis on the role of microbial degradation.

2 Factors which Influence Pesticide Mobility in Soils

The mobility of pesticides in soil is influenced by three main factors: the fundamental properties of the pesticide, the characteristics of the soil environment in question and the prevailing weather conditions. These factors have been well documented in previous reviews [e.g. Nicholls, 6] and although it is not the intention of this chapter merely to repeat previous work in this area, no review of pesticide mobility in soil could be complete without at least a brief coverage these crucial factors.

2.1 Adsorption

For most pesticides, the distribution between the solid and liquid phases of soil is one of the most critical factors in determining the potential mobility of the compound in the soil environment. The extent of adsorption of a pesticide to soil is best characterised by the simple adsorption coefficient, K_d, where:

$$K_d = \frac{\text{Concentration of pesticide adsorbed to soil.}}{\text{Concentration of pesticide in soil solution.}} \tag{1}$$

The advantages of using the adsorption coefficient in this form are that it may be determined directly by fairly straightforward experimentation [e.g. OECD, 7] and that its use does not require a knowledge, or assumption, of the mechanisms of adsorption involved for the particular pesticide/soil combination under investigation.

The extent of adsorption of different pesticides to soil may vary over several orders of magnitude: poorly adsorbed compounds such as sulphonyl urea and phenoxyacetic acid herbicides may have K_d values < 1 [Nicholls and Evans, 8] whereas K_d values of > 1000 have been reported for lipophilic compounds such as pyrethroid insecticides [e.g. Bewick et al., 9] and cationic compounds such as paraquat [Knight and Tomlinson, 10].

2.1.1 Interactions with Soil Organic Matter

The majority of all pesticides are organic compounds which are not ionised under the prevailing pH conditions of agricultural soils. If the adsorption characteristics of such compounds are studied in a range of soils, a good correlation is often obtained between measured adsorption coefficients and the organic matter contents of the soils [e.g. Nicholls, 11]. In the absence of similar correlations with the other fundamental soil properties, this indicates that the principal adsorption mechanism involves an interaction between the pesticide and the organic matter component of the soil. In these cases, it is often convenient to normalise the adsorption coefficient to take account of different soil organic matter or organic carbon contents. Thus, the K_d value may be expressed as the adsorption coefficient per unit of organic matter, K_{om}, or organic carbon, K_{oc}, where:

$$K_{om} = K_d \times 100/(\%\ \text{organic matter}) \tag{2}$$

$$\text{and,}\quad K_{oc} = K_d \times 100/(\%\ \text{organic carbon}) \tag{3}$$

McCall et al. [12] proposed a scheme to classify pesticide mobility according to K_{oc} (Table 1). This approach is perfectly valid for those pesticides which are adsorbed primarily onto organic matter but, clearly, a knowledge of the adsorption mechanism is required if it is to be used with confidence in a mobility assessment.

Soil organic matter is a relatively poorly characterised amorphous medium although it is often classified as either non-humic and humic substances. The former comprises the largely undegraded remains of plants, animals and microorganisms, whereas the latter are the transformed remains of such biota which are often intimately associated with the soil minerals and resistant to

Table 1. Classification of soil mobility on the basis of K_{oc} [McCall et al. 11]

K_{oc}	Mobility Class
0–50	Very High
50–150	High
150–500	Medium
500–2000	Low
2000–5000	Slight
> 5000	Immobile

further decay. Although soil organic matter is far from homogeneous, it shares some of the characteristics of lipids and there is often a good correlation between the lipophilicity of pesticides (as measured by the octanol–water partition coefficient, K_{ow}) and the extent of their adsorption to soil. From such correlations, empirical relationships have often been developed to estimate K_{om} from K_{ow} values. This approach has been reviewed by Lyman et al. [13] who showed a number of such empirical relationships between $\log K_{ow}$ and $\log K_{oc}$ which could be used to predict the extent of adsorption of pesticides in a wide range of soils. This approach has been further extended by workers such as McCall et al. [12] who estimated K_{oc} directly from the retention behaviour of pesticides in a simple reversed-phase high performance liquid chromatography system. The advantage of this type of approach is that pesticide adsorption to soil may be predicted on the basis of measurements which are quick and easy to perform in the laboratory, even in the absence of samples of the soils for which a mobility assessment is required and, provided the limitations of the technique are appreciated, it can often provide sufficient accuracy for at least a preliminary study.

Whilst the approaches outlined above may be suitable for neutral organic molecules, those which are ionisable require a more careful consideration. Weak acids and weak bases often have dissociation constants which profoundly affect their degree of ionisation within the normal pH range of soils, making it necessary to consider the adsorption characteristics of both their ionised and un-ionised forms. The old adage of 'like dissolves like' immediately leads one to expect that the ionised forms of such molecules will have adsorption coefficients which favour the aqueous phase to a larger extent than their un-ionised forms. All the experimental evidence supports this and large changes of adsorption coefficients at soil pH values in the vicinity of the pK_a or pK_b values are often observed. Interestingly, though, Nicholls [8] showed that the adsorption of the weakly acidic sulfonyl urea and phenoxyacetic acid herbicides was greater at 1–2 pH units above the pK_a values than would have been predicted on the basis of their pK_a values, possibly because the pH at the surface of the adsorbing soil colloids is lower than in the bulk soil solution.

2.1.2 Interactions with Soil Minerals

Although soil organic matter is the principal adsorbent of most pesticides, there are some notable exceptions. Perhaps the most dramatic examples are the bipyridyl herbicides, paraquat and diquat, both of which are quaternary salts with high water solubilities. Although these compounds have a high affinity for soil organic matter, they are even more extensively adsorbed to clay minerals. Knight and Tomlinson [10] explained the adsorption of paraquat to clay on the basis of its ability to form ion exchange complexes with the negatively charged surfaces of the clay minerals in soil. Expanding lattice clays were found to exhibit particularly high adsorption capacities for the herbicide.

Another example of a pesticide which is extensively adsorbed to soil minerals in preference to soil organic matter is the herbicide, glyphosate. Sprankle et al. [14] and Hance [15] attributed the adsorption of glyphosate to the interaction of its phosphonate moiety with soil and showed that inorganic phosphate competes for the same adsorption sites. The adsorption of glyphosate is probably more complex than that of phosphate since, in addition to the phosphonate group, it also contains carboxylic acid and amine functions and has been shown by Subramaniam and Hoggard [16] to be capable of forming chelates with metal ions.

2.1.3 Reversibility of Adsorption

If the soil adsorption coefficients of pesticides were true equilibrium constants, reversibility of adsorption would not be an issue. However, it has long been recognised that desorption of pesticide residues from soil is not achieved as readily as would be predicted from simple adsorption data. This phenomenon, often referred to as adsorption–desorption hysteresis [e.g. Swanson and Dutt, 17] may be in part an artefact arising from the limitations of the batch equilibrium techniques used for measuring adsorption. However, notwithstanding such problems, there are many examples showing that, with time, it becomes increasingly difficult to desorb pesticide residues from soil. Conceptually, multi-compartment models offer an attractive explanation for such observations. One such model, proposed by Duffy et al. [18] to explain the behaviour of the sulfonyl urea herbicides, is illustrated in Fig. 1. It was suggested that application of these compounds could result in a residue in an 'available compartment' where it is subjected to chemical and microbial degradation with concurrent transport to a 'protected compartment' where it is available only for chemical degradation.

If there is increased adsorption of 'aged' residues of pesticides in soil, with time, it is clear that the use of simple adsorption K_d values will increasingly tend to under-estimate the concentration of the compound in the aqueous phase as time progresses. As a consequence, movement within the aqueous phase of the soil will be less than expected. It is also likely that the degradation rate will be

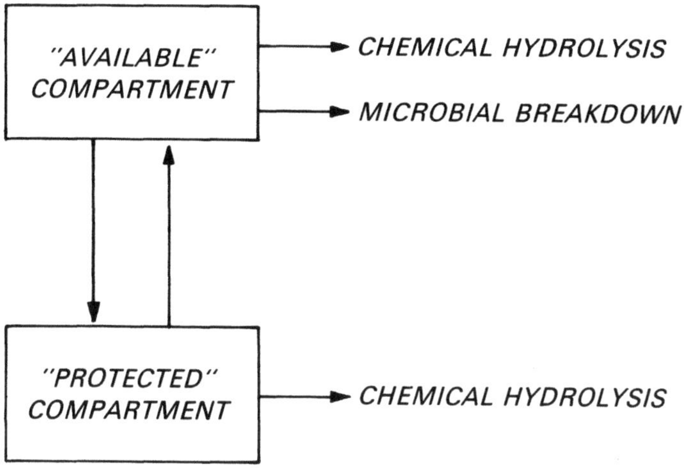

Fig. 1. The two-compartment model of soil degradation [Duffy et al. 18].

reduced but, with the exception of situations involving surface runoff of adsorbed residues, these factors will tend to have opposing, if not equal, effects in soil mobility assessments.

2.2 Degradation

When considering degradation, both the rate and route of degradation must be borne in mind. The rate of degradation of a pesticide in soil is often viewed as the key factor in determining the potential of the compound for having an adverse environmental impact and, in recent years, various persistence criteria have been introduced into regulatory schemes [e.g. Schinkel et al. 19]. Whilst it is entirely appropriate that the persistence of pesticides should be considered during the assessment of environmental acceptability, there is a danger that the importance of this factor may be over-emphasised. Logically, the persistence of a pesticide in a particular environment should be considered excessive only if it is sufficient to allow any intrinsic adverse properties of the compound to be expressed, but not merely because an arbitrary persistence criterion has been exceeded. The 'persistence' of a pesticide is a term often used as if it were an inherent property of a compound. In reality, of course, persistence relates only to the behaviour of a compound in a particular environment under a certain set of conditions and may, therefore, be very different in each of the different compartments of the soil ecosystem.

Pesticides often have precise structural requirement for activity. In general terms, then, it is reasonable to expect that any transformation they undergo in the environment is likely to represent detoxification. This is true for the majority of compounds although there are examples of environmental transformations which retain or even enhance the biological activity of pesticide residues. For example, the photooxidation of the thiophosphate, parathion, to the corresponding phosphate (paraoxon) results in an increase in toxicity [Grunwell and Erickson, 20]. Such reactions may be more subtle, however, and Bewick [21] showed that the optical configuration of the herbicidal inactive S-enantiomer of fluazifop could be inverted by soil microorganisms to yield the active R-enantiomer. These examples illustrate that a particular pesticide may undergo extensive transformation in soil without a concomitant reduction in its biological activity. A prerequisite for a complete understanding of the environmental impact of a compound is, therefore, that degradative pathways are known for all relevant environmental compartments.

2.2.1 Photolysis

Although photodegradation of pesticides may be significant on plant surfaces and in the aquatic environment, it is generally assumed to be of only minor importance in soil. Nevertheless, there are regulatory requirements to study photodegradation on soil surfaces [e.g. EPA, 22] and convenient methods to study this phenomenon in the laboratory have been developed [Klehr et al., 23; Parker and Leahey, 24]. It remains to be seen whether the photo-induced degradation of surface-adsorbed pesticides is one of the principal transformation processes, as postulated by Parlar [25].

2.2.2 Chemical Degradation

In the context of this review, the term chemical degradation is used to describe any degradative process which does not require the direct interaction of pesticides with the biota of the soil. The main reactions considered are hydrolysis and oxidation/reduction.

Hydrolysis studies under sterile conditions and controlled pH are normally conducted during the development of pesticides. These can be a good measure of the propensity of a pesticide to undergo hydrolysis reactions, although they cannot accurately predict the rate of these reactions in the environment. In certain situations, extensive adsorption in soil may inhibit hydrolysis whilst, more often, natural soil constituents may have a pronounced rate enhancing catalytic effect. Degradation studies in soil under a range of conditions are needed if this uncertainty is to be resolved. Many workers have gone to great lengths to differentiate between chemical and biological hydrolysis in such studies, often comparing degradation in fresh and sterilised soils. Beyer et al. [26], for example, studied the hydrolysis of sulfonyl urea herbicides in fresh and ethylene oxide sterilised soils and showed that chemical and microbial hydrolysis were more important at low and high pH, respectively. In order to understand the fate of pesticides in surface soils it perhaps matters little if hydrolysis is microbial or chemical in nature, merely that the overall rate and route of degradation are known. However, in sub-surface soils, an understanding of the precise mechanism of hydrolysis is crucial since microbial activity may be so low as to leave chemical hydrolysis as the main degradative reaction.

Oxidative processes are involved in the degradation of most pesticides in soil although there are few, if any, reports of this occurring by truly 'chemical' mechanisms. However, singlet oxygen may be produced photochemically at the soil surface [Gohre and Miller, 27] and has been shown to react readily with a number of pesticides. Some pesticides are also subjected to reduction in soil but, again, it is often difficult to differentiate between biotic and abiotic mechanisms. There is little doubt that reducing conditions in soil usually arise as a result of microbial metabolism, although the electron donor responsible for the reduction of pesticides may be an inorganic species.

Thus, anaerobic zones in soil, caused by microbial metabolism, may result in the formation of species such as ferrous iron which, in turn, may act as a reducing agent for pesticides. An example of this is the reduction of trifluralin reported by Williams [28].

The redox status of a particular soil environment is often characterised by its redox potential (E_h) relative to the standard hydrogen half cell. Thus, well-aerated soils have E_h values in the range 800 to 400 mV; moderately reduced soils have E_h values in the range 400 to 100 mV; reduced soils have E_h values in the range 100 to -100 mV and highly reduced soils have a E_h values < -100mV [Patrick and Mahapatra, 29].

2.2.3 Microbial Degradation

Of all the processes involved in the degradation of pesticides, microbial metabolism is generally regarded as the most important. Over ten years ago, Hill [30] extensively reviewed the range of biotransformations which had been observed for a wide variety of pesticides, but concluded that the state of knowledge at the time did not allow degradative pathways to be predicted with confidence. Since then, there have been numerous further published examples of the microbial metabolism of individual pesticides and many thorough reviews mapping the transformations of the various functional groups of such compounds [e.g. Aizawa, 31]. Arnold and Briggs [32] have recently shown how such data may be used together with expert judgement to predict the fate of pesticides in soil with sufficient confidence to at least devise a strategy for the most appropriate forms of environmental testing.

There are many examples showing how pesticides may be transformed by microbial metabolism through a variety of intermediate metabolites leading, ultimately, to extensive mineralisation to carbon dioxide. Pesticide microbiologists often find it convenient to differentiate those metabolic pathways which lead to a net benefit to the organism responsible, in terms of either energy or nutrient gain, from those which do not. The latter, arising merely from the fact that an organism may have the enzymic capability to perform particular transformations as a result of its normal physiological processes, but which do not benefit the organism, have been referred to by Horvath [33] as cometabolism. The significance of this is that unless an organism is able to derive some benefit from a particular transformation, it will never adapt to the presence of a contaminant in its environment. The potential for microorganisms to adapt to pesticides was first shown by the classical work of Audus [34] with the herbicide 2,4-D, as long ago as the early 1950s. This showed that degradation could be preceded by a lag-phase during which the organism developed the capability to metabolise the substrate. Subsequent exposure of the substrate to the acclimatised microbial population resulted in rapid degradation without a lag-phase. The full practical significance of this work was not really appreciated until the phenomenon of enhanced degradation was discovered some three decades later. Workers such as Felsott [35] reported that repeated use of certain soil-applied pesticides in the same fields could lead to an increase in their rates of microbial degradation to the extent that they lost their pesticidal activity. Clearly, this phenomenon has caused important agronomic problems for certain pesticides which are required to express their activity within the root zone. On a broader level, though, it illustrates that the adaptation of microorganisms to pesticides is perhaps more widespread than had been realised only a few years ago, and that it may be an important factor in determining the fate of many compounds in the environment.

The concentration of a pesticide in the environment may be a significant factor in determining how it is degraded by microorganisms. If a pesticide is only degraded as a result of being utilised by an organism as a growth factor, it will only be extensively metabolised if it is present at a sufficiently high concentration to at least maintain the microbial population. Alexander [36] has reported threshold values below which pesticides are not mineralised in natural waters and has speculated that similar thresholds may also exist in soil. Even though threshold values are not likely to be as important if a pesticide is degraded primarily by incidental or cometabolism, there are clearly good reasons for researchers conducting environmental fate studies to ensure that pesticides are applied at concentrations which are close to those likely occur following normal use.

2.3 Weather Conditions

The weather has a profound influence on the nature of the soil environment and, consequently, on the fate of any pesticide residues within it. The most important effects of temperature, rainfall and wind on the degradation and movement of pesticides are outlined below.

2.3.1 Temperature

Under the influence of changes in the intensity of solar radiation and air temperature, the surface layers of soils are subjected to both diurnal and seasonal fluctuations in temperature. Temperatures at the surface of agricultural soils may vary from many degrees below zero to about 40 °C, although mobility and degradation are both negligible in frozen soils. Generally, maximum soil temperatures become lower, and minimum temperatures higher, with increasing depth below the soil surface.

2.3.2 Rainfall

Rainfall and other forms of precipitation are particularly important since, together with irrigation, they have a profound impact on the levels of moisture and water fluxes within the soil system. The moisture level may influence several aspects of pesticide degradation in the soil system; this may be a direct effect associated with the availability of soil water as either a reactive species in its own right or as a solute for the substrates and products of microbial metabolism, or an indirect effect associated with soil aeration. Since soil water and air compete for the same pore space in the soil, high moisture levels are associated with low levels of aeration and this may have important consequences for both the rates and pathways of microbial degradation.

The water flux, V_d, in a soil is often estimated from the soil water balance equation,

$$P_r + L = ET + V_d + w \tag{4}$$

where, P_r = precipitation, L = irrigation, ET = evapotranspiration and w = runoff. Under normal conditions, rainfall or irrigation will result in infiltration and at least a short-term increase in soil moisture levels. If the soil moisture level exceeds field capacity, downwards movement of water will occur by percolation through predominantly the larger soil pores. Evapotranspiration will result in an upwards movement of water. Thus, although there is likely to be net annual downward movement of water from the root zone of most temperate agricultural soils, resulting in recharge to groundwater, this is not a continuous process and during any given period of time, the vertical flux of water and any associated solutes may be in either direction. Water flux and, hence, the mobility of pesticides in soil are very sensitive to both the total amount and intensity of rainfall.

2.3.3 Wind

Wind will tend to sweep vapours away from the soil surface and can significantly increase the evaporation rates of both water and volatile pesticides. The evaporation rate of water at the soil surface is important since it may result in the net upward movement of water and solutes during a given period of time. In extreme circumstances, high winds may result in the loss of surface soil and the transport of any associated adsorbed pesticide, by erosion.

2.4 Transport Mechanisms

If a pesticide is associated, even in part, with a phase of the soil which is prone to movement under the prevailing conditions, the pesticide residue will also move to certain extent. Such mass flow or convective flow processes are responsible for the leaching and surface runoff of pesticides. Pesticide mobility may also occur in soil as a result of diffusion along concentration gradients in either the liquid or gas phases. The total solute flux, J_s, may be expressed mathematically as:

$$J_s = J_{DL} + J_{CL} + J_{DG} \quad (5)$$

here, J_{DL} and J_{CL} are, respectively, the diffusive and convective fluxes in the liquid phase and J_{DG} is the diffusive flux in the gas phase. In principle, convective flux may also occur in the gas phase along pressure gradients but this has been shown to be of consequence only for extremely volatile compounds (Wagenet et al., 37). Convective and diffusive solute fluxes in the soil phase of bulk soil environment are usually considered insignificant, although slow diffusion of pesticides into organic matter complexes may be responsible, in part, for adsorption/desorption hysteresis.

Although solute diffusion may take place in both the gas and liquid phases of the soil, diffusion rates are some 3–4 orders of magnitude higher in the gaseous phase. Thus, unless the air/water distribution coefficient is very high ($> 10^3$), diffusion in the soil water phase is the more important. Diffusion of any solute in the soil results in its dispersion, and dilution, throughout a wider volume. For most pesticides, diffusive processes are likely, therefore, to result in a reduction in environmental impact but others, such as soil fumigants, rely heavily upon diffusion to reach their intended sites of action and to express their pesticidal activity.

3 Pesticide Degradation and Mobility in Surface Soils

The potential mobility of a pesticide in any soil environment is determined, essentially, by the extent of its adsorption and the rate of its degradation. The extent to which this potential is realised under a particular pattern of use depends on the water flux which is, in turn, influenced by the weather conditions, the soil physical properties and the prevailing agronomic practices.

3.1 Adsorption in Surface Soils

The root zone of agricultural soils often contains higher levels of organic matter than any other part of the soil profile (typically 0.5–10%) and, as such, provides an environment which is optimal for the adsorption of many organic pesticides. Many lipophilic compounds have been shown to have adsorption coefficients which are inversely correlated with depth. The few exceptions to this are those pesticides which are adsorbed by specific interactions with soil minerals; adsorption of such compounds may be higher in low organic matter soils and may therefore increase below the root zone.

3.2 Degradation in Surface Soils

Although all the degradative processes outlined earlier may operate in surface soils, microbial metabolism is generally the most important and will be considered in more detail here. Frehse and Anderson [38] described the three major variables which determine the rates of microbial metabolism of pesticides as:

1. The availability of the chemical to the organisms or enzyme systems responsible for metabolism.
2. The quantity of these microorganisms or enzyme systems.
3. The activity level or physiological state of the organisms.

The availability of a pesticide to potentially degrading microorganisms is related to its extent of adsorption and rate of desorption. Anderson [39], for example, increased the extent of adsorption of the herbicide diallate in a soil by the addition of various levels of activated charcoal and found a direct correlation between the concentrations of the compound in solution and the rates of degradation. It would be easy, but quite incorrect, to deduce from experiments such as these that adsorption, per se, is a barrier to microbial degradation; there are many examples of pesticides which are readily degraded in soil despite having high adsorption coefficients. For example, the insecticide lambda-cyhalothrin was found to have a half-life in soil of 25 days although it has an adsorption coefficient of over 1000 [Bewick et al., 9]. Compounds such as this must either have an extremely high inherent biodegradability or be capable of being microbially degraded in both the soil solution and the adsorbed form.

Water plays an important role as the solute for the substrates of microbial metabolism and it has long been recognised that changes in soil moisture levels may have a profound impact on the rates of degradation of pesticides. Based on laboratory studies with a number of herbicides incubated in a soil with moisture levels in the range 4–12% (25–75% of field capacity), Walker [40] developed a purely empirical equation relating half-life at 25 °C (H) to soil moisture (M):

$$H = AM^{-B} \quad (6)$$

where, A and B are constants. The effect of increased soil moisture on the rates of pesticide degradation in soil is readily accepted by most workers in the field but there are relatively few data on the mechanisms involved. In a study with diallate and triallate, Anderson [39] obtained an excellent correlation between soil water contents in the range of 2.4–19.0% and rates of degradation, attributing the increased rates of degradation at higher soil moisture levels to the increased amounts of the herbicide in the soil solution. Although variations in soil moisture clearly influence the availability of solutes for degradation, they may also affect the metabolic activity of the organisms involved. Linn and Doran [41] studied the effect of different moisture levels on the microbial respiration rates of a silty clay loam soil. Respiration rates increased linearly with increasing water content up to a maximum corresponding to 60% water filled pore space (WFP) and decreased, thereafter. This was explained on the basis that, below 60% WFP, the availability of water limits microbial activity, but above 60%, water limits microbial activity as a result of reduced aeration. It is possible that the relationship between soil moisture (almost certainly in the $< 60\%$ WEP range) and degradation rates in Anderson's study was at least partly due to the effect of moisture on the rates of microbial metabolism and was not merely due to the changes in the amounts of the herbicides in the soil solution.

The effects of differing quantities of microbial biomass on the rates of degradation of diallate and triallate were also studied by Anderson [39, 42]. In the first study, the rates of degradation of the herbicides decreased with time during the study and were accompanied by a gradual decline in the levels of soil biomass. Anderson concluded that size of the biomass had a direct influence on the rates of degradation of the two herbicides. In the later study, Anderson manipulated the size of the soil biomass by either carbon-starvation or carbon-amendment before adding the herbicides and showed, again, a clear relationship between the rates of degradation and the levels of microbial biomass. Anderson stated that "it is not unreasonable to expect that the relationship between biomass and degradation rates also applies to other pesticides". Indeed it is not, and this relationship has become one of the basic principles upon which a great deal of contemporary pesticide microbiology is based.

The soil microbial biomass comprises a diverse population of fungi, actinomycetes and bacteria. It is worth considering, at this point, the basic requirements for the maintenance of viable populations of these organisms in the soil environment. In common with all other forms of life, soil microorganisms require nutrients for the synthesis of protoplasm and for the generation of energy to drive their normal physiological processes. The synthesis of protoplasm requires a carbon source, usually in organic form; lesser amounts of nitrogen, phosphorus and sulphur, and traces of various other elements. Phototrophic organisms also require light and carbon dioxide and, as a result, are generally found only at the soil surface. In order to generate energy, heterotrophic organisms require suitable supplies of electron donors, usually in the form of oxidisable organic compounds. Obligate aerobic organisms are only able to use molecular oxygen as an electron acceptor, but facultative anaerobes are also able to use nitrate, sulphate, carbon dioxide, or reducible organic compounds if oxygen is not available. Obligate anaerobes are only able to use electron acceptors other than oxygen. Autotrophic organisms may use oxidisable inorganic substances such as hydrogen, ammonia, nitrite, ferrous iron or reduced sulphur compounds as electron donors. The nutritional requirements of many different types of microorganisms are met by the conditions which prevail in productive agricultural soils to the extent that they are able to support a total microbial biomass, in the top soil, of typically greater than 2000 kg ha^{-1} [Riley, 43]. The microbial ecology of the soil is such that the populations will tend to proliferate until they become limited by the availability of one or other of the critical nutrients or growth factors.

The effects of differing activities of soil biomass on pesticide degradation are rather difficult to study because, to a large extent, degradation itself represents the most important measure of activity. However, in a more general sense, measures of microbial activity such as respiration rates may be used. In another study with diallate, Anderson [38] studied the influence of different microbial activities on degradation. Soils were treated with the compound and incubated under otherwise standard conditions at various temperatures between 4 °C and 25 °C for up to 10 weeks. During the course of the study, the degradation of the herbicide and the levels and activities of the soil biomass were monitored. Although soil biomass decreased with increasing temperature, the rates of diallate degradation and microbial activity both increased, allowing the author to conclude that the differences in degradation rates were due to the differing microbial activities. The influence of temperature on the rate of chemical reactions, in vitro, is shown by the Arrhenius equation,

$$k_T = Ae^{-E/RT} \tag{7}$$

where, k_T is the rate constant, A is a constant, E is the molar energy of activation, R is the gas constant and T is the temperature (K). Although there is no prima facie reason why this equation should apply to the metabolic capability of microorganisms, Walker [44] found it could be used to predict biodegradation rates for eight herbicides in soil over the temperature range of 10–30 °C. The effect of each 10 °C rise in temperature was generally to increase the rate of degradation by a factor of 2–3.

The microbial degradation referred to above arises primarily from the aerobic metabolism of the soil biomass. Under conditions of high aerobic microbial activity, oxygen levels may become depleted and result in the formation of anoxic or anaerobic conditions. Anaerobicity is characterised by a reduction in redox potential and an associated switch to alternative metabolic pathways in which substances other than oxygen are the electron acceptors. This switch to anaerobic metabolism usually occurs when the oxygen concentration falls below 3×10^{-6} M [Greenwood, 45]. The substrates for anaerobic microbial metabolism become reduced in the process and this often results in quite different rates and products of degradation than aerobic metabolism. This is particularly important for compounds with nitro [e.g. Adhya et al., 46] or haloaromatic [e.g. Suflita et al., 47] substituents which are often persistent in aerobic environments but readily degraded anaerobically.

Anaerobic conditions arise in soil whenever microbial demands for oxygen exceed supply. This is most likely to occur where there are plentiful supplies of readily oxidised organic carbon and where soil aeration is limited. The latter may occur in the bulk soil environment as a result of flooding but it is more often associated with microsites which exist within a generally aerobic soil environment. Sexstone et al. [48] showed that when aggregates of a silt loam soil were incubated in air, there was often a steep oxygen gradient over very small distances from the aggregate surface. Small aggregates were generally aerobic but those above 10 mm radius often had measurable anaerobic centres. The implications of anaerobic microsites for pesticide degradation were extensively reviewed by Haider [49].

The redox status of a soil is liable to change over relatively short periods of time in response to variations in aeration and oxygen consumption. This may have a profound effect on the degradation of any pesticides which require both aerobic and anaerobic reactions for their complete biodegradation. Fogel et al. [50] showed that methoxychlor was readily degraded by reductive dechlorination in anaerobic soil, but persistent in aerobic soils. However, when the anaerobic soil was converted to aerobic conditions after the primary degradation, mineralisation was much more rapid than from the soils which had been aerobic, throughout.

To summarise, the microbial degradation of pesticides in soils is most likely to occur under conditions which support high microbial activity. Thus, rapid and extensive aerobic degradation is most likely to occur in soils with moderate levels of organic matter and which are warm, moist, and well aerated. Anaerobic degradation may also be important at the centre of aggregates in well structured soils or during periods of prolonged waterlogging.

3.3 Water Flux and Pesticide Movement in Surface Soils

The majority of recharge to groundwater, in temperate climates occurs during the winter and early spring periods. During this time, there may be significant leaching of weakly adsorbed autumn-applied pesticides, compounded by generally low soil temperatures and, hence. low degradation rates. During the late spring and summer of the following year, however, the net water movement is likely to be upwards, and may result in the exposure of a following crop to residues of pesticides applied during the previous season. This has been cited by Duffy et al. [18] as the reason for the observed damage of sugar beet by residues of sulfonyl urea herbicides which had been applied to preceding cereal crops grown in poorly drained soils in the UK. In freely drained soils, however, autumn applications of weakly adsorbed pesticides are more likely to result in leaching to ground water than late spring or summer applications of the same compound.

The mobility in soil of pre-emergence pesticides raises some interesting issues. By their very nature, such compounds need to be present in the soil during the key growth stage of the crop or target weed species, which may be several weeks after they are applied. Such compounds also need to be available for uptake by plant roots. Consequently, pre-emergence pesticides with useful biological activity are often moderately persistent and poorly adsorbed, properties often associated with mobility in soil. The leaching potential of such compounds is enhanced by the fact that, during the period following their application, the soil is likely to be bare, thus giving no scope for transpiration to limit groundwater recharge. It is perhaps not surprising that pre-emergence treatments, particularly of herbicides, are often associated with the detection of residues of pesticides in ground water.

The flow of percolation water and any associated solutes from the root zone of soil is a complex process which is heavily influenced by the physical structure of the medium. At one time, light, unstructured, sandy soils with low organic matter contents and high hydraulic conductivities were thought to be the only soils at risk from the point of view of pesticide leaching. However, there has been a realisation in recent years [e.g. Hallberg, 51; Nicholls, 6] that soil structure is a key factor in determining water flux and that, rather than flow through the small pores within the aggregates of structured soils, water tends to move, in preference, through the larger channels which surround them. Thus, 'by-pass flow' or 'macropore flow' may result in a fraction of the percolating water reaching a certain depth much more rapidly than a 'chromatographic' flow model would suggest. This phenomenon may facilitate the transport of a pesticide to a subsoil zone in which it is neither readily degraded nor extensively adsorbed and probably accounts for much of the observed leaching of 'biodegradable' compounds. The productivity of poorly drained agricultural soils is often improved by the introduction of artificial drains which remove excess water from just below the plough layer. This may be considered as an artificial form of by-pass flow and may result in the creation of a link between the leaching of solutes from agricultural soils and surface water contamination.

If the maximum infiltration rate of a soil is exceeded during a particular rainfall or irrigation event, the excess soil moisture will remain at the surface and, depending on the topography, may result in runoff. Leonard [52] defined runoff as "water and any dissolved or suspended matter it contains that leaves a plot, field or small single-cover watershed in surface drainage". Thus, if there is surface water runoff in response to intense rainfall, pesticides may undergo lateral movement and, as a result, may enter surface waters. For pesticides with moderate solubilities and adsorption coefficients, the low sediment loading of most runoff ensures that such movement will occur predominantly in the aqueous phase. If a pesticide has a very high adsorption coefficient, it will only be subject to runoff if a rainfall event is sufficiently intense to cause erosion of the soil particles with which it is associated. This may not seem a particularly important mechanism of loss, but remarkable amounts of soil can be lost from highly erodible land during heavy rainfall. It has been estimated that over 23% of the cropped land in the eastern USA suffers from an annual loss of soil of >5 tons/acre due to erosion, about 2/3 of which is removed by water [Brady, 53].

In a review of the pesticide content of agricultural runoff, Wauchope [54] concluded that, for the majority of commercial pesticides, total losses are likely to be of the order of 0.5% of that applied unless severe rainfall occurs within 1–2 weeks after application. For soil applied wettable powder herbicides, up to 5% may be lost by runoff depending on the weather and slope. From the point of view of environmental impact, however, the concentrations of pesticides in runoff are probably more important than the amounts lost from treated fields. Concentrations may vary by an order of magnitude or more during a single runoff event and may be reduced considerably by dilution or adsorption between leaving the field and entering the receiving water. Nevertheless, the concentration of pesticides in surface waters, arising as a result of runoff, may be several orders of magnitude higher than those in ground waters, arising from leaching.

4 Pesticide Degradation and Mobility in Subsurface Soils

The fate of pesticides in surface soils has been extensively studied for many years, but the level of understanding of fate in sub-surface soils is not nearly as well developed. Just to collect deep underground samples of water and sediment for straightforward chemical analysis may present severe technical problems, but even these are surpassed by the difficulty of securing uncontaminated samples of sub-soil and ground water for microbial ecology studies. However, despite these problems, there has been a steady increase in the number of publications on this subject in recent years and the state of knowledge is currently advancing rapidly.

4.1 Degradation in Subsurface Soils

In one of the first comprehensive reviews of subsurface microbial activity, McNabb and Dunlap [55] discussed the limited information available up to the mid-1970s, and considered the likely impact of various subsurface environmental factors on the suitability of the medium as a microbial habitat. Whilst recognising the scarcity of definitive experimental data, the authors suggested that the physical nature of the subsurface region, in terms of its likely osmotic pressure, temperature, hydrostatic pressure, pH and redox potential, did not generally represent an inherently hostile environment for microorganisms. McNabb and Dunlap also considered the likely availability of nutrients to microorganisms in subsurface zones and concluded that although the levels of readily utilisable carbon sources, electron acceptors, nitrogen and phosphorus were low, they were probably sufficient to maintain viable populations.

Many of the methods used to detect microorganisms in subsurface samples and to determine their biomass and activities have been adapted from soil and aquatic microbiology. Ghiorse and Wilson [56], summarised the work on subsurface microbiology reported between 1977 and 1987 which suggested that pristine sites generally had low microbial population densities (10^5–10^7 cells per gram) and activities compared to surface soils, presumably reflecting the oligotrophic nature of the environment. It is not surprising, perhaps, that shallow subsurface zones and aquifers (< 50 m) have been the most studied. Balkwill and Ghiorse [57], employed direct microscopy methods to characterise the bacteria associated with two aquifers in Oklahoma, USA, at which the water table was only 3 m below the surface. Total cell counts between 2.9×10^6 and 9.8×10^6 cells per gram were determined and the populations subsequently shown to comprise predominantly aerobic and nutritionally versatile bacteria which could survive on low concentrations of organic compounds ($< 10\ \mathrm{mg\,l^{-1}}$) without forming specialised resting cells. Microbial isolates from this site were capable of aerobically degrading a wide range of naturally occurring compounds [Swindoll et al., 58] such as acetic acid, amino acids, cellulose, cinnamic acid, glucosamine and glucose, with rate constants between 10^{-2} and $10^{-4}\ \mathrm{h^{-1}}$. Several xenobiotic compounds were also degraded, but generally more slowly, with rate constants between 10^{-3} and $10^{-6}\ \mathrm{h^{-1}}$. The organisms appeared to be capable of degrading much higher concentrations of these substances than they are ever likely to be exposed to in situ, presumably in order that they are able to make optimal use of any pulses of nutrients they encounter in their normally nutrient limited environment.

Even though the metabolism of subsurface microorganisms at pristine sites is often carbon-limited, Smith et al. [59], showed that large proportions of the available carbon are often diverted into storage products and exopolysaccharide rather than growth. Storage products such as polyhydroxybutyrate may be essential components of the survival strategies of subsurface bacteria, whereas, exopolysaccharides are thought to help the cells attach to surfaces and absorb and concentrate organic nutrients dissolved in the water phase.

Substantial microbial populations have also been found in much deeper subsurface sediments. Balkwill et al. [60], reported 10^5–10^8 viable cells per gram in sediment cores taken from depths to 265 m from a site in South Carolina, USA. The bacteria were very diverse and could assimilate a variety of carbon sources via oxidative pathways. The bacteria from different depths and, to a lesser extent, different boreholes differed in their metabolic activities although, generally, they were able to metabolise a narrower range of carbon sources than their counterparts in surface soils. Sinclair and Ghiorse [61], studied the same site and also found significant numbers (10^6–10^7 cells per gram) of diverse microorganisms in the subsurface layers. Microbial numbers did not decline with depth but, in the various layers of the profile, were positively correlated with sand content and pore water pH, but negatively correlated with clay and pore water metal concentration. The microbiology of deep subsurface environments has been studied by other workers [e.g. White et al. 62; Weirich and Schweisfurth, 63; Chapelle et al. 64] who generally reported population densities of the similar orders of magnitudes to those described above.

In sites with gross subsurface contamination, the extent of biodegradation of a pollutant may be heavily influenced by the availability of suitable electron acceptors, even if the biomass is adapted to degrade it. Since rates of oxygen transport may limit the rates of aerobic metabolism in these circumstances, it is common for anaerobic degradation to be more important here than at non-polluted sites. Suflita et al. [65] studied the microbial ecology of an anoxic aquifer beneath a land-fill site where the water table was 1–2 m beneath the surface. At least two spatially distinct zones existed within the aquifer; electron transport was predominantly by methanogenesis at one site, and by sulphate reduction, at the other. Haloaromatic compounds and tetrachloroethylene were degraded by reductive dehalogenation under methanogenic but not sulphate reducing conditions. In contrast, alkylphenols were degraded faster under sulphate reducing conditions. Factors such as these are unlikely to be important for pesticides unless the aquifer is also contaminated with high levels of other substances.

Although there are many reports of the detection of residues of pesticides in groundwater, there are relatively few concerning their degradation below the soil surface. One of the first pesticides to be found in groundwater was the nematicide, aldicarb [Cohen et al. 66] and the fate of this compound has subsequently been extensively studied in both subsoils and aquifers. Bromilow et al. [67] showed that, in aerobic soils, aldicarb was rapidly oxidised to the corresponding sulfoxide and sulfone (aldoxycarb) with concomitant hydrolysis to oximes. Smelt et al. [68] compared the degradation of the aldicarb metabolites in moist aerobic soils from above the water table with that in anaerobic flooded soils from beneath the water table. The rates of degradation of the two compounds in the subsoils were rapid with half-lives ranging from 5.1 to 131 days, but were from 8 to >100 times slower in the surface soils from the same profile. These data are rather different from those of Miles and Delfino [69] who found slower rates of degradation of aldicarb in groundwater microcosms under anaerobic conditions. A possible explanation for this was provided by Bromilow et al. [70] who showed that ferrous iron could have a pronounced catalytic effect on the degradation of the carbamoyloximes in anaerobic soils. Ou et al. [71] showed that the degradation of aldicarb in deep subsurface samples (150–300 cm) was dominated by chemical hydrolysis, although microbial oxidation was more important in surface soils.

The fate of the chloracetanilide herbicides has also been studied in soils from different depths, and in groundwater. Novick et al. [72] found that both alachlor and propachlor were microbially degraded in aquifer samples, albeit more slowly than in surface soils. Pothuluri et al. [73] studied the rates of alachlor degradation, under both aerobic and anaerobic conditions, in the surface soil,

Table 2. Degradation of alachlor in aerobic and anaerobic soils and aquifers [Pothuluri et al. 73]

Soil depth (m)	aerobic $T_{1/2}$ (d)	anaerobic $T_{1/2}$ (d)
	surface (root zone included)	
0.0–0.6	23	100
	vadose zone	
0.6–2.4	73	144
2.4–4.3	104	110
4.3–6.3	116	53
6.3–8.3	22	57
8.3–9.6	285	148
	aquifer	
11.6–12.2	324	337
13.1–15.2	320	553

the vadose zone and aquifer material from a site in Georgia, USA (Table 2). Apart from the unexplained discontinuity at the 6.3–8.3 m horizon, these data show that the rates of aerobic degradation decreased with increasing depth and were significant, albeit very slow, in the aquifer. Anaerobic degradation was generally slower than aerobic degradation. The addition of organic nutrients increased the rates of degradation in the aerobic subsoils and the upper aquifer material, suggesting that nutrient limitations may have been at least partially responsible for the slow degradation in these zones.

Several workers have reported data on the fate of the triazine herbicides in subsurface soil layers. Lavy et al. [74], studied the degradation of atrazine at 15, 40 and 90 cm depths at two sites in Nebraska, USA, and found the degradation rate increased with depth. Wehtje et al. [75] reported evidence of chemical hydrolysis of atrazine, but no microbial degradation, under simulated aquifer conditions. Wood et al. [76], studied the fate of atrazine in several groundwater and aquifer sediment samples. Half-lives of the order of 15–20 weeks were found in groundwater samples, together with extensive degradation during a 10 week incubation with the sediments. There was some evidence of a lag-phase for the degradation in sediments and microbial degradation was indicated, albeit that the concentration of atrazine employed was several orders of magnitude higher than the likely concentration in groundwater, in situ. Helweg [77], reported that atrazine was mineralised to the extent of about 13% in each of two subsoils during a 626 day incubation, although there was significant variation from one replicate to another. Moorman and Harper [78], studied the degradation of metribuzin in soil samples from the 0–10, 35–80 and 125–150 cm depths of a Mississippi delta soil. Degradation was faster, and with more mineralisation, in the surface soil. The lower degradation rates in the subsurface soils corresponded to lower microbial populations.

Walker et al. [79], reported differences in the extents of adsorption and rates of degradation of the sulfonyl urea herbicides chlorsulfuron and metsulfuron-methyl in soil from the 0–20, 20–40 and 40–60 cm depths from eight UK sites. With the exception of the site with a very acidic, organic soil, adsorption coefficients and degradation rates tended to decrease, with increased depth. Degradation rates were positively correlated with microbial biomass.

The fate of the soil fumigant, ethylene dibromide (EDB), in shallow aquifer materials was studied by Pignatello [80]. Ethylene dibromide was degraded under various redox conditions in all except sterile samples. Degradation was faster in samples from a contaminated site, suggesting that adaptation of the microorganisms to the chemical was possible. The products of aerobic degradation were similar to those found in surface soils, although the slower rates of degradation at depth (half-lives 35–350 days) reflected the lower temperatures and microbial numbers.

Helweg [81] showed the herbicide MCPA was degraded much faster in surface soils than in soils from a depth of 1 m, although degradation was still significant in the subsoil, with 15–20% mineralisation during a 308 day incubation. In another interesting experiment with MCPA, Bek Zeuthen et al. [82], showed that mineralisation in a subsoil from 1 m below a barley field with a long history of phenoxy herbicide application was much more extensive (40% $^{14}CO_2$ in 80 days) than in the subsoil from an adjacent untreated area (5% $^{14}CO_2$ in 250 days), thus providing a clear indication that microbial adaptation may take place in the subsurface environments.

Campbell et al. [83] recently reported the results of a study on the factors influencing the rates of degradation of a model arylamide compound in subsoils. *N*-(4-nitrophenyl)propanamide was chosen because it can be hydrolysed or reduced. Degradation in three soils occurred principally by microbial hydrolysis and was 10–30 times faster in surface soils than at 2 m. In some subsoils, degradation was preceded by a 3–6 day lag-phase, indicating that the microbial populations were able to adapt to degrade the compound.

There are now sufficient good examples of pesticide degradation in subsoils to suggest that the phenomenon is more widespread than had been realised until only recently. Whilst degradation rates may be slow compared to those in surface soils, they may be very significant compared to the long residence time of water in aquifers. Microbial degradation of pesticides at considerable depth has been clearly demonstrated, as has the potential for adaptation to degrade novel compounds more rapidly. However, all the examples cited above, except the last one, relate to studies with pesticides which had previously been detected in subsurface soil layers. It is possible, therefore, that whilst these data provide evidence that pesticides may be degraded in subsoils and aquifers, that they are biased towards those mobile compounds which are degraded the most slowly in these environments. There may be other examples of pesticides which are intrinsically mobile but more readily degraded in subsoils which have not yet been studied.

4.2 Water Flux and Pesticide Movement in Subsurface Soils

Once percolation water in soil moves beyond the root zone it ceases to be susceptible to the influence of evapotranspiration and is likely to move predominantly downwards, through the unsaturated or vadose zone, until it reaches the water table. The annual rates of recharge to groundwater vary enormously according to the prevailing meteorological conditions and cultural practices, but are likely to be of the order of 25 cm per year in temperate regions such as Western Europe. The rate of downwards movement through the unsaturated zone obviously depends on the water content of the medium; if this is 25%, the piston flow model indicates the rate of movement will be 100 cm per year. In the absence of significant by-pass flow, the time taken for percolation water to reach the water table may therefore vary from a few months to several decades, depending on the depth of the aquifer. In soils with high by-pass flow, a single heavy rainfall event may be sufficient to transport water from the surface to a shallow aquifer.

It follows from the above that the requirements for protecting ground water from contamination as a result of pesticide leaching are diverse. For shallow aquifers, essentially any leaching may result in contamination and any pesticides which are not at least moderately adsorbed, and rapidly degraded, in surface soils are likely to reach ground water. On the other hand, if the aquifer is several hundred meters below the surface, there may be ample time for even relatively persistent and poorly adsorbed compounds to be totally degraded before they reach the saturated zone. Even degradation rates and adsorption coefficients which are several orders of magnitude lower than those encountered in surface soils may, therefore, be sufficient to prevent significant ground water contamination as a result of pesticide leaching.

Water flux in the saturated zone depends on the hydraulic gradient, hydraulic conductivity and porosity of the aquifer. Ground water velocities vary over many orders of magnitude and may result in the transport of unadsorbed solutes for considerable distances from their point of entry into the aquifer. The residence times of water in aquifers may be many years and, again, even very slow degradation rates may be sufficient to ensure pesticide residues are never detected in abstracted water supplies.

5 Prediction of Pesticide Mobility in Use

There is an obvious need to be able to predict the mobility of pesticides before their use and subsequent movement results in the contamination of either ground or surface waters. The various methods which may be used to do this are mentioned only briefly here as they are reviewed in detail elsewhere in this volume.

5.1 Prediction of Leaching

It is quite reasonable to expect that any pesticide which is weakly adsorbed, and persistent is likely to leach to subsoils and ground water. Equally, a compound with a half-life of only a few days in soils and subsoils is never likely to leach appreciably to groundwater, even if it is only weakly adsorbed. Compounds which have intermediate persistence, moderate adsorption and low application rates pose a more difficult problem, however, and Nicholls [6] recently proposed that it might be prudent to assume that up to 0.1% of compounds with a $\log K_{ow} < 4$ and a half-life of > 10 days are capable of leaching to shallow aquifers.

Cohen et al. [84] developed a set of criteria (stated below) for predicting whether certain pesticides could leach to ground water as a result of normal agricultural use in hydrogeologically sensitive areas.

Mobility criteria:
(1) Water solubility $> 30\ \mathrm{mg\,l^{-1}}$.
(2) $K_d < 5$ and usually < 1–2.
(3) $K_{oc} < 300$–500.
(4) Henry's Law constant $< 10^{-2}$ atm–m^3/mol.
(5) Speciation – negatively charged at ambient pH.

Persistence criteria:
(1) Hydrolysis half-life > about 25 weeks.
(2) Photolysis half-life > 1 week (while pesticide is on surface).
(3) Soil half-life > 2–3 weeks.

When a pesticide met all these criteria, it was considered to have a high potential for groundwater contamination. Experts may disagree about the relevance of some of these criteria, but there is little doubt that they can be used to differentiate between those pesticides which may leach from those which almost certainly will not. According to Cohen [66], no pesticide which failed to meet or exceed these criteria has been detected in groundwater. This may be true for leaching in poorly structured soils, but these criteria cannot be expected to predict leaching due to extensive by-pass flow, especially if this occurs soon after pesticide application.

A simple scoring system which uses the degradation half-life of a pesticide in soil, $t_{1/2}$, and its partition coefficient between soil organic carbon and water, K_{oc}, but not the use rate or physical properties such as water solubility and volatility, has been proposed by Gustafson [85]. Pesticides with scores for the resulting so called "Groundwater Ubiquity Score", or "GUS index", of less than 1.8 are considered as improbable leachers and those with a GUS index of greater than 2.8 as probable leachers. The GUS index is defined as follows:

$$\mathrm{GUS} = \log_{10}(t_{1/2}) \times (4 - \log_{10}(K_{oc})) \tag{8}$$

Approaches such as these may give a good indication of whether a pesticide may leach but they cannot predict the quantities which may leach under a given set of conditions, or the resulting concentrations in groundwater. A number of sophisticated computer models have been developed in attempts to overcome this problem [e.g. Leistra, 86; Nicholls et al. 87; Carsel et al. 88; Jury et al. 89]. These all simulate leaching from a large number of input parameters, representing the characteristics of the pesticide, the intended site of use and the prevailing weather. Before a simulation model may be used, with confidence, to predict leaching, it must first be validated by comparison with appropriate field measurements. Attempts to do this have met with varying degrees of success [e.g. Wagenet, 90] but models can be valuable research tools and are being increasingly used by regulators to trigger large scale field studies. However, simulation models have not yet been developed to the point they may be relied upon to make important regulatory decisions in the absence of such field data.

Simulation models of the type referred to above may be able to predict the extent of pesticide leaching, given a scenario of a particular product used at a specific site under a defined set of weather conditions. Even if they achieve this with a high level of accuracy, the problem remains of relating the prediction to other scenarios which have not been specifically addressed by the model. However, approaches have been developed which allow the pollution potential from pesticide leaching to be considered over very large areas. Aller et al. [91] described a system which uses a combination of weighted factors to produce a numerical value known as the DRASTIC index which may be used to help prioritise the areas most at risk from leaching. DRASTIC is an acronym for the factors which are considered:

D – Depth to water
R – (Net) Recharge
A – Aquifer Media
S – Soil Media
T – Topography
I – Impact of the Vadose Zone
C – (Hydraulic) Conductivity of the Aquifer.

Hollis [92] proposed a system for mapping the vulnerability of aquifers and surface waters at the national/regional scale. Soil vulnerability classes were built up by integrating a soil and substrate hydrological classification with a soil classification based on organic matter content. This, in turn, was integrated with a classification of pesticides based on their relative mobility and persistence. Climatic and land use variables were then included to give an overall vulnerability assessment for various classes of pesticides at the national (UK) level. Various aspects of this model still need to be validated but it represents an interesting approach to the identification of those areas where water is most at risk from contamination with mobile pesticides.

5.2 Prediction of Runoff

One of the first attempts to predict the extent of pesticide runoff from treated land was made by Wauchope and Leonard [93]. The 'edge-of-field' concentrations of pesticides in bulk runoff, C_t (in ppb), were predicted from the semi-empirical relationship shown below:

$$C_t = AR(1 + 0.44t)^{-1.6} \tag{9}$$

where, A is the 'availability index' (ppb ha kg^{-1}), R is the application rate (kg ha^{-1}) and t is the time (days) between application and the runoff event. The 'availability index', A, was assigned on the basis of the properties of the pesticide and ranged from 300–10 000 ppb ha kg^{-1}. Thus, a pesticide such as atrazine ($A = 300$ ppb ha kg^{-1}), applied at 1 kg ha^{-1} would be predicted to give a runoff concentration of 10 ppm if rainfall occurs immediately after application. This prediction is close to the upper limit of a range of measured values [Wauchope, 94] and therefore represents a worst-case concentration at the edge of a treated field. The simplicity of this approach was justified by the authors on the basis that the predictions it provides are as good as the weather predictions will allow.

Simulation models have also been developed to predict pesticide runoff. Leonard [95] described the chronological development of these and reviewed their applications; he concluded, however, that no model had been developed that could give accurate predictions of pesticide concentrations in runoff. Since absolute accuracy is not yet achievable, Leonard suggested the current models should only be used to aid decisions by providing comparative estimates of pesticide losses by runoff. If used in this way, they are able to provide useful insights into the likely effects of different soil management practices, or to rank different pesticides in terms of their susceptibility for runoff.

6 Summary and Conclusions

The fate of pesticides in surface soils has been extensively studied for many years as an integral part of the environmental safety assessments carried out by the manufacturers and government regulators. The factors which affect the relative rates of degradation and movement of pesticides in surface soils have become well established during this time but, until recently, most environmental scientists had paid scant regard to the behaviour of pesticides in subsoils. Those who had considered the subject at all, probably presumed that the subsoil environment was essentially sterile and that microbial degradation, which is so important in surface soils, was not feasible at depth. The detection of residues of pesticides in ground water, from the late 1970s onwards, has stimulated research and prompted a major re-evaluation of this presumption.

Although the sheer inaccessibility of the subsoil environment poses some difficult technical problems, methods have been developed to obtain experimental samples from the deep vadose and saturated zones. Adaptations of the techniques of soil and aquatic microbiology have revealed significant populations of diverse microorganisms (typically of the order of 10^6 cells g^{-1}) in such samples and there is an increasingly convincing body of evidence showing they are able to metabolise a wide range of natural and xenobiotic compounds, including pesticides. Although degradation rates are often much slower in subsoils than at the surface, they may be sufficient to reduce significantly pesticide residue levels in ground water, particularly in deep aquifers.

In the majority of the subsoil degradation studies which have been conducted with pesticides, only the disappearance of the parent compound and, occasionally, mineralisation rates have been determined. Clearly, further work is needed if we are fully to understand the mechanisms involved and the nature of the degradation products formed. However, it is important that any new work should be conducted with realistic concentrations of pesticides as the nutritional balance of the normally oligotrophic subsurface environment may be easily perturbed by the demands of high substrate concentrations. Most of the existing studies on pesticide degradation in subsoils have been conducted with those compounds which have already demonstrated their potential to leach in normal use. Such compounds are likely to be only slowly degraded, at best, and there may be other intrinsically mobile pesticides which have not yet been studied, but which are more readily degradable and which may provide additional useful insights into the processes involved.

The prediction of pesticide mobility beyond the root zone is still largely an imprecise science, but there is little doubt this will improve along with our understanding of the basic transport and degradative mechanisms. With better quality input data, further refinement, and validation, sophisticated computer simulation models are bound to play an increasingly important part in this process.

7 References

1. Ritter WF (1990) J Env Sci Health, Part B 25: 1
2. Leistra M, Boesten JJTI (1989) Agric Ecosystems Environ 26: 369
3. Fielding M (1991) Pesticides in ground and drinking water. Water Pollution Research Report, 27. Commission of the European Communities
4. World Health Organisation (1984) Guidelines for drinking-water quality. vol 1. Recommendations. WHO, Geneva, Switzerland

5. Council of the European Communities (1980) Directive of the Council of 15th July 1980 relating to the quality of water intended for human consumption. The Official Journal of the European Communities L229, 30th August 1980.
6. Nicholls PH (1990) In: Jones KC (ed) Organic Contaminants in the Environment. Elsevier Applied Science, London and New York, p 87
7. OECD (1981) Guideline for Testing of Chemicals No 106 "Adsorption/Desorption"
8. Nicholls PH Evans AA (1985) Proc Brit Crop Prot Conf Weeds p 333
9. Bewick DW, Bharti H, Vickers JA (1987) 11th International Congress of Plant Protection, 5–9 Oct, Manilla
10. Knight BAG, Tomlinson TE (1967) J Soil Sci 18: 233
11. Nicholls PH (1988) Pestic Sci 22: 123
12. McCall PJ, Swann RL, Laskowski DA, Unger SM, Vrona SA, Dishburger HJ (1980) Bull Environ Contam Toxicol 24: 190
13. Lyman WJ, Reehl WF, Rosenblatt DH (1982) Handbook of Chemical Property Estimation Methods. McGraw-Hill, New York
14. Sprankle P, Meggitt WF, Penner D (1975) Weed Sci 23: 229
15. Hance RJ (1976) Pestic Sci 7: 363
16. Subramaniam V, Hoggard PE (1988) J Agric Food Chem 36: 1326
17. Swanson RA, Dutt GR (1973) Soil Sci Soc Amer Proc 37: 872
18. Duffy MJ, Hanafey MK, Linn DM, Russell MH, Peter CJ (1987) Proc Brit Crop Prot Conf Weeds p 541
19. Schinkel K, Nolting HG, Lundehn SR (1986) Guidelines for the Official Examination of Plant Protection Products, Part IV, 4–1. Biologische Bundensanstalt fur Land und Forstwirtschaft. Braunschweig, Germany
20. Grunwell JR, Erickson RH (1973) J Agric Food Chem 21: 929
21. Bewick DW (1986) Pestic Sci 17: 349
22. United States EPA (1982) Pesticides Assessment Guidelines, Subdivision N
23. Klehr M, Iwan J, Reimann J (1983) Pestic Sci 14: 359
24. Parker S, Leahey JP (1988) Proc Brit Crop Prot Conf Weeds p 663
25. Parlar H (1990) In: Hutson DH and Roberts TR (eds) Environmental Fate of Pesticides (Progress in Pesticide Biochemistry, Vol 7). Wiley, Chichester, p 245
26. Beyer EM, Brown HM, Duffy MJ (1987) Proc Brit Crop Prot Conf Weeds p 531
27. Gohre K, Miller GC (1986) J Agric Food Chem 34: 709
28. Williams PP (1977) Residue Rev 66: 62
29. Patrick WH, Mahapatra IV (1968) Adv Agron 20: 323
30. Hill IR (1978) In: Hill IR and Wright SJL (eds) Pesticide Microbiology. Academic, London, p 137
31. Aizawa H (1982) Metabolic Maps of Pesticides, Academic, New York
32. Arnold DA, Briggs GG (1990) In: Hutson DH and Roberts TR (eds) Environmental Fate of Pesticides (Progress in Pesticide Biochemistry, Vol 7), Wiley, Chichester, p 101
33. Horvath RS (1972) Bacteriol Rev 32: 146
34. Audus LJ (1950) Nature 166: 365
35. Fellsott A, Maddox JV, Bruce W (1981) Bull Environ Contam Toxicol 26: 781
36. Alexander M (1985) Environ Sci Technol 19: 106
37. Wagenet RL, Hutson JL, Biggar JW (1989) J Env Qual 18: 78
38. Frehse H, Anderson JPE (1983) In: Greenhalg R, Drescher N (eds) IUPAC Pesticide Chemistry, Human Welfare and the Environment, Vol 4, Pesticide Residues and Formulation Chemistry, p 23
39. Anderson JPE (1981) Soil Biol Biochem 13: 161
40. Walker A (1976) Pestic Sci 7: 41
41. Linn DM, Droan JW (1984) Soil Sci Soc Am J 48: 1267
42. Anderson JPE (1984) Soil Biol Biochem 16: 483

43. Riley D (1991) In: Walker A (ed) Pesticides in Soil and Water: Current Perspectives. BCPC Monograph No 47, p 11
44. Walker A (1987) Weed Res 27: 143
45. Greenwood DJ (1961) Plant Soil 14: 360
46. Adhya TK, Barik S, Sethunathan N (1981) J Agric Food Chem 29: 90
47. Suflita JM, Horowitz A, Shelton DR, Tiedje JM (1982) Science 218: 1115
48. Sexstone AJ, Revsbech NP, Parkin TB, Tiedje JM (1985) Soil Sci Soc Am J 49: 645.
49. Haider K (1983) In: Greenhal R, Drescher N (eds) IUPAC Pesticide Chemistry, Human Welfare and the Environment, Vol 3, p 351
50. Fogel S, Lancione RL, Sewall AE (1982) Appl Env Microbiol 44: 113
51. Hallberg GR (1989) Agric Ecosyst Environ 26: 299
52. Leonard RA (1990) In: Cheng HH (ed) Pesticides in the Soil Environment: Processes, Impacts, and Modelling. Soil Science Society of America, Madison, Wisconsin, USA, p 303
53. Brady NC (1990) The nature and properties of soils, 10th edn, Macmillan New York.
54. Wauchope RD (1978). J Env Qual
55. Mcnabb JF, Dunlap WJ (1975) Groundwater 13: 33
56. Ghiorse WC, Wilson JT (1988) Adv Appl Microbiol 33: 107
57. Balkwill DL, Ghiorse WC (1985) Appl Environ Microbiol 50: 580
58. Swindoll CM, Aelion CM, Dobbins DC, Jiang O, Long SC, Pfaender FK (1988) Environ Toxicol Chem 7: 291
59. Smith GA, Nickels JS, Kerger BD, Davis JD, Collins SP (1986) Can J Microbiol 32: 104
60. Balkwill DL, Fredrickson JK, Thomas JM (1989) Appl Env Microbiol 55: 1058
61. Sinclair JL, Ghiorse WC (1989) Geomicrobiol J 7: 15
62. White DC, Fredrickson HF, Gehron MJ, Smith GA, Martz RF (1983) Dev Ind Microbiol 24: 189
63. Weirich G, Schweisfurth R (1983) Geomicrob J 4: 1
64. Chapelle FH, Zelibor JL, Grimes DJ, Knobel LL (1987) Water Resource Res 23: 1625
65. Suflita JM, Miller GD (1985) Environ Toxicol Chem 4: 751
66. Cohen SZ, Eiden C, Lorber MN (1986) In: Garner WY, Honeycutt RC, Nigg HN (eds) Evaluation of Pesticides in Ground Water, ACS Washington DC, p 170
67. Bromilow RH, Baker RJ, Freeman MAH, Gorog K (1980) Pestic Sci 11: 371
68. Smelt JH, Dekker A, Leistra M, Houx NWH (1983) Pestic Sci 14: 173
69. Miles CJ, Delfino JJ (1985) J Agric Food Chem 33: 455
70. Bromilow RH, Briggs GG, Williams MR, Smelt JH, Tuinstra LGMT, Traag WA (1986) Pestic Sci 17: 535
71. Ou LT, Rao PSC, Edvardsson KSV. Jessup RE, Hornsby AG (1988) Pestic Sci 23: 1
72. Novick NJ, Mukherje R, Alexander M (1986) J Agric Food Chem 34: 721
73. Pothuluri JV, Moorman TB, Obenhuber DC, Wauchope RD (1990) J Environ Qual 19: 525
74. Lavy TL, Roeth FW, Fenster CR (1973) J Environ Qual 2: 132
75. Weijte GR, Spalding RF, Burnside OC, Lowry SR, Leavitt JRC (1983) Weed Sci 31: 610
76. Wood M, Harold J, Johnson A, Hance R (1991) In: Walker A (ed) Pesticides in Soil and Water Current Perspectives. BCPC Monograph No 47, p 1750
77. Helweg A (1990) 7th Congress of Pesticide Chemistry, IUPAC, Hamburg, 07B–56
78. Moorman TB, Harper SS (1989) J Environ Qual 18: 302
79. Walker A, Cotterill EG, Welch S (1989) Weed Res 29: 281
80. Pignatello JJ (1987) J Environ Qual 16: 307
81. Helweg A (1987) Tidsskr. f. Planteavl, Beretning S 1881, 78
82. Bek Zeuthen S, Madsen T, Helweg A (1988) 5 Danske Plantevaernskonference, I, 23

83. Campbell AT, Nicholls PH, Bromilow RH (1991) In: Walker A (ed) Pesticides in Soil and Water: Current Perspectives. BCPC Monograph No 47, p 155
84. Cohen SZ, Carsel RF, Creeger SM, Enfield CG (1984) In: Krueger RF, Sieber JN (eds) Treatment and Disposal of Pesticide Wastes, ACS Washington DC, p 297
85. Gustafson D (1989) J Environ Toxicol Chem 8: 339
86. Leistra M (1978) Plant Soil 49: 569
87. Nicholls PH, Walker A, Baker RJ (1982) Pestic Sci 13: 484
88. Carsel RF, Mulkey LA, Zorber MN, Baskin LB (1985) Ecol Modelling 30: 49
89. Jury WA, Focht DD, Farmer WJ (1987) J Env Qual 16: 422
90. Wagenet RJ, Rao PSC (1991) In: Cheng HH (ed) Pesticides in the Soil Environment: Processes, Impacts, and Modelling. Soil Science Society of America, Madison, Wisconsin. p 351
91. Aller L, Bennett T, Lehr JH, Petty RJ, Hawkett G (1987) National Technical Information Service, Springfield, Virginia, USA, EPA 600/2-87-035
92. Hollis J (1991) In: Walker A (ed) Pesticides in Soil and Water Current Perspectives. BCPC Monograph No 47, p XXX
93. Wauchope RD, Leonard RA (1980) J Environ Qual 9: 665
94. Wauchope RD (1978) J Env Qual 7: 459
95. Leonard RA (1991) In: Cheng HH (ed) Pesticides in the Soil Environment: Processes, Impacts, and Modeling. Soil Science Society of America, Madison, Wisconsin. p 303

Strategies to Minimize the Pollution of Water by Pesticides

A Critical Review to Plant Protection Measures in the Past, Present and Future

Christian Guyot

Rhone-Poulenc Agro, Centre de Recherche de la Dargoire, 14–20, rue Pierre Baizet, P.O. Box 9163, 69263 Lyon Cedex 09, France

Strategies aimed at minimizing water pollution by agricultural chemicals are reviewed. They involve all possible participants in the areas of agriculture, industry and regulations. Contribution of low input policies calling for optimized use of products is discussed. Significant decrease of pollution incidents in catchments is expected from careful handling of products and farmers education. In the long range, development of modern formulations containing new active ingredients effective at low dose rate and more environment-friendly will also limit non-point source contamination in conjunction with the development of site-specific management practices to protect surface and groundwater. Complementary to the strategies regarding the use of pesticides, possible measures designed to limit contaminant transfer in the watersheds such as implementation of buffer zones are outlined. Modern water treatment processes required to comply with stringent quality standards for drinking water are described. Applicability of multifaceted strategies is considered in a context of reinforced pesticide regulations, evolving farm agro-economic setting and emotional reactions of public opinion to environment protection issues.

1 Crop Protection and Protection of Water Resources

Systematic use of agricultural chemicals in intensive agriculture has resulted in contamination of environmental compartments not directly exposed to treatments and particularly water. Since the detection in the late 1970s of certain pesticides in groundwater, organic microcontaminants from agricultural chemicals have been found in increasing number in natural waters, particularly in developed countries where intensive agriculture makes extensive use of crop protection products. Until the early 1980s most people thought that soil and rock formations protected groundwater from contamination by pesticides. By 1988, the U.S. Environmental Protection Agency had detected 74 active ingredients in the groundwater of 38 states. Of these, 21 were herbicides that the agency could confirm as due to normal agricultural use. Once contaminated with molecules which do not degrade rapidly, groundwater is extremely difficult and expensive to clean up. Because it is slow to recharge, some contaminations are likely to remain for long periods, typically several years, even after remedial action is taken. The degree of dependence on groundwater as a source of drinking water varies widely from country to country but is always significant in proportion to total needs. The presence of agricultural chemicals in drinking water has also aroused considerable concern. Pesticides may also cause other adverse environmental effects, including surface water contamination. The increasing awareness of potential problems associated with surface and groundwater contamination has led to the establishment of monitoring programs. Water surveys in Northern America [1,2], Europe [3,4,5,6,7,8] and continuous monitoring of raw waters by water companies show that contamination of surface and groundwater occurs in some circumstances.

As problems became more patent, important information was gained on pesticide behavior in soil and water through many laboratory and field studies conducted or sponsored by industry, universities and regulatory agencies. Basic mechanisms of pesticide degradation and transport are well understood now.

At the same time, stringent regulations aimed at maintaining high quality standards for drinking water, probably to an extent far more stringent than for any other food product, were established. For instance, limits set by the European Directive 80/778/EEC (parameter 55) on potable water reflects that pesticide residues have no place in drinking water. These extremely high standards on drinking water quality will have an important impact on the use of some agricultural chemicals. Regulatory actions against certain pesticides clearly identified as water contaminants or against crop protection products as a whole have been taken in the United States and in Europe. Despite the fact that pesticides are already considered as the most stringently regulated products among chemicals commonly used, the trend is a general movement towards strengthened regulations regarding registration and use of agricultural chemicals. Product ban or restricted use are the short and medium term consequences of these regulations.

Paradoxically, pesticides in drinking water do not pose severe human health problems. Conversely, exposure of aquatic organisms to higher chemical concentrations in surface water has led to a significant degradation of biota in certain areas. Crop protection products are certainly not the only agents responsible for this alteration. Probably because of the negative way these products are perceived and their possible effects on living organisms, they are considered as more dangerous than other chemicals. Public opinion is very receptive to information largely conveyed by media on environmental issues but remains globally unaware of reality. Public knowledge is generally based on a few spectacular, distorted and excessively emphasized cases. However the influence public opinion is having on pest control technologies is certainly increasing and channelled through ecologist pressure groups. Regardless of the emotional reactions that contamination of natural waters may induce, protection of the water resources is nevertheless an urgent need, not only to ensure a sufficient supply for a continuously increasing demand in high quality drinking water, but also to maintain intact a precious natural heritage.

The setting for discovery of new agricultural chemicals is now bound by severe regulations, leaving only limited freedom for pesticide manufacturers. Legal actions are likely to be extended to pesticide end-users regarding product and land use. Many steps have been already implemented, are being prepared or could be imagined to minimize the impact of pesticide use on water quality. There are several areas where improvements can be made including the discovery of new active ingredients, various methods of formulating agrochemicals, application systems, correct and optimum use of pesticides by farmers, modified management practices and users' education. Strategies can therefore be developed in many areas to improve the situation and will involve various actors: pesticide manufacturers for discovery of new active ingredients and formulation development, farmers for product use, public authorities for land and water management, water companies for raw water treatment.

Possible strategies are related to different activities and fall upon the corresponding participants. They are not independent of each other and there is little chance that one could bear fruit on its own if relevant steps are not simultaneously taken in other areas. This is the consequence of complex interactions in environmental mechanisms which forces us to act at all necessary levels in order to avoid the mere replacement of one problem by another. From a technical point of view, strategies are applicable and can be considered at different spatial scales. This allows for a better understanding of their complementariness and applicability in various local conditions. Strategies designed to minimize water contamination must include preventive and curative means. Actually, contaminant transfers to water are inevitable even when an important decrease of losses from treated areas is achievable. Absolute protection of natural waters is not workable and high quality standards for drinking water require specific treatments to remove contaminants.

There are not many examples where consistent strategies have involved all possible participants at the different levels in order to minimize water contamination by pesticides. Based on the knowledge of pesticide environmental fate and regulations, it is in the scope of this chapter to discuss the usefulness of solutions already existing or likely to be introduced. Identify factors that limit the efficacy of actions is also a primary objective. A good strategy should then do nothing else than strive to improve the consistency amongst possible actions at various levels. It is evident that they are not restricted to technical considerations and have also to deal with education of actors and public information.

2 Pesticides and Water Directives

There is no technically acceptable technology available to replace pesticides at the moment. In addition to the primary goal of crop protection, chemical pest control is aimed at eliminating unnecessary pesticide use and minimize any adverse impact on the environment. In the early years of the pesticide era, the long term environmental impact was not sufficiently assessed and unacceptable environmental damage was caused by persistent pesticides entering the food chain and aquatic systems. The awareness of water issues associated with the use of agricultural chemicals, firstly in the United States in the early 1980s, and in Europe more recently, triggered the move towards regulations aimed at safeguarding water quality: regulations have become more stringent with increased requirements for pesticide registration. Furthermore, legislative action was taken to restrict the use of products identified as water contaminants.

Registration packages of many pesticides have been reviewed by the U.S. Environmental Protection Agency over the past years. The new regulations resulted in registration cancellations for a few products and restricted uses for some others. A similar review of all registered active ingredients is to be started in the EC using the recent registration directive and the process will take several years.

2.1 Directives on Water Quality

The European Community has produced various directives, relating to the quality of water intended for human consumption (80/778/EEC), and to the protection of groundwater against pollution caused by certain dangerous substances (80/68/EEC) or concerning the quality required of surface water intended for abstraction of drinking water (75/440/EEC). The Maximum Admissible Concentration (MAC) set by the parameter 55 of the 1980 EC Drinking Water Directive is 0.1 $\mu g\,L^{-1}$ for individual substances and 0.5 $\mu g\,L^{-1}$ for total pesticides. The establishment of the 0.1 $\mu g\,L^{-1}$ limit for pesticide concentration represents the first time in the evaluation of pesticide risks that a fixed limit value has been introduced that does not have any bearing on the data requested from registrants [9]. This unscientific approach reflects the political view that no pesticide residues should be found in water, as the 0.1 $\mu g L^{-1}$ is still considered as an analytical detection limit.

The World Health Organization (WHO) has proposed drinking water guidelines for individual chemicals based on their toxicity. WHO limits for pesticide residues in water are maximum tolerable concentrations based on toxicological data and cannot be compared with EC limits which are aimed at maintaining high quality standards for drinking water. The surrogate zero residues EC limit reflects the idea that pesticides have no place in drinking water. Therefore any cooperation with international bodies such as the World Health Organization or the Codex Alimentarius Commission in the interests of international harmonization has been ruled out. The industry view that the arbitrary 0.1 $\mu g L^{-1}$ limit should be replaced with limits reflecting the toxicology of individual pesticides has been steadily presented and supported by some European governments, France and the United Kingdom for instance [10, 11]. At the opposite, other countries are continuing to press for an EC-wide ban of some pesticides. No advisory limit for total pesticide content has been proposed and the view taken widely is that the concentration of individual pesticides is of primary importance. Nevertheless, compliance with the EC total pesticides limits will become increasingly difficult as the range of pesticides analyzed extends. If human exposure to pesticides induced directly by consumption of residues in food is easily traced and well understood, the pathway via drinking water is very complex and difficult to describe and predict. There is a need for health-based guidelines on the safe level of pesticides in water.

In this context of high quality standards for drinking water, the use of several pesticides, mainly herbicides applied on major crops, has created a relatively uncomfortable situation for the regulators who have to take steps to protect water on one side, and the manufacturers and the farmers blamed of polluting and poisoning the water and the environment on the other. The statement that pesticides do not belong to drinking water may be acceptable but cannot be extended to all water. With regard to pesticides in general, rejection of their presence in all types of water constitutes an utopian view that fails to consider a scientific and economic evaluation of all the issues. Integrity of water resources must be achieved in a realistic manner, using all means that the most recent technologies provide. Surface water and even groundwater cannot be pure in the absolute sense.

2.2 Pesticide Registration Requirements

The current registration systems for pesticides take into account a large set of standard tests designed to generate useful information on pesticide behavior in soil, plants, water and air. Many countries have their own national guidelines for risk assessment of plant protection products (U.S. EPA, EC). There are also several general international guidelines produced by organizations such as EPPO (1992), FAO (1989), OECD (1981), GIFAP (1990), IUPAC (1988) and the Council of Europe (1989). Each of these bodies takes a slightly different approach, and there have been few attempts to establish consistent and explicit decision making schemes [12]. Finally uniform principles at the European level are expected to be adopted in 1993 and will be applicable for new active ingredients and for commercial products as well.

Important factors affecting the movement of pesticides are well understood and enough is known to set down basic requirements for registrants to produce relevant information in support of approval applications. Hopefully, a clear set of data requirements is becoming established and more or less accepted worldwide for regulatory purposes. However, methodology is of primary importance in evaluating the environmental fate of pesticides and caution must be exercised in interpreting some results. For instance, certain regulatory authorities request lysimeter studies as a minimum data requirement because of the trigger values set to determine whether they should be conducted. These results are certainly very useful but will not predict the likely transport of a chemical in all field conditions. Similarly, not all field studies will be relevant to all conditions met in agricultural practice. Pesticide use will result in some contamination of water albeit at low levels.

2.3 Regulations to Minimize Impact on Water Quality

When an individual product is identified as a water contaminant, all that can be done at the regulatory level is to reduce the number of approved uses to those less likely to be used in sensitive conditions. For instance use of leachers could be permitted only at times of the year when there is no downward movement recharge to the aquifer. On the other hand, restricting the use of or banning one chemical without reviewing the whole subject of use could simply be exchanging one problem for another. Risk/benefit analyses, as those used by the U.S. EPA in the special review process, are central to decisions regarding use or cancellation of certain pesticides [13].

2.3.1 Restricted Use of Pesticides

According to the U.S. EPA, restricted use of a pesticide provides a means to maintain the benefits of use by imposing a variety of restrictions that will protect health and the environment and avoid the need for cancellation. The Agency has classified certain pesticides for restricted use if unrestricted use may have unreasonable adverse effects on the environment. Classifying a pesticide for restricted use is considered if any of its ingredients, including any component of a pesticide product (active ingredient, inert ingredient, impurity, metabolite or degradation product of any component) has the potential to reach groundwater on a widespread basis if it is shown, through laboratory studies, that it is mobile and persistent.

Criteria based only on a pesticide's potential for reaching groundwater are appropriate for considering restricted use classification of products to reduce the likelihood of groundwater contamination. Contamination data above certain toxicologically based levels are also one criterion. The criteria used by EPA for selection of pesticides classified for restricted use are either the measured persistence and mobility of an active ingredient of the product or the detection of the ingredient in groundwater at least three times in distinct locations. EPA decided to consider restricting a pesticide's use without regard to the specific level of contamination that may occur, or has occurred, for a number of reasons. While one can predict with some certainty whether a pesticide may reach groundwater, there is limited ability to predict the specific level of groundwater contamination which may occur.

Restricted-use classification can result in risk reduction because restricted pesticides may only be sold to, or used by certified pesticide applicators or individuals under their direct supervision. Such certified applicators are more likely to follow pesticide label instructions than non-certified users because they are better trained in pesticide mixing, loading, application, and disposal techniques and are more aware of the adverse environmental consequences of not following the label instructions than non-certified users.

One particular aspect of restricted use is the drinking water well setback, prohibiting use of a product within a certain distance of the wellheads. In Germany, a positive list of active ingredients authorized in the vicinity of wells was established for the different states (Länder).

2.3.2 Ban on Pesticides

Product ban or voluntary withdrawal by the manufacturer is an extreme measure, usually when the regulatory context does not allow for a correct balance between benefit and risk. Under some circumstances, management practices such as those described further may not provide adequate protection of the water resources and prohibiting the application of such agricultural chemicals may be necessary. The prohibition of the use of several different pesticides in the highly vulnerable region of Long Island in the state of New York is an example of such restrictions.

A complete ban of a chemical or group of chemicals recognized as being dangerous to the environment is a uniform policy. Such a ban does not discriminate between situations where the elimination of a chemical would result in major or minor cost increases. It usually means a loss of knowledge and technology in the agricultural and industrial aspects of crop protection. In many cases, a substantial share of the environmental benefits associated with a complete ban can be preserved by introducing a partial ban or a restrictive-use policy. In such cases, pesticide use is allowed only in situations where substitutes are not efficient or nonexistent.

2.4 Ambit for Possible Actions

The governments of many developed countries strive to reduce inputs of agrochemicals as a result of the growing concern for environment protection, mostly expressed by environmental pressure groups. Regulation of the crop protection industry is expected to be tightened on many aspects of chemicals use, with increasing periods and costs in registering new products. As political pressure increases, governments could make use of the "polluter pays" principle. Pesticide fees or taxes can have effects similar to those of a partial ban or limited use policies. Fees can restrict environmental risks below target levels at the least cost whereas uniform pesticide regulations may be much more costly than fees in attaining policy targets [14]. These considerable changes in the pesticide legal setting are relatively recent and we are too close for a proper view on their usefulness and their consequences on agriculture and industry.

3 Low Input to the Environment

The relationship between the presence of residues in surface water or groundwater and the use of pesticides in a given area is trivial. Decreasing the overall load of xenobiotics to the environment is often proposed as an obvious and necessary step to minimize the pollution.

The European Parliament (12 countries) and the Council of Europe (22 countries) have clearly expressed their will to limit to a minimum the use of agricultural chemicals. A number of countries have manifested their intention to reduce significantly the pesticide input to decrease the damage to the environment, Denmark, Sweden, The Netherlands, for instance. In the U.S., the LISA program (Low Input Sustainable Agriculture) started in 1988 at the request of the US Congress to promote the concept of integrated crop management through demonstration farms. More stringent regulations are partly intended to reduce this input which results both in a decrease of the pesticide consumption and a lower number of products available on the market. In Denmark, the total use of pesticides has to be reduced by 50% before 1997 compared to the average use from 1981 to 1985. The decrease will include the tonnage of products and the way they are used, as reflected by a treatment intensity index. This index is calculated every year on the basis of the total product sales and represents the number of times arable land could have been treated with the recommended dose of a product [15, 16]. A reduction in the treatment intensity index is obtained by reducing the treatment number or by applying reduced doses. A 25% reduction of active ingredients was obtained but decrease of frequencies of treatment failed. A similar policy of low input is followed in the Scandinavian countries. Sweden managed to halve pesticide use mainly through reducing dose rates and partly by shifting to highly active products [17].

Low input to the environment will be achieved primarily through reduced application rates, introduction of new molecules with higher activity, progress in formulation and better use of products. Most of the actions are now driven by legislation which tends to impose regulations of increasing stringency to existing products and also to new products which are certainly less prone to contaminate water. The strategies intended to minimize water contamination will take place within the limits set by these regulations.

3.1 Reduced Application Rates

In the present regulatory context, a possible strategy may be to cut chemical rates for some commercial products. Manufacturers recommend doses that are designed to give high levels of control in all situations. The effective dose rate is determined by experimentation for commercial and environmental considerations. Some areas may not need this level of control, giving scope to drop rates. With reduced rates, control might be variable and there is a risk of resistance building up. Incomplete pest control affects primarily farmers income so that none of them takes a chance overlooking necessary treatments. All these elements must be considered when striving to minimize pesticide application rates.

3.2 Effects on Water Quality

The cost effectiveness of pesticides has always been an important factor in the quantities used by farmers fully aware of the cost of crop protection. Except a small number of cases that could be remedied with appropriate education, systematic overdosing is scarcely encountered nowadays and cannot be reasonably considered as a major cause of non-point source pollution.

Important amounts applied every year on major crops have made herbicides good candidates for reduced application rates. Triazines have been on top of the black list because of the relatively frequent detections of residues in both surface and groundwater and their high potential for movement in soil. Use of atrazine is actually extremely controversial in Europe. This active ingredient was banned in Germany in 1991. In the Netherlands, registration of specialties containing atrazine was maintained until 1992. In Belgium, non-agricultural uses of atrazine are forbidden. In Denmark, atrazine can be applied once every three years on maize at a rate of 750 g ha^{-1}. Applications are prohibited within 10 meters of drinking water wells. In Italy, atrazine was re-registered after one year withdrawal and its use is allowed at the rate of 750 g ha^{-1} in combination with other active ingredients. In Austria, the authorized application rate is 1 kg ha^{-1}. In France and Switzerland, the dose rate was limited to 1.5 kg ha^{-1} in 1990. In addition to decreasing the authorized dose rate, recommendations have been made to minimize the amounts applied: 750 to 1000 g ha^{-1} on heavy soils and at low weed infestation level; replacement of broadcast application on bare ground after sowing by pre-sowing spraying followed by soil incorporation or post-emergence application. As a result, the average dose rate dropped from 1340 g ha^{-1} to 1240 g ha^{-1} between 1990 and 1991 with positive effects on water quality. A few other major herbicides will probably be affected by similar use restrictions [18].

Based on its excellent tolerance in maize and because of lack of adequate products to solve difficult weed problems, rates up to 7 kg ha^{-1} of atrazine were formerly applied on this crop [19]. Autumn applications of atrazine for control of couch grass (*Elymus repens*) were dropped due to the high application needed and the winter conditions conducive to important leaching. Due to high rates needed for weed control in orchards and industrial weed killing, certain atrazine uses have been or will be withdrawn. Today, maximum rates of 1.2 kg ha^{-1} are recommended. Results of water monitoring in Switzerland show a decrease of atrazine level in groundwater as a result of reduced rates. Two years after the application rate of atrazine was reduced in France, improvement in the quality of surface water remains difficult to demonstrate. The principal reason is that water surveys are not specifically designed to evaluate the effect expected from reduced application rates and most of the farmers had already decreased the dose rates when this limit was set. Adapted monitoring procedures may not be sufficiently accurate to assess correctly the effectiveness of this theoretically beneficial step since large variations in contamination levels may originate from uncontrolled environmental factors. As relatively small amounts of material are able to contaminate water courses on a large scale, and significant efforts have already been made to adjust the application rates, further reductions of application would probably result in effects of limited amplitude on water quality. The example of atrazine is self-explanatory: this active ingredient has a recognized potential for movement in soil and is likely to contaminate surface and groundwater. By reducing the treatment rates, contamination is likely to be decreased accordingly but not eliminated.

3.3 Product Use and Biological Efficacy

The agronomic goal of pesticide plant protection can be achieved using strategies adapted to the factors of the cropping system: crop, pest, soil characteristics and climatic conditions. Chemical weeding provides a good example of various agronomic approaches taking advantage of very different biological properties of molecules. As they also differ in a wide range of environmental fate characteristics, the extent of possible impacts on aquatic compartments may vary significantly from one molecule to another. Some representatives of the principal herbicide families are effectively major surface or groundwater contaminants.

The use of pre-emergence herbicides in autumn is dictated by the presence of weeds that would be more difficult to eliminate at a later development stage. The efficacy of pre-emergence weed-killing is controlled by diffusion barriers and transfer kinetics in both soil and plant. For example, a rough estimate of the critical concentration of atrazine in the soil solution likely to induce the herbicidal activity is about 10 $\mu g L^{-1}$. Several weeks are needed to insure a sufficient loading of the target PS II protein for inhibition. Furthermore, germination of the various weed species is also spread over a similar period of time. Hence, a minimum atrazine concentration in soil should be maintained during this period [20]. The biological activity of the product is obtained when a critical concentration of several $\mu g L^{-1}$ in soil water is reached and maintained during several weeks. It is necessary to ensure a sufficient concentration of herbicide in soil for a given period of time to compensate for product losses due to degradation and leaching in the root zone. This scenario is applicable to other molecules having a similar mode of action and used as pre-emergence herbicides. Corresponding requirements are expressed in terms of persistence and mobility for the molecule. This situation might not be compatible with environmental concerns in certain conditions. A similar situation is experienced with the relatively highly water soluble alachlor although its biological activity is observed before an important transpiration flow takes place. Very little is known about the critical concentrations in the soil solution for other molecules which do not require high transpiration flows in their modes of action (e.g. pendimethalin).

The strategy for post-emergence herbicide treatment is quite different and is based upon other principles. There are some possibilities to eliminate developed weeds without affecting the crop using selective molecules. In a few hours, the herbicide must penetrate through the leaf cuticles and reach the target in sufficient amounts [21]. Consequently, the biological activity is not strongly dependent on the persistence and mobility characteristics of the molecule. Chemical or biological degradation rate in soil has little impact on the treatment efficacy.

At first glance, molecules used in post-emergence treatments appear more environmentally friendly than those used in pre-emergence applications. However both strategies are not equivalent. For instance, pyridate is used in post-emergence treatment when pre-emergence applications of atrazine and/or alachlor failed to fully control resistant weeds like lambsquarter, *Amaranthus* and *Polygonum* species. Furthermore application has to be made at a sensitive stage. Despite environmental drawbacks of some active ingredients, pre-emergence herbicide strategies cannot be systematically replaced by safer post-emergence treatments although a partial shift to spring applications is likely to occur in certain conditions. However pre-emergence application rates can be decreased according to site-specific conditions and by controlled released formulations. The organic matter content determines soil retention for most non-ionic active ingredients. Adjusting application rates to the organic matter content is relatively simple and can prevent crop injury. This is particularly important in low organic matter soils where the low sorption potential facilitates high pesticide efficacy even at low application rates.

3.4 Applicability of Low Input Strategies

By reducing the herbicide dose rate, efficacy of weed control becomes strongly dependent on weed spectrum, weed growth stage and environmental conditions. A range of environmental factors such as temperature, air moisture and rainfall will affect the efficacy of foliarly applied herbicides whereas soil moisture will be the main factor to consider for soil applied herbicides. How much the dose rate can be reduced is a difficult question to answer. In Northern Europe, there is a political expectation of a transition towards integrated cultivation systems, combining preventive measures using crop rotation, tillage methods, choice of crop varieties, time and density of sowing and fertilizer strategies. Weed control thresholds, reduced pesticide dosage, split application and mechanical weeding call for long term research in weed science and the development of advisory systems.

In Denmark, a computer advisory system using the concept of factor-adjusted doses is based on the assumption of parallel dose-response curves at the different growth stages of a weed. Knowing the dose response relationship for each weed species, the dose rate is adjusted according to the weed density. Many herbicide products used in Europe contain two or three active ingredients with different modes of action. A weed can be primarily controlled by one of the active ingredients whereas another is controlled by others. Reducing the dose rate is likely to affect each weed differently. Therefore predicting the dose rate reduction for adequate weed control remains difficult. The Danish advisory system has been tested on winter cereals with possible further implementation on major crops. The purpose of this system is to provide help to farmers in selecting relevant herbicides, adjusting the dose rate on the base of weed density and growth stage, and making information on weed control readily available. A major hurdle to overcome in using the advisory system efficiently is the ability of farmers to identify weeds at an early growth stage [22].

Better use of thresholds for weed treatment and application of appropriate rather than recommended dose rates are opportunities for adjustment but total suppression of herbicide treatment is a mistake. Studies conducted over a period of three years in France demonstrated that missing one year of herbicide treatment makes weed control difficult in the following years [23]. Even at a low infestation level, overlooking herbicide treatment is an unrealistic practice. Seeds of self-propagating species can be transported from adjacent fields, fallow land or set aside where they are not thoroughly controlled. A single seedling of wild oat can produce 500 seeds, and a poppy 50000. The arable soil of each field constitutes a reservoir for weeds that is difficult to deplete as seeds survive for long periods: up to 15 years for wild oat and even 40 years for poppy with all seeds able to sprout. Anti-grass weed treatment could be skipped for two successive years because of low foxtail infestation (5–6 seedlings per m^2) without detrimental effects on the crop but 500 to 600 seedlings per m^2 emerged on the third year. In another experiment, 15 seedlings of foxtail survived the 1986 treatment. Despite total destruction of this weed in the following years to avoid additional production of seeds, 80 seedlings per m^2 still emerged in 1990 from the 1986 seed stock [24].

These studies show that extensive agriculture does not lend itself to cheap weed killing. Low density crops receiving lower amounts of fertilizers compete less with weeds that are more difficult to eliminate as late sowing dates favor the development of a late flora which is not controlled by early treatments. Minimum soil preparation tends to enhance the development of new weeds more difficult to control. Management of seed stock could be critical in the long run if herbicide applications are significantly reduced. In contrast, optimized (rather than reduced) application rates belong to intensive farming for economic and environmental reasons. Early post-emergence herbicide treatment is largely used, providing efficient elimination of weeds before they compete too much with the crop and allowing the right selection of active ingredients. In these conditions, dose-rate reduction is workable as an element of weed control optimization.

Weed effects are twofold: direct resulting from competition with the crop and indirect such as difficulties in harvesting, moisture and purity of the crop. Economic objectives in weed killing call for new management methods including harmfulness thresholds. Many attempts are being made to optimize pesticide use with this concept. This is a relatively recent trend with environmental implications but mainly driven by economical considerations. Rough estimates range between 10 to 35% of variable costs for crop protection. It is not surprising that considerable attention has been paid to maximizing the cost-effectiveness of agrochemicals. Systems of managing herbicide inputs, based on assessment of weed populations and estimation of their likely effect on grain yield have been proposed. Weed management systems based on weed threshold models have been tested only on a limited scale and on short-term sites. They show the possibility of reducing the herbicide dose rates without negative impact on grain yield [25]. As the weed population per unit area increases, the gain in crop yield from chemical control becomes greater than the cost of the chemicals and their application. The threshold density is where the cost of control is equal to the net benefit from control. In the short term, the cheapest strategies are the most cost-effective inasmuch as crop yield remains unaffected. Potential influence of treatments applied one year on yields and weed population in subsequent years must be investigated to reliably identify the best long range strategies. Little is known about the long term effects of repeated applications at reduced dose rates, particularly in terms of evolution of weed seeds in soil.

A practical alternative widely used by farmers is to adopt a prophylactic, insurance, approach, but to reduce application rates below those recommended on the product labels. Dissatisfaction with control thresholds has led to put more emphasis on the prophylactic use of targeted doses. Although it does not minimize herbicide use, this approach reduces risk in both short and long term and, above all, is easily communicated to farmers who are often willing to cut herbicide rates but not withhold weed control completely [26]. Defining costs and benefits of crop protection in all situations is not straightforward. For weed control, benefits derived from the ratio weed density over yield loss are unreliable. Long term effects of weed seed return from not controlling weeds are impossible to determine with any certainty. The relevance of an approach based on a cost/benefit analysis to help decision on pesticide use is doubtful given that availability of reliable data is too scarce. It is unlikely that operational threshold-based systems will be proposed in the near future because of the difficulty of modeling these extremely complex situations. Therefore, the prophylactic use of targeted doses for broad-leaved weed control offers a practical solution. The principal factors to take into account in optimizing herbicide rate are discussed in the last reference. The paper calls for simple assessment of weed species and density to allow the correct choice of herbicide and select the dose rate which will minimize the risk of poor weed control.

Repeated fungicide applications on cereals in absence of pathogens or at a low level of infection are also considered as insurance treatments that are probably not necessary in all cases. They take advantage of the preventive properties of the molecules rather than the curative ones. Up to now, the cost of preventive treatments in a strategy of maximum yield has been exceeded by the possible losses of crop quality and yield resulting from a minimum curative treatment. The situation may change if the price of commodities continues to drop, pushing farmers to adopt strategies of economic optimization. Better predictions of the disease development should increase the treatment profitability at two levels: phase the treatments with the start of the disease; predict the potential for disease damage as a function of climatic conditions for better adjustment of the number of treatments. Thus, effective management of plant diseases and pests requires that early and accurate information on threats to the crop becomes available.

3.5 Integrated Crop Protection Management

The choice of a treatment can be based on pest or disease identity, distribution and density. Most of the time, observation of external symptoms is the only method of plant disease diagnosis and this may lead to many mistakes. Conventional methods for detection of pathogen infection are time-consuming and expensive. Significant progress in the detection of plant diseases or pests at an early stage is expected from new diagnostic tools. For instance, immunoassays employing the specificity of antigen-antibody affinity reactions enable the rapid and early detection of diseases when symptoms are not evident or have not fully developed.

Integrated pest management calls for diagnostic tools to assess infection levels triggering chemical treatment. Development of such tools for direct use by farmers is limited by economic and technical reasons, particularly preparation of representative samples and test operation in controlled conditions. Because of the technical level required, only qualitative or semi-quantitative tests could be made available to farmers, whereas accurate measurements by immunodiagnostic or other promising techniques such as PCR would be conducted in specialized laboratories.

The quantitation of pathogens and pests levels may provide some indication of the potential economic loss to the farmer, influencing decisions on the choice of preventive measures and on the economic viability of taking such action. Techniques for risk assessment of potential damages caused by crop enemies are primarily intended for optimization of cost effectiveness in crop protection. The technical reliability and economic profitability of these tools need to be demonstrated at a large scale before implementation. Positive impact on the environment by reduced dose rates appears only as a secondary objective and a consequence of the primary economic goal. Expensive active ingredients in insecticides and fungicides recently introduced on the market, that do not exhibit a high potential to contaminate the water compartments, are primarily concerned by these techniques whereas herbicides known as water contaminants will remain outside their scope.

In the future, models of pest propagation associated with early detection of pests and diseases could help us to assess the level of crop attack and to adapt chemical control in proportion to the threat. Similarly, expert systems to determine which methods of weed control to use, taking into account herbicide composition, dosage and integration of weed control with other cultural practices could be used to select the most economical and effective herbicide doses and combinations, to recommend control at economic thresholds and to vary rates according to the weed spectrum [27]. Any means allowing for a better adjustment of the dose rate appears as a contribution to environment protection [28, 29]. The technical reliability and economic profitability of these tools need to be demonstrated on a large scale before implementation. Positive impact on the environment by reduced input is nevertheless a secondary goal in comparison to the expected benefits due to the optimization of plant protection.

3.6 Alternative Methods

Mechanical weed control has been unsatisfactory up to now but, when combined with reduced rates, it may be another possibility of reducing pesticide contamination in water. Possibilities of alternative weed control methods for several crops have been discussed, including preventive measures (crop rotation, seedbed preparation) along with mechanical control by plowing, harrowing and hoeing and reduced herbicide application. Strategies for herbicide application in integrated control systems may lead to economical weed management using mechanical control in conjunction with reduced herbicide rates but there are no practical applications at present [30, 31].

Non-chemical weed control strongly depends on crop rotation, cultivation, mechanical weeding and prevention of seed dispersal. Cultivation techniques rely on burial, cutting and desiccation to remove weeds from the crop, or before the crop is planted. Ploughs, harrows and cultivators are used for land preparation either with or without agrochemicals. Hoes can be used selectively within row crops. Cultivating and hoeing may be effective in spring sown row crops but are of limited interest in autumn. Small seeded crops and winter cereals are much less amenable to weed control by hoeing, yet they are major crops in Europe [32]. Research into alternative methods of weed control must continue if they provide additional or complementary management tools for the grower. One has to recognize that there is no alternative to chemicals for killing the toughest weeds.

New practices such as treatment placement in the rows and mechanical weeding by inter-row hoeing could decrease the application rate and the cost of weed killing at the same time. Although not attractive because of the extra workload, this method sounds workable in conjunction with the use of new molecules for economical reasons. Similarly, whenever possible, soil incorporation of some commonly used herbicides could prevent their entering surface water but may also increase erosion [33].

In some circumstances, split herbicide applications at reduced rate give a similar weed control as standard application at regular rate [34]. As some of the more aggressive weed species occur in well defined locations, the concept of targeting application to weed patches in crops has been developed so that sprays can be applied only to the areas of interest which are usually perfectly well known by farmers. Alternatively a low dose rate can be applied over most of the field with a higher dose rate of a different mix of herbicides used to treat the identified weed patches [35].

3.7 Product Replacement

Not all commercial pesticides are problem products with respect to water quality so that replacement of a product by another existing one with more favorable environmental fate is often considered as an easy solution. Irrespective of water quality problems, atrazine is still technically and economically attractive for maize weeding. It provides an excellent control of certain weeds such as umbellifers and pansies despite the availability of new molecules which are more expensive, require more skills in weed control strategy and exhibit a narrower efficacy spectrum. Weed control of maize has become more difficult since more stringent restrictions were set on triazines [36, 37]. An atrazine ban would probably induce important farming changes in the countries where its use is still permitted. A part of the maize acreage would shift towards other crops or be set aside with corresponding effects on animal husbandry.

The approach often used to assess the impact of banning pesticides is based on results of experimental studies to estimate cost and yield effects per unit of land. These effects are then summed across crops and regions to provide aggregate cost estimates. This approach tends to ignore possible price and land-use changes. It also tends to overestimate effects on growers, while underestimating consumer effects [14]. Estimation of effects of canceling a product should take into account various technical and economical factors such as the availability of substitutes, the capacity for additional supply and the responsiveness of demand to prices. The immediate effect of canceling a pesticide is a shift to chemicals that are usually more effective or less effective. When differences in cost and efficacy are small, the economic impacts are small. However, when effective substitutes are not available, the impact may be large. Another effect of banning a pesticide is a possible shift of production to different regions in response to increased product prices. The ban can also redistribute income among producers. As demand becomes inelastic, pesticide bans are likely to have a greater price effect on commodities. Another economic aspect that must be considered is the reduction in government price-support expenditures due to price increases associated with the pesticide ban. When a family of pesticides is banned, substitution possibilities are reduced more than when just a single pesticide is banned. A recent study found that there were no substitutes for 30% of the pesticides that would have been banned by a proposition aiming at eliminating pesticides known to cause cancer or reproductive damage in California.

Some of the negative impacts of a pesticide ban may be mitigated by new technologies. Increased commodity prices are likely to stimulate research efforts but the trend is currently the opposite. In the long run, research efforts may provide substitutes which will alleviate the effects of the ban but the capacity to develop such alternatives should not be overestimated because of the time and money necessary to discover and develop new molecules. The broader the imposed restriction, the lengthier and more arduous the process of adjustment or replacement [14].

3.8 Stalling Tactics Versus New Crop Protection Methods

Old active ingredients applied at a high rate are primarily affected by reduced rate strategies, and particularly the major herbicide families, i.e. phenoxies, triazines and substituted ureas. Reduced dose rates and elimination of unnecessary treatments are not likely to produce spectacular effects on water quality since inputs to the environment remain important. Reduced input should nevertheless remain a general objective in the future because of the expected positive impacts on the environment as a whole. Significant progress has already been made but has showed little effect with respect to the ambitious targets to be reached. This strategy of water protection is probably not likely to be one of the most effective when implemented solely through regulatory actions. Further efforts in this direction could even be very disappointing in terms of technical results and certainly detrimental to the business of certain pesticide companies.

There is a range of costs and benefits which relate to decisions about the use of chemicals for pest control which do not impact on farm budgets and which are not taken into account by farmers when making economic decisions. The costs of reduced yield, which may result from lack of appropriate control, will be felt at various levels. Many estimates have been made about crop losses due to weeds and pests. They show large regional and seasonal variations. Economic losses primarily relate to reduced farmer income. They also involve certain aspects of rural and regional development such as impact on regional employment and income, and on external balance of payments at the nation level, currency exchange rate and inflation. On the other hand, a set of externalities needs to be taken into account and these relate to the environmental consequences of chemical crop protection. The cost of contamination of water supplies is extremely difficult to determine in an objective way. A clear cause and effect linkage between pesticide application by an individual farmer and any subsequent environmental impact cannot be established, except in case of accidental pollution. Therefore, a polluter-pays system would be hard to implement.

As modern agriculture and the pesticide industry are together facing tremendous changes in a constantly evolving context of reinforced regulations and economic crisis, the final picture of pesticide use in terms of consumption and number of products registered by the turn of the century remains uncertain.

4 Pesticide Formulation and Water Quality

Significant progress in the protection of water has already been made and will be pursued in the area of new products development to achieve important goals: reduced inputs with the discovery of highly active molecules and reduced environmental impact through more specific and better targeted crop protection products. Given that an important proportion of any pesticide applied misses the target, and is likely to move away from the zone of interest, more attention should be paid to better targeting of pesticides. The areas where improvements are needed are those where the risks are the greatest, i.e. safety and the environment.

Consequently, modern formulations are developed with less active ingredient than older products. In the past two decades, dose rates have decreased from several kilograms to about 10 grams per hectare thanks to the discovery of new molecules which are much more biologically effective. If smaller amounts of chemicals are put into the environment, contamination of the water compartments should reasonably decrease accordingly [38]. Contrary to an opinion frequently voiced, higher biological activities for new molecules do not result in increased risks to human health or the environment. In addition to lower inputs with modern formulations, the potential for water contamination can be reduced above all with less persistent and less mobile molecules.

Beside enhancing the efficacy of active ingredients on the target pest, reduction of the user's exposure to toxic agrochemicals has been an important objective for formulators. New formulation technologies are being developed which indirectly reduce the environmental impact of pesticides in combination with new packaging and new delivery systems. Effects of pesticides are not totally specific to the target pest and the potential for harmful residues, damage to non-target organisms, spray drift, leaching or runoff into water exists. When formulating a new molecule or reformulating an existing compound, minimal or zero effect on the environment has become a primary objective.

Many commercial products have been on the market for many years and are still applied using conventional spraying or application equipment. Their efficacy would be undoubtedly significantly enhanced by reformulating using modern additives and formulation technologies, thus contributing to the low input policy. New delivery systems must result in improved productivity and also be more environmentally friendly. The new active ingredients must not only satisfy the requirements of lower input and safe environmental characteristics but also allow for new and safer formulations. Manufacturers must also recognize the need for continuing research and development of new methods of packaging, handling, application and disposal of agrochemical products.

4.1 Reduced Application Rates

As application rates decrease with the discovery of new active ingredients, a parallel innovative progress must be made in formulations since delivering a molecule to the target must be optimal because of the smaller amount to be applied. In addition to the development of modern formulations containing new active ingredients with higher biological activity, reduced inputs can be achieved in several ways, e.g. use of optically pure enantiomers by eliminating 50% of the inactive isomer, and replacement of some formulations additives, organic solvent particularly, by safer alternatives. Many marketed products still use outdated materials. Reformulating a product is certainly costly but not as expensive as developing a new active ingredient. Replacement of old formulations by safer alternatives is limited by the time-consuming effort of reformulation and the costly industrial investment.

Wettable powders and emulsifiable concentrates have become less popular due to safety and environmental concerns. Wettable granules are a possible alternative to dusty wettable powders despite higher manufacturing costs. Emulsifiable concentrates are the main type of formulations used for water insoluble active ingredients because they are easily developed and manufactured at low cost. Many of these products are still on the market today. Agrochemical formulations used 150 000 tons of solvents in 1989 and the trend is to use the less toxic ones. Spillages are not easily cleaned up and disposed of. In contrast, water is the solvent used in suspension concentrates which are considered as being more environmentally friendly although they leave contaminated packages if not carefully rinsed. From the environmental standpoint, spillages can be readily rectified and packaging disposed of easily with water dispersible granules.

4.2 New Formulations

In addition to better safety and efficacy, new emerging formulation technologies like microemulsions, suspoemulsions and effervescent tablets are likely to reduce the environmental impact by allowing a safer handling of the product and a better adjustment of the dose rate. Controlled release formulations, e.g. slow release granules, were expected to decrease the risk to groundwater contamination as they provide a better distribution of the active ingredient in time. Biodegradable polymers exist that can be used as a slow release matrix for active ingredients [39, 40]. Unfortunately, the technique is not applicable when a high concentration in soil water is required to obtain the biological effect. Microencapsulation is a slow delivery system allowing release of a chemical in a controlled manner depending on environmental conditions. Despite high development and manufacturing costs, microencapsulation techniques could come again in favor with the environmental pressure as they significantly reduce the use of solvents and may decrease the potential for groundwater contamination.

New formulations are developed to maximize the benefits of a new active ingredient. The biological activity can be enhanced by combination of surfactants and various formulation additives making the active ingredient more available to the target, e.g. enhanced adhesion to the plant surface and increased penetration for foliarly applied products [41]. Spray foliar retention is influenced by several factors such as droplet size. Spread and foliar penetration are also important parameters controlled by formulations. New formulations using adequate additives can also result in a reduction of spray drift. Decrease of dose rates and of wash-off from crop canopy will probably lead to a reduction of transfer to water courses. In contrast, there is little chance that formulations could have a significant direct effect on the soil mobility of the molecules.

4.3 Formulations and Pesticide Mobility

The properties of the active ingredient and, to a lesser extent, the formulation and the treatment application mode are the product-related factors which are mainly responsible for the potential for movement of a crop protection product. The mobility of the molecule is assessed on the molecule itself in laboratory tests using the adsorption/desorption method [42, 43]. The role of formulations or certain additives in soil mobility and possible transfer to groundwater is still questionable. Studies conducted on special slow release formulations (granules and microcapsules) demonstrated a tendency for the active ingredient to be less mobile in some cases. Influence of formulation additives on leaching was studied in the field and in the laboratory, on soil columns and soil thin layer chromatography plates. Most of them demonstrated how little influence the formulation additives have on mobility or gave inconsistent results, either in the direction of greater mobility or the opposite [44, 45]. If a significant influence of the additive can be shown, the concentrations used are far higher than those encountered in normal conditions (several orders of magnitude). Only a few publications described the influence of formulation on runoff [46].

Despite the small amount of information available, there is little hope that formulation additives could significantly influence leaching of an active ingredient to offer sufficiently attractive prospects for practical applications. Alginate stickers have been developed to reduce pesticide mobility by trapping it in the top soil layer. The role of soil organic matter and natural colloids on contaminant mobility has been suspected many times and is likely to overshadow the effects of the formulation additives. At the moment, there is no clear information regarding relationships between formulation type and the ability for a chemical to move through soil macropores. Providing a sufficient product efficacy is obtained, entrapping the active ingredient in a polymer or granules for controlled release could restrict fast downwards movement through soil macropores shortly after application.

4.4 Formulation Strategies

By reducing the use of a few active ingredients spread on vast acreages, growers could prevent pesticide contamination of water courses. In Europe, complex formulations or tank mixes of agricultural chemicals could be one way to keep pesticide levels in water below legal limits. For instance, some herbicides are used on a very wide scale. Spraying weed killers in spring rather than autumn could help avoid overwinter leaching or runoff due to low ground coverage by the crop. But these products tend to be most effective on small weeds. Most growers are therefore compelled to maintain autumn treatment for efficacy reasons. Adopting complex formulations with several active ingredients to prevent the domination of a single active ingredient is playing the legal system to keep within the law but might be a way round the problem.

A decline in the number of pesticide companies involved in the research and development of new active ingredients will probably continue. With more stringent tests required for registration, the number of new molecules put on the market every year is likely to decrease. Several active ingredients will not survive the re-registration process to be started in Europe, hence the need to extend the life and use of existing active ingredients with formulation. One of the most effective ways of extending their use is by reformulation which also requires re-registration. In this field the chemist also faced with regulatory constraints and must develop delivery systems that have to comply with legal requirements. This process is limited by technical feasibility, research and registration costs. For instance, only a small number of optically pure active ingredients are manufactured (e.g. mecoprop). Most often manufacturing expenses are a deterrent. However, in comparison with the discovery and development of a new active ingredient, the development of a suitable formulation is a cheap task.

4.5 Pesticide Packaging

Most media and public attention has been focussed on the environmental issues associated with container disposal and contamination. Disposal of empty containers has been a puzzling challenge for many farmers anxious about environment protection. The problem of packaging waste is also a concern for the pesticide industry since residues in the containers may be a major source of contamination. In recent years, the pesticide industry has started to consider packaging as an integral part of a delivery system with possible product benefits via packaging design. Until recently the focus was mainly on user's safety aspects, however, trends in pesticide packaging are now taking into account disposal. Agrochemical companies are encouraged to launch new products in environmentally friendly packages. Disposal and recycling of all types of packaging have also been matters of discussion in the last few years.

Collection of used agrochemical packaging is difficult and expensive. Many agricultural chemicals are classified as toxic and the used containers are scattered over large areas. Laws are being introduced to regulate packaging disposal, reuse or recycling of materials. They are not solely directed at pesticide industry but at packaging in general. Germany has introduced a new packaging regulation according to which manufacturers and distributors are required to recover certain types of used packaging form their customers for reuse or recycling.

Where possible, large size containers tend to be used in the United States where large farms predominate. Containers can be reused, partly solving the problem of packaging collection and minimizing possible contamination by formulation residues. Herbicides are applied in very large quantities, making them good candidates for bulk or minibulk systems. The technique is only practical for licensed agrochemical suppliers and the large farms, which practice large scale monoculture. Repackaging is simply the transfer of pesticide from one bulk container to another and a retention tank minimizes environmental harm due to spillages. For bulk herbicides, many manufacturers favor the use of cheap reusable packaging designed to last for a few uses.

In contrast, the most widely used small volume pesticide packs are polyethylene bottles containing between 1 and 10 liters. Residues are present in used containers and rinsing is a prerequisite for any reuse or before disposal. Residues can be reduced with packs easier to rinse (wide necks and isolated handles). A major initiative for nonreturnable packaging involves the standardization of container necks and closures (63 mm). This effort has been led by GIFAP to facilitate the adoption of closed systems for transferring and metering the product out of the container that allows proper container rinsing, while minimizing pesticide exposure to the end-user. The United States have already set a stringent standard for pesticide removal from containers before disposal with municipal waste and will be followed soon by the European Community.

In most countries, the only way to dispose of used agrochemical packaging is by careful rinsing and dumping in landfill sites or by burning. Recycling is expected to increase in the future. Prior to recycling, containers must be thoroughly rinsed to reduce any risk of contamination. The rinsing standard required in the United States is known as the "six nines" i.e. 99.9999% of the product must be removed from the container. This level can be reached after triple rinsing or, preferably, pressure rinsing. Pressure rinse nozzles are introduced through the bottom of containers whilst they are inverted over the spray tank. 30 seconds of rinsing is normally sufficient to remove the residues. The characteristic hole left by pressure rinsing is a convenient way of detecting whether or not a container has been rinsed. Manual triple rinsing involves filling a container one fifth with water, replacing the lid, shaking it and draining the liquid into the spray tank. Even after this procedure has been repeated three times, there will be up to one hundred times more residues remaining than with pressure rinsing. The Dutch rinsing standard also called "four nines" requires that 99.99% of the product is removed. This rule is rigorously enforced and it also states that all sprayers must be equipped with a rinsing apparatus. With such stringent requirements, residue removal from the package will become an important design parameter for both formulation and package development. Recycling associated with careful rinsing is an obvious way to reduce waste and decrease water contamination by pesticides. Unfortunately, like other recyclable materials, it costs more to process than the eventual resale value. Apart from contamination which makes manufacturing of new containers from old ones a complex problem, there is a limit to the use of recycled polymers since these materials have worse performance characteristics than virgin plastic.

Small volume returnable containers have been studied by pesticide manufacturers. These containers could be used in Europe where a relatively large number of products are used in small quantities compared with American farms. There are still many factors to investigate such as size, material for cost-effective reuse, need for immediate rinsing by farmers, etc.

The idea of water soluble packs for agrochemicals is not new but development of the technique has been hampered with technical problems: quality of water soluble polymers, compatibility of the pesticide with the film, specifications allowing use of conventional spraying equipment, manufacturing technology and storage restrictions. This technique has a good future since it leaves no contaminated packaging. Some formulations such as flowables have a tendency to be sticky and are therefore difficult to rinse. Reformulating products as wettable powders and packing them in water-soluble film seems to be an acceptable alternative. The balance between pesticide chemistry and packaging is a delicate one. Altering formulations to suit specific packaging techniques is not always a good approach since it may reduce product efficacy, necessitating higher dose rates. A range of safe and environmentally friendly packaging systems rather than reliance on a single technology seems to be more effective. Use of water soluble packs is expected to grow since this system solves the problem of formulation residues and packing disposal in addition to minimizing user's exposure. The main drawback in the context of low input is that packages deliver a single dose which cannot be easily adjusted to recommended rates in local conditions. This, coupled with the fact that farm acreage varies considerably, could limit the popularity of any measured dose packaging. It is considered that single dose packs of any type could prevent farmers from applying a sufficiently small dose of product. This would mean that a considerable amount of product is wasted or stored, defeating the objective of safer packaging.

Depending on their own markets and product lines, pesticide manufacturers may adopt very different policies with regard to packaging, some preferring large size, refillable containers or small size one-way packs that are easy to rinse and recycle or dispose of. Irrespective of the options retained, modern packaging systems are aimed at minimizing the amount of chemicals likely to contaminate the environment. They are an important element in the strategies for water protection inasmuch as adequate education is also provided to the pesticide end-users.

4.6 Spraying Equipment

Many of the new formulations may require more sophisticated spray equipment and more demanding application techniques. For example, agitation systems will need to be more intense with suspensions to prevent settling in the spray tank. Long range objectives in this area include the development of new application technologies for efficient delivery of modern formulations minimizing application rates and environmental exposure.

New equipments provide electronic control of the spraying with a better control of delivered volumes (nozzle calibration, spray rate proportional to the speed). Special systems are able to limit the spray drift [47]. These new techniques are expensive and mostly applicable to major crops. Unfortunately costly equipment is not likely to be used by many individual farmers at the moment except on large farms.

An important area for improvement is the design and correct use of spray equipment. Pipes with reduced dead volumes, tanks with minimum residual volume, systems for automatic package rinsing, tank filling control devices (water metering and non-return valves to prevent back-siphoning of the spray liquid), closed systems for formulation transfer to the tank [48], and systems for decontamination of rinsing waters [49] are some of the features found on modern spray equipment. They are directly aimed at minimizing point source contamination of water. Training farmers on the proper use of spray equipment and new generation formulations will require a lot of effort [50].

5 New Agricultural Chemicals

The efforts of the pesticide industry into research for the discovery of new active ingredients include the selection of molecules on the basis of their physical, chemical, biological, toxicological and ecotoxicological properties. Laboratory and field studies are designed to provide general information on pesticide properties and behavior in various environment compartments. With regard to environmental concerns, high biological activity minimizing application rates and environmental fate characteristics leading to low potential for movement outside the treated areas are features of primary importance for a new molecule. Based on these characteristics, the scientist and, in a second step, the legislator are able to make appropriate decisions regarding the safe use of the product or ask for further information.

5.1 New Active Ingredients

Only large companies can afford to invest in long term research and development, and remain innovative in the area of new molecule discovery. The development of a new agrochemical costs roughly $ 125 000 000. Testing of up to 30 000 molecules is necessary and it takes an average 8 year period before launching onto the market once the right molecule has been discovered. Low input to the environment resulting from the use of new generation molecules are therefore part of long term strategies without immediate results in a highly uncertain context. Manufacturers must continue to research and develop new products which are effective, economical and perceived as environmentally friendly. Industrial research is already targeted this way. Today, triazole fungicides and sulphonylurea herbicides are prominent examples of the dose rate reduction. These products also illustrate the outstanding progress made by the pesticide industry research to conceive and create new molecules aimed at highly specific targets.

5.2 New Products Selection Criteria

Decisions regarding development of new pesticides are obviously based on efficacy against pests and potential market considerations. After a chemical is synthesized and starts down the path towards development, the success of the compound and ultimately the usefulness of this product to agriculture depends upon several critical areas. One of these areas is environmental protection, especially protection of the water resources. Whatever the target of the treatment may be, pesticides ultimately reach the ground where they are affected by transfer and degradation processes. Soil is the primary path for contaminants on their way towards ground or surface waters. Therefore the pesticide fate in soil must be assessed with considerable attention, mainly using standard tests requested by regulatory agencies.

Based on laboratory, semi-field (lysimeter) and field tests, persistence and mobility in soil allow identify potential environmental concerns associated with new products [51, 52]. The basic processes affecting pesticide fate, mainly degradation and movement in soil are now well known. Moreover, the factors contributing to these processes are correctly understood. Screening assessments should serve primarily to classify the behavior of a compound and to indicate the possible need for additional investigations. Screening methods should not serve as cut-off criteria upon which regulatory decisions are based. Depending on the results of initial exploratory studies, additional laboratory or field testing may be needed to further clarify the environmental fate of a molecule and support product registration. For the researcher these criteria are only indicators that will trigger further investigations during the development phase of the product. Certain new compounds may be ruled out if these criteria are strictly followed. The persistence criterion may conflict with a sufficient residence period of the molecule necessary for its efficacy. Similarly, the criterion directed towards identifying potential for groundwater residues will almost always be violated for nematicides and systematic soil insecticides.

Preliminary experiments have proved to be very useful before starting the studies for registration purposes. Performing preliminary environmental assessments of new active ingredients can help identify potential impediments to registration and minimize instances were considerable resources are expended on new product development, only to recognize later that environmental considerations preclude registration. Environmental assessments should therefore be started very early at the time of preliminary evaluation of efficacy in the field, or even earlier if a key environmental characteristic is searched among several possible candidates in the same class of compounds. Early screening tests for environmental behavior assessment require considerable efforts in analytical method development when radiolabelled material or immunoenzymatic techniques are usually not available.

Irrespective of other features decisive in the successful development of a new product, it appears that an adequate compromise should be established between product effectiveness and environmental behavior, in soil particularly. Biological activity is positively correlated with persistence as a sufficient residence time is necessary to manifest the desired effects. For systemic molecules, physical properties affecting soil mobility like sorption coefficient and water solubility, are of primary importance in the absorption and translocation processes within the target organism. Often, a critical concentration in soil water must be reached which means that a potential for movement in soil and into groundwater may exist for the product.

There is certainly no such thing as an ideal or green pesticide. It is essential that the positive aspects of a product outweigh the negatives. If the positive aspects can be rather easily demonstrated, the negative are difficult to assess. But these products can be handled in such a way that the risk for water contamination is significantly reduced.

5.3 Registration Requirements and Risk Assessment Uncertainty

Regulatory agencies use criteria to scrutinize new chemicals and decide whether to classify a new chemical, without waiting for it to reach groundwater. It is out of the scope of this paper to discuss the methods for assessment of pesticide mobility in agricultural environments. For reasons of legal certainty, decisions for pesticide authorization are based on unequivocally defined trigger and cut-off values of appropriate criteria determined using internationally harmonized and accepted test guidelines. Various procedures are used, relying on scientific rationale for decision trees and trigger values [53].

Laboratory, semi-field and field studies are designed to provide general information, qualitative and to some extent quantitative, on pesticide properties and behavior. Based on the data available, the legislator is able to make appropriate decisions or asks for further information. Results contained in the registration package provide an overall view of the pesticide environmental behavior and justify the claim that the use of a product is environmentally safe.

The protection of soil and water is one major point of the recent regulations. This raises the question as to whether or not data produced in the laboratory are able to reflect the field situation, and whether these data are suitable for realistic assessment of environmental hazards of chemicals. The combination of laboratory experiments, which examine few parameters with a relatively high statistical reproducibility of results, with field experiments, with their inherent stochastic variability, might help overcoming the shortcomings of both approaches [54]. However, due to the considerable variability of natural conditions, the stringency of standards for water protection and the technical difficulty to make accurate predictions of contaminant concentration at very low levels, it is impossible to assert that legal limits will never be violated in particular conditions. This is particularly true for some systemic soil-applied insecticides. The many reasons for the difficulties can be found in the manifold interactions between chemicals and abiotic and biotic components of soil, the regulation of these interactions by microbial conditions and agricultural management practices and the considerable variability of environmental conditions at the different geographic scales.

There are currently no definitive studies available that have correlated the concentrations determined in leachates from laboratory columns or lysimeters with the actual concentrations in shallow groundwater using piezometers. In the absence of data, some regulatory agencies are tempted to apply groundwater standards to leachates exiting from lysimeters. This approach is inappropriate since it ignores the continued degradation and dispersion of residues below the root zone and in the saturated zone. The contamination of waters by registered pesticides has manifested the extreme difficulty faced when assessing the risk for water contamination at low concentration levels and particularly at the maximum acceptable concentration in drinking water set by the European Community to 0.1 $\mu g\, L^{-1}$.

One important result of the uncertainty to predict the behavior of pesticides in soil is the difficulty of determining the possible contamination level in drinking water when legal values are set extremely low, and extended to groundwater, for instance in the European Community. In this particular case, this low level means zero residues and was chosen to prevent any contaminant in groundwater. Conversely, zero input into groundwater can be only achieved in cases where total degradation or retention of pesticides occurs in the unsaturated zone of soil [55].

Approving new pesticides is therefore fraught with difficulty as the authorities have to deal with increasing environmental concerns. In particular, registration procedures have to take into account various EC directives on water whereas risk of drinking water contamination cannot be assessed correctly in all conditions. Some factors affecting pesticide movement to water are largely outside product-specific parameters. When evaluating a new product, it is difficult to predict if it will reach water sources in levels above the statutory limit, because pesticide movement at trace level is controlled by other factors, irrespective of pesticide properties. Therefore, regulations are difficult to set down because predicting in a quantifiable way the likely concentration of a pesticide in water will depend on many factors outside the properties of an individual chemical and cannot be done satisfactorily at present [56, 57].

Some old active ingredients will probably not pass the re-registration process using stringent environmental requirements mostly defined following the detection in water of compounds highly mobile or recalcitrant to degradation. New active ingredients selected on new registration criteria will offer better guaranties for environment protection. They are supposed to progressively replace former ones but several generations of active ingredients are likely to coexist, particularly in Europe where complex herbicide formulations containing several active ingredients are largely used. As previously mentioned, this allows for the use of old cost effective molecules while decreasing the input to the environment.

6 Factors Affecting the Transfer of Contaminants to Water

The behavior of pesticides in both the unsaturated and saturated zones is a complex process influenced by product physical and chemical properties, soil and groundwater properties, climatic conditions, and management practices. The various ways pesticides can reach the aquatic environment has been reviewed in many papers [58]. Those include spills, careless use of pesticides (e.g. overspraying water courses or drainage ditches, heedless disposal of empty containers), leaching and runoff.

The main factors affecting the movement of pesticides through or on soil are largely outside product-specific parameters. Assessment of potential for water contamination by a chemical and decisions regarding its use cannot therefore be solely based on the intrinsic product properties. Movement of contaminants is strongly dependent on local soil and weather conditions. Potential for surface or groundwater contamination should be assessed in site-specific conditions so that critical instances for the use of a particular chemical could be identified. Recognition of discriminant factors related to product use (application rate and timing) and site vulnerability is a prerequisite to the possible implementation of practices aimed at reducing contaminant transfer to water. Modern techniques are available, including mathematical modeling, databases on environmental parameter values and geographical information systems, for quantified land evaluation and environmental risk assessment. In the future, information and decision-support systems will help to identify vulnerable combinations of cropping, soil and climate conditions within catchments and to indicate the range of pesticides likely to contaminate water [59, 60]. Rather than adding to the number of black-listed pesticides, such sophisticated expert systems should become precious tools in selecting and evaluating the relevant management practices to implement at the field and watershed levels or to assess the economic impact of product restriction or cancellation.

6.1 Assessment Procedures of Pesticide Mobility

A variety of criteria and indexes have been used to identify pesticides and land areas where potential exists for pesticide residues to enter groundwater [61, 62, 63, 64, 65, 66, 67]. Instead of considering cut-off criteria, the combined influences of compound mobility and dissipation can be estimated using a screening assessment such as the Gustafson Ubiquity Score (GUS) to identify the compounds that have the potential to enter groundwater. Important efforts have been devoted to compiling pesticide properties, particularly persistence and mobility [68, 69, 70, 71]. However, most of the time, only one value is provided for each chemical while the actual value in the field may vary significantly due to climatic or soil properties. The more comprehensive USDA data base helps overcome this limitation by providing field half-life data and sorption measurements for all locations or soils for which data has been provided. However, not all registered active ingredients are included in the database. Persistence and mobility parameters may considerably vary, causing sharp differences when ranking the leaching potential if the GUS index is calculated using published data [72]. Because of the complex interactions between pesticides, climatic conditions and soil properties, no screening method has been accepted by the scientific community as adequate, although DRASTIC has been widely used by state and federal regulatory agencies in the United

States. The best screening methodology for indicating which pesticides have the potential to reach groundwater under some conditions probably consists of correlations between chemical properties such as persistence and mobility [73].

No simple criterion can usefully account for potential contamination of surface waters. Transfer of residues to surface water may arise regardless of mobility and soil retention properties if the compound is not soil incorporated. For more mobile compounds, residues may be present in runoff water from recently treated fields, while less mobile compounds may be present in suspended material and eroded sediments [74].

6.2 Methods for Risk Assessment

A wide range of experimental approaches is used for assessing pesticide mobility in soil. In addition to laboratory and field tests, computer models play a significant role in the development and registration of products in helping rationalize the results of studies, in providing estimates of likely leaching behavior in various conditions and in assessing the probability of groundwater contamination from normal product use [75]. Models of different complexity are available to describe pesticide behavior. There are still some shortcomings in modeling namely validation in the field including the key problem of parameter identification, influence of spatial heterogeneities of soil parameters, and effects of pesticide transformation. Most of these models are valid for homogeneous soils. More research is needed in the field of preferential flow and spatial variability.

A lot of emphasis is put on new model development or model improvement. Surprisingly little effort is devoted to minimizing or even better understanding the main sources of variation which constitute the limiting steps in good use of modeling. There is no pressing need for new models to be developed. On the contrary, there is a plethora of models which can be objectively tested and evaluated and there is a need for a critical overview and validation of models currently in use [73, 76]. A clear distinction should be made between research and operational models or their use.

Mathematical models are being proposed by registration authorities (e.g. the United States, Germany, the Netherlands) and various groups working on risk assessment as a tool for estimating environmental concentrations. Current models are often poorly documented, not correctly validated and relevant databases for input parameters not readily available. Modelers tend to work independently, adding to the number of models not sufficiently validated. A limitation associated with the use of models for regulatory purposes is the frequent lack of model evaluation prior to adoption of a particular modeling scheme, leading to unacceptable misuse.

For instance, predicting contaminant concentration at the 0.1 $\mu g\,L^{-1}$ level in shallow groundwater using mechanistic models and laboratory data as input parameters is largely conjectural. Leaving aside the considerable variability of input parameters used, leaching models can actually handle contaminant quantities that are only two or three orders of magnitude below the amount of pesticide applied. Furthermore, real concentrations in groundwater near the water table may be significantly higher than 0.1 $\mu g\, L^{-1}$ because of possible macropore transfer in the unsaturated zone, irrespective of the pesticide mobility properties. It is imperative that legislators and industry agree on appropriate models and modeling practices to ensure that the results obtained reflect the best understanding that is achievable with a reasonable compromise of computational intensity, numerical accuracy, physical reality, input data requirements and user friendliness. Regulatory decisions should not be based solely on the results of simulation modeling but should take into account the use of models as one additional means to understand the fate of a pesticide in the environment [77].

6.3 Vulnerability Assessment

Most research activities focus on specific processes at a specific scale. This scale is not always compatible with the scale at which the environmental risks are situated. Many of the present environmental issues related to the fate and mobility of pollutants are at field or regional scale. However most of the ongoing research for regulatory purposes study the processes on the laboratory or microplot scale. This scale is very suitable for identifying mechanisms and critical parameters [78]. Care should be exercised in applying these parameters to a higher scale, without considering the inherent variability at these scales. A clear need exists to integrate present knowledge from different fields obtained on various scales, particularly the values of environmental parameters to be used on different scales, and to establish the links between them.

The ability to predict the fate of a pesticide under a specified set of conditions is not sufficient for environmental fate assessment. Rather, one must predict the likelihood of pesticide contamination throughout an entire geographical area, taking into account variation in climate, soil properties, pesticide properties and usage practices. In addition, managing pesticide use requires an understanding of which environmental and chemical factors have the greatest effect on contamination potential throughout that entire region [79].

The effectiveness of these methodologies depends largely on the availability of locally relevant data and validated models [59]. Unfortunately, the use of distributed, stochastic, mechanistic models at the field and regional scale require that a large number of parameters and input variables be known at a detailed level. One way to account for spatial variability of soil is to use input in the form of probability distributions to arrive at a stochastic model output. Several simpler procedures have been proposed to enable regulators to identify those areas where assessment of the magnitude of pesticide contamination in surface or groundwater is needed [80].

The U.S. Soil Conservation Service (SCS) conducted a program for estimating the potential for residues of agricultural chemicals to occur in surface or groundwater [81]. A screening procedure was developed which farmers could use to evaluate the relative loss of pesticides from soils and to guide their selection of products. The technique used considers the soil and pesticide properties separately. Based on GLEAMS simulations [82], soils were divided into high, medium and low vulnerability. Pesticides were divided into similar categories based on their soil half-lives, water solubilities and organic carbon partition coefficients, the groundwater ranking using the Gustafson mobility index (GUS). This procedure does not consider actual rainfall patterns so that arid regions are considered equally vulnerable as areas with high rainfall. The procedure ignores site-specific characteristics which are also extremely important. For example, subsoil properties and depth to water table are ignored in estimating vulnerability of groundwater. Surface water predictions ignore field characteristics such as slope and proximity to ditches or streams [73].

A soil classification based on hydrologic properties has been developed in Great Britain by the Soil Survey and Land Research Centre to allow predictions of the hydrologic response of a catchment as a function of soil profiles. To assess the vulnerability of water sources a methodology based on soil and substratum hydrological classification and top soil organic matter content was set up. Separate vulnerability classes are defined for each type of water source. Using simple mathematical models, soil vulnerability and pesticide classes are combined to predict the likelihood of pesticide compounds reaching ground or surface waters in concentrations greater than 0.1 $\mu g\,L^{-1}$ [83]. Climatic variations which strongly affect pesticide leaching are introduced with the duration of field capacity period and the amount of excess of rain during this period. Critical factors for surface water vulnerability are the length of time between application of a pesticide and a significant rainfall event and the amount of rain in that event. By overlaying the spatial distribution of soil vulnerability classes on important climatic variables, maps showing aquifer and surface water vulnerability assessments for different pesticide classes can be produced.

Screening procedures to minimize adverse water quality impacts of pesticide application may become extremely complex to match several pesticide parameters to site-specific soil ratings (e.g. relative leaching potential index, relative runoff potential index, U.S. EPA lifetime health advisory level, aquatic toxicity). These procedures may produce recommendations against important portions of pesticide application which would mean significant changes in agricultural practices and pesticide use [84, 85]. Besides the need for procedure validation against pesticide concentrations found in water sources, another criticism is that the assessments based on extremely simplified hydrological models may be inaccurate at a low contamination level such as 0.1 $\mu g\,L^{-1}$. Systematic and comprehensive studies should be designed to evaluate and validate the different techniques for assessment before they are used for regulatory purposes.

The soil profile exerts a strong influence on hydrologic processes and the response to precipitations of a hydrologic unit. The succession of strata in the various horizons may induce a different response of a watershed as a function of time: fast with intense runoff or slow with underground flow. Intermediate situations occur with sub-superficial flow when soil permeability decreases suddenly between two successive soil layers. In the same way, tile drains are additional vulnerability elements. The significance of non-conventional transfer mechanisms for contaminants that may originate from physical or anthropogenic singularities has not been sufficiently investigated up to now. Risk or vulnerability assessment procedures are probably not relevant in circumstances where these mechanisms account for an important part of the water pollution.

Regulatory steps on product authorization will have poor effects if not accompanied by other dispositions related to the handling of products by the end-users and to agricultural practices. Small amounts of agricultural chemicals, out of proportion with the total amounts used, can significantly alter water quality or induce effects that cannot be overlooked with regard to the effects observed.

6.4 Pollution Incidents in Agricultural Catchments

Occurrence of pesticides in water are typically grouped into two categories. Non-point source is used to describe those instances where residues reach surface or groundwater as a result of normal movement on soil surface or through the soil profile. Point source is used to describe residues resulting from spills, handling, mixing, equipment rinsing, bad disposal practices or direct river or well contamination.

Pesticide activities include such activities as mixing, loading, application and disposal. Both misuse and failure to exercise good judgement in using pesticides can contribute to the potential for surface and groundwater contamination. Strategies are aimed at making recommendations to farmers regarding operations such as handling of products, preparation of spray solutions (tank filling), application of pesticides (buffer zones along water courses), equipment cleansing and disposal of rinsates. High concentrations found in ditches and attributed to point sources illustrate the importance of careful practice by operators when handling pesticides. It also suggests that some relatively large inputs to water could quite easily be reduced. For example, certain pesticides which are applied from a tank into irrigation water drawn from groundwater, direct the user to avoid allowing "back siphoning" of the tank of pesticide into groundwater. Similar precautions must be taken when filling tanks to avoid contamination of the drinking water distribution network.

The documented pollution incidents are generally those which cause visible changes in water quality or have other identifiable impacts, such as fish deaths. It is likely that a large number of cases of accidental pollution in rivers remain unknown inasmuch as the macroscopic effects of the pollutant are not observable nor observed, unless the contaminant is identified and determined in a sample. If readily apparent problems can be identified, traced and corrected, diffuse and chronic inputs are not seen nor easily identified. The severity of pollution impact usually depends on the concentration of the material. The degree of dilution of a particular chemical depends on the water flow through soils, aquifers and water courses. A point source can have locally devastating effects but these effects may be diluted with time and distance. Diffuse pollution is more difficult to track and control, and may have smaller but continuing and pervasive effects across the whole catchment area.

Determining whether residues in a drinking water well result from point source or non-point source contamination is difficult since the well may be located near treated fields or near storage areas. Phase II of the EPA National Pesticide Survey [1] has found statistical associations between factors that support several of the commonly accepted theories concerning the presence of chemicals in water from drinking wells. The amount of a pesticide applied for non-agricultural purposes was linked to the chance of nearby contaminated well. A number of factors affecting transport of chemicals to groundwater, including precipitation, the presence of surface water close to the sampled well, and other wells operating near it, were found to be related to the presence of pesticides in well water. Older wells and shallower wells were also found to be more likely to contain detectable amounts of residues. The probability of detecting pesticides was found to be greater in wells with lower temperature or pH. Pesticides with longer half-lives were more likely to be detected.

Point source contamination can be attributed to spills near wells or water courses, therefore increasing emphasis has been placed on the proper use of agricultural chemicals. Many programs have been initiated by regulatory agencies, agricultural organizations and pesticide manufacturers to educate farmers and applicators on appropriate practices. Documents outline basic rules for safe use of pesticides. Any spills which occur during the transport or handling of products should be immediately cleaned up. Application equipment should be washed away from any well or from water courses. Excess spraying volumes and rinsates from tanks should be collected and properly disposed of, preferably onto an appropriate crop and away from water courses. These practices to prevent point source contamination apply to all agricultural chemicals regardless of their chemical properties. As information regarding the proper use of pesticides is more efficiently conveyed to farmers, diffuse inputs, often the result of changes in land use practice or the accumulation of many minor problems, may become more significant.

A comprehensive assessment of the scale and nature of the current and possible future problems of pesticide contamination in water cannot be made yet. This is because of the limitations in present monitoring programs, the absence of adequate analytical techniques for many of the active ingredients and lack of knowledge on pollution pathways, particularly in relation to effects on groundwater. Some pesticides are widespread at low levels in surface waters but their presence is often characterized by strong seasonality and variability due to incompletely understood environmental factors [86]. Significant pesticide contamination of groundwater to date has been limited to triazines and, where sources can be identified, generally from non-agricultural activities [87]. The future extent of pesticide contamination of deep groundwaters is very difficult to assess.

As previously mentioned, many factors affecting the movement of residues to water are largely outside product-specific parameters. Therefore strategies solely based on pesticide selection and use precautions would be inefficient without relevant action in the catchments to minimize the inputs to water by reducing the transfers of contaminants. Sound development of agriculture in respect to conservation and protection of soil and subsoil, surface and groundwater environments from degradation and contamination calls for modifications of agricultural practices. The objective is to combat the pollution problem by preventing contamination at the source, i.e. before it reaches the water compartment, without overlooking any other possible means. Except for a few products, only little work has been done to prove whether different agricultural practices can reduce contamination of the environment to a minimum.

7 Management Practices for Groundwater Protection

Agricultural practices have elevated various pesticides and plant nutrients to unacceptable levels in groundwater throughout North America and Europe. Groundwater pollution by pesticides in agricultural regions has been cited as one of the main problems leading to non-sustainable use of groundwater in the EC. Management practices are becoming increasingly important tools in protecting water supplies from contamination by pesticide residues. Most of the experience in this area has been gained in the past decade with products well-known for their leaching properties and their ability to contaminate groundwater. The detection in 1979 of agricultural chemicals in groundwater in certain areas of the United States (Central California and Long Island) has induced monitoring and research programs by universities, regulatory agencies, and pesticide manufacturers [88, 89]. As sustainable use of groundwater can be achieved only if pesticide leaching mechanisms are understood, the objective of these programs was to assess the degree of contamination of water resources and better understand the fate of agricultural chemicals in soil and groundwater.

In addition to conducting research and monitoring programs, regulatory agencies and pesticide manufacturers began to develop practices to protect drinking water wells from residues that might result from future agricultural chemical applications. Measures have ranged from complete prohibition for some active ingredients to adoption of management practices for others [90]. Management practices can be tailored to the use of a specific chemical under a prescribed set of circumstances. Potential management practices include optimization of product application and irrigation procedures, establishment of buffer zones around drinking water wells and water courses, and use restrictions in limited areas.

In many cases, it is difficult to determine whether the presence of pesticide residues in groundwater is due to accidental spillage, direct well contamination or genuine leaching as a result of normal agriculture practices. Identification of the cause of contaminant detection in groundwater is a necessary prerequisite to the development of effective management practices. Controlling diffuse pollution is perceived to present some difficulties. By its very nature, it will obscure the source of individual contributions, so limiting the use of legal sanctions.

Because of the many factors that can influence contamination of groundwater and drinking water wells and because groundwater contamination is mostly irreversible, an approach that concentrates on pollution prevention is the most effective means of protecting drinking water wells from contamination [1]. A variety of environmental conditions and human activities combine to affect the occurrence of pesticides in drinking water wells and no single factor can explain the presence of pesticides. Among the steps that need to be considered are appropriate reductions in the use of pesticides, site-specific assessments to accurately target vulnerable groundwater, identification and protection of groundwater recharge areas and protection of wellhead.

The ultimate goal in groundwater protection research is to develop management practices that prevent agricultural chemicals from reaching groundwater while still maintaining the benefits of continued chemical use. Development of effective management practices for agricultural chemicals includes several steps [90]:

- determine the groundwater to be protected and to what extent. In the EC, the surrogate zero residues limit for drinking water is also applied to groundwater, which means that no residues should reach groundwater,
- determine why residues may potentially reach groundwater,
- determine the degradation and mobility characteristics of the chemical so that the potential for groundwater contamination can be evaluated,
- develop, when necessary, chemical-specified management practices based on the properties and uses for the chemical.

Ideally, preventive measures should be tailored to provide the optimum of groundwater protection without undue restrictions on agricultural productivity. The potential for groundwater contamination at a given location depends on site characteristics, specific pesticide properties and use rates, and agronomic and management practices [91].

7.1 Product Application

The adverse impact of agrochemicals on the long term quality of shallow groundwater resources can be minimized or eliminated by effeçtive management of fluxes of these substances in the various soil and aquifer zones. Local groundwater quality can be maintained if the rate of agrochemicals input into soil is balanced by the rate of degradation to innoxious end-products, i.e. zero net flux.

One type of management practice is to optimize product application: timing, placement, dose rate. Delaying aldicarb applications to potatoes from planting to emergence in certain areas of the United States reduced the amount of residues likely to reach groundwater by avoiding heavy spring rainfall and by enhancing degradation rates as a result of warmer soil temperatures [92].

The placement of the product may also be important since applications that place the material near the soil surface may result in faster degradation rates compared to deeper applications. Rate reductions usually have less effect on the residue level than changes in other practices or differences in environmental conditions. Reducing the application to the minimum needed to provide the desired control does contribute to reduced residues.

In areas where water supply by rainfall must be supplemented by irrigation practices, the amount and type of irrigation can also affect the movement of residues in soil. Rainfall and irrigation water being the driving force for downward movement of residues, the amount of irrigation must be adjusted to compensate for evapotranspiration losses. Excessive irrigation increases downward movement of water and, hence, residues. The type of irrigation may also affect the potential for movement of residues. Unevenly applied irrigation (furrow irrigation for instance) may interact with the mode of product placement and influence residue movement [93].

Management practices should be developed for a specified product, crop and set of environmental conditions. Such practices may not be applicable to other circumstances. Under abnormal circumstances (unusually heavy rainfall, high groundwater velocity), management practices that are usually effective may be inadequate.

7.2 Protection of Drinking Water Wells

For a long time water suppliers have been aware of the potential dangers to both abstraction wells and aquifers from point source pollutions, such as leaking sewers or waste disposal. For example, guidelines for protecting against such pollution have been in force in Germany since 1953 and similar measures were passed by France in 1964. All adopted the stepped zoning of the drainage catchment around a wellhead, with potentially polluting activities being increasingly restricted as the well is approached. Although principally protecting the borehole from horizontal movement of pollutants, the methods also recognize the possibility of downward movement directly affecting the resources. The zone immediately around the wellhead covers an area in which only water abstraction activities should be permitted and can extend from 10 to 30 m from the borehole. The next zone, sometimes called the inner protection zone, is intended to be far enough from the borehole to allow time for destruction of pathogens by the natural cleansing capability of the sub-surface layers. Estimates of this time have varied from 10 to 400 days, but the most generally used figure is 50 days. Methods of estimating the speed of groundwater have been proposed and allow for converting the 50 day line into a distance on the ground.

The integrity of boreholes can be threatened by a range of persistent chemicals for which the 50 day line is irrelevant. Although defining an outer protection zone can be helpful in estimating time and space between the origin of pollution and the borehole, the longevity of some chemicals is such that their downward movement threatens even deep aquifers.

The threat posed by diffuse sources is more complicated to deal with as there are many starting points for contaminations. Agricultural nitrate is the first and more obvious example of diffuse pollution. Important advances have already been made towards understanding the problem and providing methodologies for both combating pollution incidents and protecting the most vulnerable aquifers.

Prohibiting applications within a specified distance of drinking water wells may be an appropriate management practice for those chemicals that continue to degrade in groundwater [94]. Restrictions requiring buffer zones around shallow wells have been adopted for aldicarb applications in certain areas of Canada and the United States. The basis for the distance specified in these restrictions has been a combination of field research, potable well monitoring and computer modeling [95].

Current EC legislation does not adequately protect groundwater from a number of known pollution threats. Proposals on future legal requirements will be made by mid-1993. It is moving towards a comprehensive policy of zoning around abstraction wells and designation of much larger areas of land where industrial and agricultural activities may be restricted to protect aquifers. Regarding restrictions of pesticide use, the extent of the protection zones will probably be subject to many discussions as the risk of groundwater contamination at the 0.1 $\mu g\, L^{-1}$ level remains technically extremely difficult to assess.

An important aid will be provided by the publication of national and regional vulnerability maps. Several groundwater studies have shown that triazine herbicides tend to reach groundwater more under cool climates and shallow soils of moderate permeability, or warm climates and shallow depth to groundwater and very permeable soils, or in areas where fractured media provide conduits from the soil surface to the aquifer [96]. Detailed maps could help growers choose the right agrochemicals to avoid polluting water. For instance, the SEISMIC project (Spatial Environmental Information System for Modeling the Impact of Chemicals), a joint project between the British Ministry of Agriculture, the British Agrochemicals Association and the Soil Survey and Land Research Centre (UK), aims to highlight areas that are more vulnerable than others in relation to use of specific pesticides on particular crops [97].

Geographical information systems have been used for mapping vulnerable aquifers, nitrate pollution or contamination pressure. Mapping systems aimed at identifying areas of agricultural land which are more vulnerable to groundwater contamination must be developed [98,99]. Combination and integration of different kinds of information resulting from different sources and experiments having a different resolution in time and space can only be done in an economic and feasible way using versatile information systems. Soil, crop and climatic databases must be combined with the properties and the use pattern of particular chemicals to help draw up appropriate management strategies, i.e. to predict which chemicals are safe to use in a particular location. Management strategies can be drawn up for individual chemicals on local and regional scales.

At the moment, combination of variability of parameters in modeling and use of geographical information systems require so many data that extrapolation to larger areas is not achievable [100]. The important objective of extrapolating data results to large areas remains conjectural. This significantly restricts practical applicability to environmental management.

In the long run, procedures for assessment of land suitability, soil vulnerability, sustainability of farming and environmental policies should be developed, using modern technologies: remote sensing for surveying regional information with direct coupling of geographical information systems containing static and time-variant site information to the integrated mechanistic modeling framework, advanced statistical procedures for interpretation of spatial and temporal variations of soils, and decision support systems in agricultural and environmental policy.

7.3 Use Prohibitions

Under some circumstances, management practices may not provide adequate protection for groundwater and restrictions prohibiting application of a chemical may be necessary. The prohibition of the use of several agricultural chemicals in the highly vulnerable region of Long Island (State of New York) is an example of such a restriction.

8 Management Practices for Surface Water Protection

Similar strategies, based on the understanding of pesticide fate and the development of management practices minimizing contaminant transfer, should be implemented for the protection of surface water. However, the complexity of the mechanisms involved in an extremely large range of environmental conditions and the urgency of remedial actions call for the short term testing of possible modified practices without gaining a useful but time-consuming experience. An important background lies in the considerable work done on problems of soil erosion and water contamination by nutrients. Since agricultural chemicals are transported to surface water using the same vectors, i.e. runoff water and suspended soil particles, environmentalists should take advantage of the experience in this area. Pesticides certainly exhibit a wider range of physical and chemical properties than nitrate or phosphate. Implementation of management practices to reduce water contamination by fertilizers proved to be effective and should be adapted to organic contaminants as a general rule.

In the United States, action against non-point source pollution is closely related to soil conservation. The objective is to limit pesticide losses (nutrients and heavy metals as well) which occur during rainfall events through runoff. Preventive and curative steps called Best Management Practices are taken at the level of individual farms on a contract basis. These steps are defined according to the results of scientific investigations and empirical observations. For each volunteering farmer, a plan for soil conservation is established by the Soil Conservation Service, with the objective to implement best management practices. Conservative practices are recommended, for instance contour plowing, crop rotation plans to shorten the period when the soil remains bare, careful tillage and establishment of buffer zones, filter strips and grassed waterways for instance.

One of the major problems is controlling the efficacy of these steps at various levels: field, farm and catchment. Implementation of best management practices are associated with research programs involving water quality monitoring in the field, the water courses and the aquifers. They are often conducted on a contract basis with universities (Vermont, Maryland, Oregon). Real size evaluation of best management practices are performed in farms or in small watersheds. Specific field experiments may be designed to accurately study certain aspects. Relevance of these practices in specific conditions is assessed through computer simulations using the CREAMS model [101].

8.1 Reduction of Contaminant Transfer

Complex mechanisms are involved in pesticide losses by runoff at different geographical scales. In addition to understand these mechanisms better, the objectives of many studies are to evaluate preventive and curative methods suitable for each scale. Human action to limit contaminant transfer can be conceived at each of these levels, providing that adapted methodologies exist.

As previously mentioned, pesticide mobility and persistence are critical factors. Small scale rain simulation and relevant modeling provide basic information of transport capability of a chemical in specific situations and allow for establishing mobility classes. Field experiments provide a way to estimate runoff concentration associated with the formation of rills and gullies. Impact of agricultural practices and other dispositions devised to retain microcontaminants should also be evaluated at the catchment scale. A potential exist in the watersheds for self-elimination of organic contaminants involving their retention and their transformation. The methods proposed below have been developed for soil conservation and applied to the reduction of contamination by fertilizers.

Non-cultivated buffer zones located near cropped areas are known to modify the soil hydraulic properties and are per se beneficial to fauna. They also contribute to restore the quality of water. Efficacy of grassed strips in restricting the movement of suspended particles and contaminants increases with the strip width but decreases with the number of flooding events. 70% of the concentration in solid particles and 30% of the phosphorus concentration is dropped when filtering runoff water through a 5 meter width grass strip.

The ability of buffer zones to filter and retain agricultural pollutants lies on basic soil properties. Well-known physical, chemical and biological mechanisms are involved, e.g. sedimentation, adsorption or precipitation, and degradation. Whereas chemical and biological processes require a relatively long period of time and therefore a long distance transfer, water movement in a field or at its boundaries occur rapidly on a short path. Irrespective of the size of a watershed, action planning to limit runoff and erosion should take into account where the transfers in liquid and solid phases take place and how intense their relative amplitude can be. Measures of different natures are probably necessary to limit the movement of mobile phases along their way downstream. Action at the source, i.e. in the field or nearby, is likely to be more effective because of the limited possibilities to reduce runoff as it tends to concentrate. The applicability to crop protection products of the techniques developed to limit fertilizer dispersion in a watershed has not been demonstrated on the basis of a sufficient number of relevant studies. Some techniques experimented with at the very beginning by the Soil Conservation Service may be adapted to efficiently trap pesticides in runoff water or eroded soil, i.e. use of buffer zones, including filter strips, channel terraces and grassed waterways.

8.2 Soil Conservation Practices

The possible effects of rate reductions have already been discussed with the low input strategies. Transport of pesticides occurs soon after application and usually averages less than 2% of the amount applied [38]. As surface water can be contaminated by small proportions of the products applied, rate reductions have less effect than changes in other practices.

The conservation program in the United States generated many results, sometimes inconsistent. Several studies demonstrated the influence of no- or minimum tillage with a significant decrease of runoff (up to 50%) and suspended particles transfer (10 to 20 times less). Reduction in the amounts of nitrate losses through runoff is related to increased infiltration rate due to no-tillage. When comparing the effects of modified practices with a control plot, the largest differences are observed in situations of heavy rainfall, with small soil coverage by the crop or in crusting and capping soil situations.

Field studies have demonstrated the possibility of reducing runoff losses of pesticide residues using appropriate tillage practices [102, 103, 104]. For instance, the influence of tillage systems and row orientation on runoff of soil applied pesticides (alachlor, carbofuran) was studied using a rainfall simulator [105]. No conservation tillage system completely eliminated pesticide runoff but contoured rows plus strip-till or no-tillage minimized losses most effectively by controlling both water and sediment movement. Significant changes of the ratio between concentrations of adsorbed and dissolved material were observed depending on hydrological and pedological conditions [106]. The relative importance of transport vectors were linked to local hydrological conditions. Metolachlor appeared to accumulate in footslope soils and then to be washed out by rainfall.

Based on the Wischmeier equation (Universal Soil Loss Equation) or other approaches describing changes in the soil surface, empirical procedures have been developed to assess the risk of erosion by agricultural practices. Runoff can be reduced by agricultural practices but to a limited extent. Additional specific management practices are necessary to control runoff and erosion under European climatic conditions [107, 108].

8.3 Proximate Protection of Water Courses

It is evident that the ability to farm right up to the river bank has made pesticide transfer to surface waters a lot easier. In addition to traditional soil conservation practices in the field and on hillslopes, different measures exist for the proximate protection of water courses against contaminant transfer. Steps to limit soil erosion and runoff have also been proposed to diminish the nitrate pollution. To some extent, they are likely to be applicable to organic pollutants. They are designed to limit the formation of runoff and protect the areas where it tends to concentrate. Therefore they are introduced in the immediate proximity of the permanent hydrographic network. These techniques draw their inspiration from those developed by the U.S. Soil Conservation Service to limit the effects of runoff and erosion. They have been studied and used principally in the United States [109] and adapted to other situations, Europe particularly [110, 111].

8.3.1 Buffer Zones

The transfer of solid particles near water courses can be significantly reduced through buffer zones that are able to filter the sediments and to favor runoff reinfiltration. Their ability to retain the water flow and make it less destructive is supported by numerous observations. The transfer rate of water and contaminants is strongly influenced by the hydraulic rugosity of the waterways. In the past years, cultivating crops up to the river bank has resulted in an increase in transfers of sediments and contaminants. Conversely, they can be lowered if the hydraulic rugosity is raised along water courses through natural and artificial obstacles such as hedges, brushes or vegetated strips.

8.3.2 Riparian Woods

Riparian woods are particularly good buffer zones. In addition to regulating the temperature and oxygenation of surface water, their beneficial effects include the improvement of the biological quality of water and the decrease of nitrate content in shallow alluvial aquifers. They also protect against bank erosion and non-point source pollution [112]. They tend to be more effective under European temperate climate than in harsher Amercian conditions.

A particular problem is posed by the possible exchanges between surface water and adjacent shallow groundwater. According to the season, water may move from ground to the river and vice-versa, with a similar transport of contaminants. Riparian woods have been used for efficient extraction of agricultural nitrate and phosphate, in addition to preventing direct transfer of contaminants to water. Regarding organic contaminants, this latter direct effect on transfer is probably efficient but little is known about possible filtering action on residues transported in the soil profile by sub-superficial flow [113].

8.3.3 Grassed Filter Strips

The loss of agricultural chemicals with runoff water should also be reduced by increased infiltration, reduction of sediment carrier and adsorption on organic matter using grassed filter strips. Little information is available about the efficiency of such buffer zones in various conditions.

Vegetative buffer strips have been proposed as an effective management practice for the control of non-point source pollutions from cropland or animal production facilities [114, 115]. A few publications have documented the ability of vegetated buffer strips to decrease significantly the pesticide content in runoff water [116]. The efficiency of buffer strips has been demonstrated primarily in experiments of short duration with simulated rain. They are designed primarily for the control or erosion and work reasonably well with coarse soil particles. Larger strip widths seem necessary for smaller elements or soluble contaminants [117, 118]. Particle size selectivity during transport and dispersion plays a crucial role in the total amount of residues exported. As pesticides bind more readily to small size particles having a higher adsorption capacity (concentration factor), the width of buffer zones should be sufficient to retain the larger part of fine particles. As retention of the highly soluble nitrate ion has been clearly demonstrated, it is likely that contaminants with low sorption coefficients will be also significantly retained. Due to the variation in particle distribution during the transport, partition conditions of a chemical are modified as a function of the distance from the starting point. Sorption coefficients derived from runoff-erosion measurements are highly variable. Important discrepancies with laboratory experiments are commonly observed. Phase equilibrium is probably never reached in the latter dynamic circumstances contrary to laboratory conditions. However, the evolving partition conditions along the transport path probably account for the major differences. Determination of sorption coefficients suitable for assessment of contaminant movement through runoff is critical. Description and modeling of important mechanisms cannot be solely based on a single value of the partition coefficient but rather on a dynamic distribution of this parameter.

Contaminant losses during rainfall events start with runoff formation and take place in a very short time. Transport duration is also a function of hydraulic conditions of overland flow and particularly of natural or anthropogenic contaminant trapping systems. Pesticide persistence is also an influential factor inasmuch as the period between product application and the time of rain can significantly affect the amounts of residues likely to be lost by runoff and erosion [103]. Pesticide persistence in deposited materials with different moisture contents must also be taken into account. As physical and chemical conditions in sediments differ from those of a cultivated field, pesticide degradation rates are likely to be modified. Little information is available on pesticide persistence in these conditions and comparisons with field measurements are difficult.

The retention capacity of grassed buffer strips is probably significantly dependent on local soil properties, infiltration rate as a function of moisture status, soil organic matter content and vegetation coverage. Complete elimination of contaminants might not be achievable in certain situations if prohibitive buffer widths have to be implemented. In this case, avoidance of excessive fluxes of contaminants to the water bodies in the weeks following pesticide application should be a primary goal complemental to other practices.

The efficacy of buffer zones must be evaluated on the field scale with a wide range of soil, topology and crop conditions, and pesticide environmental characteristics. Buffer strips should be designed according to local factors such as slope, hydrology and vegetation. Determination of strip width is empirical. Using different methods, the width can be calculated for the retention of a certain fraction of sediment or for a given infiltration capacity. Avoiding the formation of concentrated runoff in rills or gullies and filtering the contaminants in the solid and liquid phases are two important objectives that should be correctly balanced. The usefulness of facilitated infiltration is also a difficult matter of discussion as buffering effectiveness might be reduced or even be useless in saturated or flooding conditions. Nothing is known about the filtering performances of aged buffer zones and the maintenance they need to maintain the required level of performance.

8.3.4 Dispersion of Concentrated Runoff

If uniformly spread over the soil surface, runoff has less transport capacity and smaller mechanical effects than if it were concentrated in rills. Consequently, the filtration capacity of buffer zones is seriously depleted when the vegetation is flooded or the runoff flow is concentrated. Ideally, the flow should be uniform along the upper edge and laminar through the grassed strip for efficient filtration. Concentrated runoff should therefore be dispersed prior to entering grassed buffer zones. Small grassed terraces across the thalweg can intercept and disperse the concentrated runoff. Although their retention and infiltration capacities are spatially limited, they operate like channel terraces (water and sediment control basins) spreading the flow over a larger width. They should contribute to maintaining a diffuse runoff flow along the slope for appropriate filtration conditions of the buffer zone located downstream.

8.3.5 Grassed Waterways

Grassed waterways have been used for many years as a means of controlling water movement and soil erosion from farmlands. By diverting runoff from cropped areas through adjacent vegetated areas, loss of suspended soil particles may be reduced. The results indicate that the vegetation is able to intercept a large portion or the contaminants that enter the grassed waterways. They operate exactly the same way as long grassed filter strips. Their location at the point where runoff concentrates and their important filtering length make them very effective in removing contaminants in both the solid and the liquid phases. They appear as possible complementary steps where buffer zones cannot be efficiently implemented.

8.3.6 Location of Buffer and Infiltration Zones

Once the buffering potential of various elements conducive to reducing contaminant transfer has been experimentally determined in specific field conditions, another important problem is the location of buffer zones throughout a catchment to improve water quality at the outlet. One can hardly imagine that grassed strips could be installed at the bottom end of each field because of the excessive and uneconomic cropland reduction this would induce. It is probably wiser to implement buffer zones in critical spots where direct contaminant transfer to permanent water courses occurs.

Most of the time, temporary water flow resulting from concentrated runoff during rainfall events reaches the river at specific locations. For proper operation, buffer zones must be located at such critical points. A good understanding of the relationships between farmland and the aquatic compartment is gained by spotting the preferential runoff paths, the areas of runoff production, the existing buffer zones along the temporary waterways (meadows, hedges, etc) and other singularities likely to affect the transfer of water and sediments to the river. Some traditional elements of the countryside landscape such as hedges and embankments that disappeared with modern farming in the past two decades could also be restored at critical locations inasmuch as they constitute natural obstacles to water and contaminant movement.

The strategy to adopt hinges on both local and regional conditions as they directly influence runoff intensity. Rolling or hilly areas with short slopes and small size catchments may accommodate light measures at the field scale. In contrast, large watersheds with weak but long slopes collecting rain over an important acreage are prone to induce high runoff flow rates. This situation calls for more complex steps to limit the transfers.

8.4 Implementation of Best Management Practices

Increasing management can certainly result in minimizing non-point source pollution but is facing enormous technical and social difficulties. For practical and political reasons, the concern for environment protection leads to action being taken before the situation is fully understood and the possible solutions comprehensively studied. Despite the poor knowledge of the performances expected in multifaceted conditions, corrective measures based on soil conservation practices are a good principle to follow. A more difficult question is how to motivate farmers to increase management intensity? It has been learned from the nitrate experience that a limitation of contaminant fluxes implies the reduction of inputs and losses in the field. Adjusting the supplies to the plant needs is an insufficient rule of thumb which explains why the strategies for better use of nitrogen fertilizers have been so difficult to determine and implement. Strategies for optimized use of agricultural chemicals are even more difficult to adopt.

Implementing management practices may enforce farming practices that favor environmental quality at the expense of agronomic productivity. This will probably make it difficult to get the farmers' agreement because of a possible negative impact on their income and the additional workload they entail. Maintaining uncropped zones along water courses or beside fields is negatively perceived as they constitute an important reservoir for pests and weeds. Minimum maintenance of buffer zones must be planned, for instance using mechanical means, providing they are not detrimental to wildlife, mainly in spring. Incentive steps must therefore be taken by public authorities to facilitate their implementation, compensate for the loss of farmland and the maintenance expenses. Implementation of buffer zones along the river banks implies that a strong consensus exists on separating the water course and the cropland. Politicians will determine if recommended solutions will be implemented voluntarily or otherwise [119]. In the United States, intensive management practices referred to as the prescription planning process are based on the voluntary participation of farmers.

Another limiting factor is that officials sometimes seem reluctant to promote these steps because of the difficulties encountered with their enforcement. Demonstrating the benefits of modified practices and the efficacy of investments is of primary importance from the political point of view. The scientific recognition of benefits resulting from management practices is a difficult task.

Effects of controlled factors are not easily discriminated from those due to uncontrolled conditions such as weather. Observing biological indicators is often a way to report the improvement of water quality which becomes visible only after a relatively long time because of the environmental inertia.

As soon as corrective measures are undertaken in a watershed, correlations between water quality and farming practices should be established, based on a detailed survey of the products used and the monitoring of water. Ascertaining whether the implemented steps are beneficial is assisted by the acquisition of numerous analytical data at the watershed outlet and, if possible, comparison of results between twin catchments. Long term observations are fundamental to assess the impact of modified management practices. Modeling water and contaminant transfers at the field and watershed scales is a privileged instrument for data integration throughout a large range of environmental conditions.

Starting with the existing experience on the reduction of water contamination by management practices, it is extremely difficult to generalize the results with hindsight in different situations. Neither a clear nor simple relationship is established between land use and agricultural pollution because of the large number of determinant parameters, their inherent variability and the complexity of land organization. Despite this considerable uncertainty, introduction of best management practices has proved to be efficient in improving the quality of surface water. Results are sometimes contradictory as control of erosion and runoff is conducive to enhanced leaching if the American experience is followed too closely in a different set of environmental conditions. In the United States, a long tradition of anti-erosion practices has favored certain strategies. Fighting pollution at the start is not the only measure since economic feasibility is questionable. The objective is a control of transfers in respect of the pollutant properties, as illustrated by different strategies for phosphate and nitrogen, at the field or large watershed scales, respectively. As demonstrated for phosphate, strategies based on trapping contaminants during their transfer are relevant in the case of low solubility pollutants. Losses are spatially concentrated, thus making buffer zones very effective. Combatting the transfer at the start cannot be overlooked. Buffer zones are just one way to restrict pollutant movement which does not excuse us from minimizing the losses using modified agricultural practices and an increase of soil coverage by crops.

8.5 Non-Conventional Contaminant Transport

8.5.1 Macropore Flow

Significant contamination of surface water by various pesticides, irrespective of their potentials for movement in soil, has suggested the possible role of soil macropores in sub-superficial contaminant transfer [120]. In some circumstances, classical methods for assessment of pesticide movement in soil seem irrelevant. Flow processes determining the behavior of agrochemicals in homogeneous soils are well known but some situations exist where conventional flow theory does not apply [121, 122]. Preferential flow paths in soil are widespread. Two important causes are the presence of shrinkage cracks or other macropores and the occurrence of wetting front instabilities. Water repellency of soil is indeed an important factor in inducing unstable wetting fronts and preferential flow. As a result, water and solutes may move and reach groundwater faster than expected on the basis of their average residence times. Characterization of conditions conducive to preferential flow using water monitoring and modeling is important as this mechanism is responsible for inducing losses of agrochemicals to groundwater and surface water.

Experiments are needed to make a clear distinction between macropore flow and regular infiltration but corresponding methods are not readily available nor easily applicable. Preferential flow of water and solutes to groundwater and surface water is one of the important processes contributing to the loss of agrochemicals that has not been adequately modeled yet. Modeling solute transport is fissured soils, especially in the case of nonconservative compounds, is still inadequate. For soils that are moderately dry, ca.20% of field capacity, pesticides can be transported through macropores to drains and to streams by lateral flow. Identification of areas where important macropore transfer may occur is based on soil properties (texture, hydrology) and presence of drainage. Oriented surface water monitoring, for instance at exits of tile drains, may help to discriminate between contaminant transport by direct runoff or through macropores and drainage.

8.5.2 Colloid-Facilitated Transport

Colloids dispersed in the soil solution typically increase the apparent solubility and transportability of pesticides by adding another mobile phase to the soil system [123, 124]. Soil colloids' mobility depends on colloid-size distribution. The different modes of colloid immobilization depend on soil chemical properties that affect colloid flocculation and dispersion within the soil solution and flocculation or adsorption to mobile soil solids. Without mobile colloids, pesticide leaching depends on its organic carbon partition coefficient. When mobile colloids are present, pesticide leaching depends on its chemically-driven distribution between stationary soil colloids, soil solution, and mobile colloids [125, 126]. The importance of suspended colloids increases as the pesticide water solubility decreases or the pesticide binding strength to mobile colloids increases. Colloid facilitated transport is an important process that enhances pesticide movement into groundwater or subsurface drains in agricultural regions [127, 128]. Preferential flow of colloids through larger pores causes particle movement to be faster than the average water movement, thus enhancing sorbed contaminant mobility. Colloid-facilitated transport will depend on regional climate, soil properties, particularly organic matter characteristics and nature of binding between pesticides and soil components. Theoretical calculations suggest that the more strongly a contaminant binds to natural soil organic colloids, the more significant colloid-facilitated transport becomes relative to dissolved contaminant transport [129]. The amount of potentially mobile colloids in soil, the strength and stability of pesticide binding and the mobility of colloids are factors that affect this non-conventional transport.

At the moment, little information is available about macropore flow, colloid-facilitated contaminant transport and their contribution to water pollution. Once again, the limiting factor is the lack of adapted methodologies to study these mechanisms in the field and to correctly represent their spatial variability for purposes of land vulnerability assessment.

8.5.3 Adapted Management Practices

Strategies can be aimed at reducing the movement of residues to drains and hence surface water by preferential flow. They include soil management practices and pesticide application methods.

Deep plowing or adapted soil preparation techniques may disrupt macropores and provide a slower downward movement of residues, allowing a better degradation in surface soil. The decrease of soil organic matter content has been a concern with intensive monoculture. Incorporation of crop residues or techniques likely to increase organic matter content is known to favor biomass development and improve soil structure. This should also provide greater soil adsorption capacities for pesticides and enhance degradation potential. Soil incorporation of pre-emergence herbicides in surface soil could lead to a more even distribution of the chemical in the soil top layer and minimize its availability for washoff through macropores. Unfortunately not all soil applied active ingredients which are effective through root absorption are suitable for soil incorporation.

Soil moisture is expected to play an important role in macropore formation and fast water flow versus regular infiltration. Little information is available on the magnitude of transfer as a function of soil moisture status. Application timing could be adapted to precipitation forecast if acceptable statistical relationships can be established between the soil moisture level and the transport capacity through macropores. Obviously such practices require a high level of sophistication for field practical application and are not likely to be implemented before comprehensive field studies have demonstrated their effectiveness.

9 Non-Agricultural Use of Pesticides

Herbicides are largely used in non-agricultural situations by various organizations: local authorities (city and regional), transport (railway, airport and port authorities), power and industry (maintenance of industrial sites and storage), leisure areas (golf courses, recreation areas).

There are significant differences in the types of herbicides applied in non-agricultural conditions according to the users, primarily influenced by their weed control requirements. Except for leisure areas which require skills for selective weed control, total weed killing is obtained with simple inexpensive formulations often containing one single active ingredient among the more common chemical families: triazines, phenoxies and ureas. When authorized, triazines represent a large proportion of the volume of active ingredients used. Persistence of action is sought to keep the treatment frequency at a minimum. Different products may be applied by contractors with rotating treatments on a cycle of several years, a different active ingredient being used each time.

A movement towards best practices takes place with non-agricultural use of pesticides. The debate has moved away from the simple question of "spray" or "not to spray" towards a concern to develop best practices which can balance the need for effective chemical weed control and the desire to reduce environmental impact. This is a learning process, as organizations evaluate the consequences of modifying practices, including costs, environmental impact, level of weed control and possible implications for herbicide use in the future. Changes can be made at a number of stages: specification of chemicals as suitable and not suitable, restriction of the number of chemicals that can be used according to the situation and specification of time of year when selective or non-selective herbicides can be applied. Instructions for non-agricultural use of products may differ significantly from those given for normal agricultural conditions. The target area requiring weed control is somewhat smaller than the sprayed area because of soil impermeabilization, imposing thus an adjustment in application rates (paved areas, railroad, parking lots).

Practices adapted to non-agricultural uses have focussed mainly on products rather than the way they are applied. Many water surveys have highlighted the role of urban and industrial pollution by pesticides. Following the prohibition of certain active ingredients, the replacement of a product by another, aminotriazole for instance, has made the water monitoring difficult so that subsequent improvement of water quality remained questionable. Following the example of agriculture, similar efforts should be devoted to the right use of pesticides in non-agricultural situations by integrating site-specific land and weather conditions.

10 Land Use Strategies

To efficiently implement best management practices in proportion to the threat, the risk of contamination must be assessed over large cropping areas, taking into account all relevant information available on products, land and weather. Recent studies have indicated the importance of spatial variability of hydraulic properties and physical and chemical properties of the solid and solution phases on the fate and transport of agricultural chemicals and dissolved substances. Research is needed to characterize the spatial vulnerability in terms of hydraulic properties and critical parameters influencing the reactivity of the systems, their effects on transport, and to define parameters upscaling strategies. For this purpose, experimental data at various scales are required for each site.

10.1 Environmental Risk Analysis

Land suitability, soil vulnerability, sustainability of farming and environmental policies cannot be assessed without the development of environmental risk analysis systems for field, farm and regional levels, providing a rational basis for the evaluation of environmental risks from different agricultural systems and land use practices. The objective of several research programs is to develop or improve tools to evaluate the impact of different farming systems on soil and quality of water and particularly predict the likelihood of exceeding concentrations defined as critical in water. They use data derived from soil maps to provide regional impact assessments with respect to risk from product use.

Developing risk analysis systems to use at the farm level is also an important goal. Prototypes of integrated farming systems have been designed to make the necessary comparisons between conventional farming and integrated farming systems, in terms of emission of pollutants. At the moment, the state of the art lacks the adequate linkages between field, farm and regional models. Actually most of the existing models are too-short term, too restricted either spatially or commodity wise and fail to place soil and water related technological risks in their agro-economic setting.

10.2 Structural Change

Agriculture in developed countries and particularly in the EC is facing enormous economic and environmental problems so that future policies should be aimed at changes in land use. Changes in farm management practices or technologies are expected in response to environmental regulations, codes of good conduct, or specific market evolution.

Embarrassingly large supplies of agricultural commodities, with production well ahead of effective demand, led policy makers to seek ways to control production. European governments have introduced measures to bring production of the major crops in line with demand. There are also external pressures through the General Agreement on Tariffs and Trade (GATT) to operate a substantial and progressive reduction in support and protection for agriculture. All these changes and proposals are taking place in a climate of growing concern for the environment. But surpluses, which are seen by many to be based on too high a price structure and the overuse of pesticides and fertilizers, have been greatly reduced. The relative fragility of the balance between surplus and shortage cannot be ignored as demonstrated by recent droughts in some parts of North America and Europe. Natural catastrophes recur at unpredictable intervals and few of the critics against modern farming methods recognized that production might fall and surpluses become strategic reserves [130]. In the near future, structural change will reduce cropping areas which account for a major proportion of agrochemicals and fertilizers. Farmers will have the option of reducing output by cultivating a smaller area more intensively or switching to less intensive production practices, with opposite effects on technological risks.

The major thrust to removing land from farming use lies under the set aside policy in the EC. Set aside land may make a small contribution to reducing agricultural production and environmental problems but it can create serious difficulties for farmers. Set aside will provide sites for overwintering pests and diseases and some options for the management of set aside conflict with the principles of weed control. For instance, without herbicide treatment, it will be difficult to prevent seed return from weeds, even with regular mowing. The use of set aside to create buffer zones where intensive agriculture may be suspected of having deleterious effect on water quality has been suggested. Inclusion in the rotation of fields selected for set aside should be also considered so that uncultivated fields would also contribute to positive environmental effects.

10.3 Farming Systems

Most studies to-date have considered the negative impacts of production and land use from an essentially technical and static viewpoint. Only limited attention has been given to examining future changes in the nature or magnitude of the risks as a consequence of structural change in agriculture and other socioeconomic pressures. Experiments are being carried out towards integrated farming systems or other changes in land use which must lead to sustainable agriculture. These systems use less pesticides and less fertilizers without loss of profit by maximizing farm income and minimizing the inputs. But their actual impact on the environment has not been fully investigated. Much more work is required to evaluate the necessary changes in land use for a sustainable agriculture and identify farming systems that are commercially and environmentally feasible.

Changes affecting commodities mostly out of balance with market demand, or in dispute with international trading partners will strongly impact the major users of arable land, pesticides and fertilizers. This is a strategic key point but out of the scope of a technical approach. Because of the swift and for the most unexpected changes at the end of the 20th century, strategies based on availability of or access to basic food resources are not likely to be abandoned just for environmental concerns. Therefore, risk analysis systems at the farm and regional levels, including predictions and their probabilities, need to be developed in the light of necessary changes in land use to meet environmental directives and other socioeconomic aspects.

11 Water Treatment

In developed countries, availability of high quality drinking water has become so common that one can hardly imagine that sophisticated techniques are necessary for its production. Because water is sensitive to pollution, its quality is nowadays intensively monitored and subject to drastic regulations. Analytical techniques for the detection of organic pollutants, purification processes and protection of the water resources are facing tremendous challenges. Therefore, the needs for drinking water supply can be conflicting with those of agriculture, not only quantitatively but also qualitatively.

11.1 Raw Water and Drinking Water Quality

Raw water is generally not sufficiently pure to be used directly, irrespective of its origin. Contamination by natural or artificial substances occurs during its movement in and on the surface of soil. Among many possible organic contaminants such as fuels, solvents and other chemicals, pesticides are in no case the more quantitatively important ones detected. For instance, long-term leakage, spillage or illegal disposal of engineering degreasing agents (chlorinated hydrocarbons) have caused extensive contamination of aquifers. Therefore treatment of water is necessary to make it suitable for use by man. Often considered as a symbol of purity, water has progressively become the most regulated food, subject to the most stringent quality standards. Consequently, sophistication of the techniques used for the production of drinking water has been increasing and is likely to continue to progress.

Most of the time, water quality rather than human safety is an issue where pesticide residues are concerned. In most of the cases, residue levels are far smaller than the concentrations which are likely to have a toxic effect. In the EC, the responsibility of keeping drinking water below extremely low limits falls upon the water companies who are having to spend important sums of money monitoring supplies and removing pesticides which are in any case not threatening health [10, 11].

11.2 Treatment of Contaminated Water

Water treatment is the last possible curative step to meet quality standards before potable water is delivered to the consumers. Enactment of the EC directive on drinking water has led to the implementation of industrial processes for removal of pesticide residues. Another reason is the need for elimination of natural compounds present in water at trace levels and responsible for bad tastes or bad smells. Treatments for pesticide removal from water are therefore not specific. They must handle a large variety of molecular structures of xenobiotics and be efficient on natural organic matter which exhibits also a considerable chemical diversity.

Studies at the water plant, together with bench-scale studies, on pesticides commonly monitored in surface waters from agricultural areas have demonstrated poor control by conventional treatment processes. The relatively high adsorption capacities of these chemicals indicate that granular activated carbon can be cost-effective for their elimination from raw waters [131, 132]. Powdered activated charcoal applied at rates used for the control of tastes and odors can also be effective in some conditions [133]. Many water companies were concerned about the performance of the treatment technology to achieve the 0.1 $\mu g\,L^{-1}$ limit for a wide spectrum of compounds at a very low concentration. Research by French water companies in the field of pesticide removal directed largely towards optimizing granular activated charcoal (GAC) filtration with respect to carbon type, contact time and regeneration intervals has proved that the treatment strategy calling for GAC and ozone is the correct one.

The elements of a potable water plant may include a mere chlorination step in the case of groundwater treatment or involve a set of different steps such as clarification, oxidation, adsorption, etc. Removal of pesticides commonly found in water can be low if the treatment is not specifically intended for elimination of microcontaminants and organic matter. Efficacy of clarification including coagulation, flocculation, sedimentation and sand filtration, is strongly dependent on the nature of the contaminant and the quality of raw water, and usually not sufficient. The most efficient treatment is adsorption on granular activated carbon used as a second filtration, following sand filtration. GAC efficiency is variable according to pollution characteristics (concentration peaks) and declines with time but its lifetime is lenghtened with previous ozonation. Addition of hydrogen peroxide in the ozonation step significantly improves the oxidation. An ideal water treatment plant should therefore include optimized clarification and oxidation steps and use of activated charcoal for proper removal of chronic organic pollution. Modern water plants only, including preozonation, clarification, oxidation with ozone and hydrogen peroxide, and GAC filtration, can produce drinking water in compliance with present regulations at a reasonable cost [134].

Seasonality of pesticide occurrence, primarily in surface waters, is a major obstacle for efficient industrial treatment of water. Apart from the extra production costs resulting from the new regulations, treatment plants must be scaled up to be large enough to cope with swift contamination peaks following pesticide application periods, with a considerable increase in investment and running costs. Water companies have different strategies to supply drinking water in compliance with standards, irrespective of raw water quality. Warning stations located upstream from large treatment plants are designed to detect contaminants before they reach the plant inlet. If the treatment capacity is not sufficient to take out the contaminant, usually of unknown nature, the plant may stop producing drinking water for several hours, with considerable economic impact. Potable water supply can be maintained during crisis periods with a temporary reserve of finished water stored beforehand. Raw water may also be stored as a precaution allowing continuous operation of the plant. In populated areas, several water works using different raw water supplies can be interconnected, alleviating the impact of the discontinued operation of one plant.

Drinking water production from surface waters may be more difficult to achieve for small cities in rural areas because the necessary investment for adequate water treatment has not yet been made. Promising results are expected from small treatment units using reverse osmosis [135] or nanofiltration [136]. A nanofiltration membrane has the sieving properties of an ultrafiltration membrane as well as the diffusion properties or a reverse osmosis membrane. Rejection of organic contaminants by nanofiltration is dependent on their molecular weight. Total or partial elimination of certain synthetic organic chemicals using this technique was demonstrated at the pilot plant scale.

Using modern treatment processes, the quality of drinking complies with European legal limits for the most widely occurring pesticides in surface waters, providing concentrations in raw water do not exceed roughly 3 $\mu g\,L^{-1}$ [137]. Where pollutant concentrations are higher (above 5 $\mu g\,L^{-1}$), more extensive GAC treatment is required and, even at 30 minutes contact time, occasional exceeding of the 0.1 $\mu g\,L^{-1}$ limit may occur in the finished water.

11.3 Cost of Water Treatment

In several countries, it has been estimated that it could cost several billion dollars to meet the European Community legal limit for pesticides. The question as to whether bringing pesticide residue levels within such low limits was the most cost-effective way of improving water quality has been raised many times. For the water companies, more promising returns are available on investment in modernizing water mains, which lose almost a quarter of the water they carry through leakage. Metering is almost universal in Europe and is an effective way of maintaining the inexorable rise in demand for water within acceptable limits. The price of tap water will increase as a consequence of a better quality. However, what is expensive is the service provided to the consumers, not the raw material itself and the treatment represents less than 20% of the service total cost. The treatment of sewage is often more costly than the production of potable water.

11.4 Water Treatment and Protection of Water Resources

Water treatment and protection of the water resources are not conflicting strategies. A suitable treatment is necessary for surface water to meet quality standards and is required to eliminate natural or synthetic chemicals, should the starting material be free of pesticide residues or not.

Irrespective of the raw water quality, water treatment is mandatory to achieve quality standards and is just one piece of the puzzle. Hence the need for close cooperation between participants from agriculture and water industry to minimize the contaminant occurrences in raw and finished waters and protect the resources using all available means. Actions tending to make water monitoring and water treatment more effective by sharing knowledge and techniques between pesticide manufactures, users and the water industry should be encouraged.

12 Communication on Pesticides and Water Quality

The public view that all water is contaminated, any trace of pesticide in drinking water puts people's lives in jeopardy, and all pesticides are evil urgently needs addressing. Pesticides definitely pose a problem but it is not as bad as in many press reports. The pesticide industry has been one of the most conspicuous to the public eyes. Therefore it has to demonstrate that it is devoting a considerable effort towards making the world a better place to live for everyone. The pressure on the agrochemicals industry tends to overshadow the increasing need for food due to world population growth. More intense demand on the world's natural resources is required to keep pace with the already inadequate per capita nutritional intake, thus justifying intensive cropping systems under temperate climates. However environmental problems associated with intensive farming have not been fully anticipated by agricultural policies based on yield increase. Water contamination issues should be set at their right place among other environment protection concerns. Complex strategies to minimize water contamination by pesticides are unlikely to reach their objectives without proper education of all the participants involved, primarily farmers. Furthermore, communication with politicians and the public should emphasize the improvements in environment quality and balance them neatly with the probably questionable economic profitability of the measures they require.

12.1 Public Concerns for Environment Protection

Among other concerns like global change, chemical residues in food or the ozone layer, the idea that the environment is endangered by chemicals, especially pesticides, is perceived worldwide and frequently raised by the media. Public opinion now considers the use of fertilizers and pesticides in modern agriculture as a plague for the environment. 94% of European people think that excessive use of agricultural chemicals is one of the major sources of pollution.

Whereas food shortage was still fresh in the minds during the post war decade, people are no longer concerned by the need for food to survive. Evidently, the satisfaction of basic needs with sufficient quantities of commodities is largely due to the outstanding progress in land management practices, seed breeding and use of fertilizers and crop protection products. Individuals now focus their attention on quality of life for themselves and for the future generations. Along with this concern, the excess of which is an unjustified food scare due to the presence of residues in crops and water, chemicals and particularly pesticides are perceived as responsible for poisoning the environment on a planetary scale. Aquatic compartments appear to be most vulnerable because of the ability for contaminants to be transported and disseminated over large distances throughout water bodies or persist in groundwaters for extremely long periods in comparison to human lifetimes. Furthermore, the public is extremely receptive to water contamination, making it one of the most emotional issues.

The public has little understanding of pesticide use, benefits and risks. The general perception is negative with the feeling that natural solutions are far better than artificial ones. The media tend to reinforce this perception, highlighting the worst aspects despite the overwhelming evidence that risk to human life is extremely small in comparison to other human activities. The public does not distinguish between different chemicals, concentrations and toxicity, so pesticides are perceived as universally bad. Public knowledge is based on a few spectacular facts often presented in a simplistic and emotive way. Agriculture is viewed through the drawbacks of commodity surpluses, pesticide contamination and government subsidies. Public perception of risk lacks the minimum expertise required to correctly interpret technical data. Performances of new generation active ingredients remain totally unknown whereas the increasing number of commercial products is anxiously retained without proper judgment. Technical and economical constraints of farming are not taken into account. Conversely, false alternatives, unsuitable for the supply of food demand are proposed. Past problems with pesticides such as DDT have not helped but there is still a strong distrust of officials and experts who have been proved wrong before. Unfortunately, water standards for pesticides not based on health risks but set by politicians as a surrogate zero residues are prone to feed public distrust.

The influence public opinion is having on pest control technologies is certainly increasing. Pesticide regulations have led to the cancellation of products such as daminozide and dinoseb. The agrochemical industry has also received increasing attention from various environmental pressure groups. This trend is expected to continue as better information becomes available. Up to now, environmental problems have been publicly addressed and responded to by ill-informed, unscientific pressure groups who have been allowed to exercise a dominant role in a time of agricultural surpluses. Agriculture and its dependent industries must not continue to fail in communicating the scientific truth.

12.2 Public Education

As most of the criticism against the widespread use of pesticides in modern agriculture is not presented in a rational way, considerable efforts should be devoted to educating people about the real benefits of agricultural chemicals. Positive progress towards reducing the impact of agrochemicals on the environment is certainly one way to say something new in this area and re-balance the ratio between risks and benefits of pesticide use. Benefits of crop protection are easy to convey, evaluation of risks much more difficult. It becomes urgent to restore a certain common sense about agriculture and crop protection in order to valorize in the eyes of the public the outstanding progress made by modern farming and related industries. Educating the public and addressing their fears about health and safety-related aspects of pesticides is identified as a major hurdle to jump over [138].

12.3 Farmers Education

Good pesticide use has to be improved constantly through appropriate education of farmers. Campaigns aimed at promoting careful use of pesticides in agricultural and non-agricultural situations in the context of water quality have been launched by associations of pesticide manufacturers in many countries. The goal of such campaigns is to make pesticide users aware of the need for careful handling of products in the vicinity of water courses and drinking water wells. This applies to selection, storage, preparation and application of the product and to the disposal of spray tank washings and containers.

In order to help reconcile economic objectives of modern farming and protection of the environment, best management practices are promoted by organizations in various countries, for instance FIP (Födergemeinschaft Integrierter Pflanzenbau) in Germany and LEAF (Linking Environment and Farming) in the United Kingdom. As responsible use of pesticides grows with appropriate farmer education, and the number of major incidents decreases, the challenge of diffuse pollution will become more apparent. In the long range, educational effort will have to focus on management practices at the field and catchment scales. As previously mentioned, best management practices call for intensive follow-up and convey important advisory information to farmers, irrespective of the system adopted in implementing these practices. Farmers must promote the image that they apply agrochemicals responsibly, in a professional manner, and communicate effectively with local people and visitors explaining the advantages of agrochemicals use.

12.4 Confidence Crisis Versus Science-Based Communication

The agrochemical industry, farmers and the food industry share a common interest and it is one of education and management. It is necessary that the pesticide industry ensures that its customers properly understand how to use their products and have the right equipment to do so. At the same time, it is absolutely vital that the general public knows and understands why pesticides are being used, what regulatory systems are in place to control their use and receive a meaningful assurance that protection of water resources from the environmental and human health points of view is taken into account. For successful efforts in communication and education, scientists and experts must recover the full trust of public opinion so that measures can be calmly proposed and expressed in terms of regulations by public authorities without any press overdramatization.

13 Conclusion

Pollution of water in the present legal setting has created an uncomfortable situation for farmers, pesticide manufacturers and water authorities. All agree on the need to improve the quality of water and protect the water resources. Despite growers following recommendations from public authorities and manufacturers, pesticides are still reaching the water. Agricultural practices would have to change and there is already a move in the right direction.

Future improvements of water quality in agricultural catchments are likely to depend on the significant progress in a few key areas [139]:

– Research

The pesticide industry is committed to devoting all the necessary efforts to discover and develop new plant protection products which minimize the risks to health and the environment. There are deficiencies in our understanding of processes of pesticide transfer in soil and pollution pathways. Better monitoring, predictive modeling and analytical techniques are required to allow for better judgments of the impacts of agricultural practices and land-use change on the water quality in combination with economic profitability. The research also needs extents to the area of social policy to give a more-balanced account of the decisions in relation to environmental and social pressures.

– Best Management Practices

A combination of different pest and weed control methods should be based on adequate crop rotation and tillage practices, optimized crop planting and pesticide application rates, using monitoring techniques, weather forecasting and information and decision support systems. In addition, land management should call for adapted means to limit the transfer of contaminants out of the treated areas.

– Information and Education

All people involved in land management, from farmers to land-use planners, should be aware of the possible impacts that their activities might have on the water environment. Adequate information of different forms should be made available to guide various activities: risk or vulnerability maps, soil and product databases, etc. The public should also be better informed to avoid excessively emotional reactions. Otherwise politicians will be pushed into making silly decisions.

– Cooperation

As many participants in different disciplines are directly or indirectly involved in water management, cooperation between product manufacturers and users, industry research and universities, water companies, water agencies, regulatory bodies is urgently needed to seek economic and optimum solutions consistent with environmental objectives and modern agriculture. Future strategies to protect vulnerable water sources from contamination by pesticides will require considerable cooperation between agriculture, industry and government. The concept of responsible care must be applied globally and the industry will continue to make an important contribution to this task.

There is a more difficult question to answer. Who will pay? This problem is particularly crucial for farmers when the price of commodities drops inexorably and production costs need to be adjusted. From the consumer point of view, the effects of stringent potable water standards on water bills cannot be ignored. At the time EC drinking water standards were set, cost implications were not fully recognized. Economic considerations should not be allowed to jeopardize health or the environment but they cannot be left out of any consideration on risk and benefit. Maybe this question has been raised rather late. Are we willing to pay for these benefits?

14 References

1. EPA (1992) National pesticide survey–Update and summary of Phase II results. U.S. Environmental Protection Agency, Office of Water, EPA 570/9-91-021
2. Ritter WF (1990) J Environ Sci Health, B25: 1
3. Fielding M, Barcelo D, Helweg A, Galassi S, Torstensson L, Van Zoonen P, Wolter R, Angeletti G (1991) In: Pesticides in ground and drinking water. Commission of the European Communities – Directorate General for Science, Research and Development. Water Pollution Research Report 27
4. Iwan J (1988) Gesunde Pflanz 40: 208
5. Legrand MF, Costentin E, Bruchet A (1991) Environ Technol 12: 985
6. Leistra M, Boesten JJTI (1989) Agric Ecosyst Environ 26: 369
7. Marcomini A, Perin G, Stelluto S, Da Ponte M, Traverso P (1991) Environ Technol 12: 1127
8. OECD (1986) Water pollution by fertilizers and pesticides. OECD Publications No 43705, Paris
9. GIFAP (1990) Pesticide residues in water. Position Paper. Revised Version, June 1990, Brussels

10. GIFAP (1987) Toxicological evaluation of pesticides in drinking water. Position Paper, Nov 1987, Brussels
11. Ohnesorge FK (1991) Gesunde Pflanz 43: 26
12. European and Mediterranean Plant Protection Organization (1992) Decision-making scheme for the environmental risk assessment of plant protection products. 15 Sep 1992, Paris
13. Bailey JE, Tilmon HD, Phipps PM (1991) Plant Dis 75: 1077
14. Zilberman D, Schmitz A, Casterline G, Lichtenberg E, Siebert JB (1991) Science 253: 518
15. Kudsk P (1989) In: Proc Br Crop Prot Conf Weeds p 545
16. Haas H (1991) Phytoparasitica 19: 322
17. Thonke KE (1991) In: Proc Br Crop Prot Conf Weeds p 1183
18. Blanc D (1992) Phytoma 467: 38
19. Seiler A, Brenneisen P, Green DH (1992) Water Supply 10: 31
20. Tissut M, Nurit F, Ravanel P (1992) In: Proceedings of the Phyt'Eau Symposium, 21–22 Oct 1992, Versailles
21. Haddad G, Ponte-Freitas A, Ravanel P, Tissut M (1992) Plant Physiol Biochem 30: 173
22. Baandrup M, Ballegaard T (1989) In: Proc Br Crop Prot Conf Weeds p 555
23. Decoin M (1992) Phytoma 443: 31
24. Curé B (1990) Perspect Agric 146: 89
25. Proven MJ, Courtney A, Picton J, Davies DHK, Whiting AJ (1991) In: Proc Br Crop Prot Conf Weeds p 1201
26. Whiting AJ, Davies DHK, Brown H, Whytock GP (1991) In: Proc Br Crop Prot Conf Weeds p 1209
27. Green JM (1991) Weed Technol 5: 894
28. Edwards CA, Regnier EE (1989) In: Proc Br Crop Prot Conf Weeds p 585
29. Lebaron HH (1991) Phytoparasitica 19: 326
30. Doll JD, Proost R, Visocky M, Buhler D (1991) Proc North Cent Weed Sci Soc 46: 48
31. Klaus M (1992) Gesunde Pflanz 44: 251
32. Barrett DWA, Sutton PB (1991) In: Proc Br Crop Prot Conf Weeds p 523
33. Fawcett RS (1992) Proc West Soc Weed Sci 45: 4
34. Pallut B (1992) Gesunde Pflanz 44: 255
35. Miller PCH, Stafford JV (1991) In: Proc Br Crop Prot Conf Weeds p 1249
36. Miesner H (1992) Gesunde Pflanz 44: 107
37. Feser M, Fischer HS (1992) Gesunde Pflanz 44: 117
38. Battaglin WA, Goollsby DA (1992) Abstr Pap Am Chem Soc 204: 150
39. Gorski SF (1991) In: Proc Br Crop Prot Conf Weeds p 343
40. Mehltretter CL, Roth WB, Wealey FB, McQuire TA, Russell CR (1974) Weed Sci 22: 414
41. Stock D (1991) In: Proc Br Crop Prot Conf Weeds p 315
42. Calvet R (1988) In: Methodological aspects of the study of pesticide behaviour in Soil. INRA p 119, 16–17 Jun 1988, Versailles
43. Kuhnt G (1988) In: Methodological aspects of the study of pesticide behaviour in Soil. INRA p 151, 16–17 Jun 1988, Versailles
44. Gaynor JD, Volk VV (1976) Weed Sci 24: 549
45. Street JE, Wehtje G, Walker RH, Patterson MG (1987) Weed Sci 35: 422
46. Wauchope RD (1978) J Environ Qual 7: 459
47. Dombrowski N, Hibbitt I, Strachan JR (1989) In: Proc Br Crop Prot Conf Weeds p 663
48. Lavers A (1989) In: Proc Br Crop Prot Conf Weeds p 649
49. Harris DA, Johnson KS (1991) In: Proc Br Crop Prot Conf Weeds p 715
50. Bailey JD (1992) Rev Environ Cont Tox 129: 17

51. Leake (1991) Pestic Sci 31: 363
52. Behl E, Eiden CA (1990) In: Agrochemical residue sampling design and techniques: Soil and groundwater. ACS Symposium Series No 465; Nash R, Leslie A Eds, Washington DC
53. Rijtema PE, Leeuwangh P, Leistra M (1992) In: Environmental criteria for assessing agricultural pesticides. Report 73, Feb 1992, DLO Winand Staring Centre, Wageningen
54. Schuster E (1991) Tox Environ Chem 30: 159
55. Van Haasteren JA, Van den Berg R (1992) In: Pesticides and groundwater: Policy and situation in Europe and guidelines for protection. Council of Europe, 28–29 Apr 1992, Paris
56. Tooby TE (1989) In: Proc Br Crop Prot Conf Weeds p 1165
57. Tooby TE, Mardsen PK (1991) In: Pesticides in soils and water: Current perspectives. BCPC Monograph No 47 p 5, 25–27 Mar 1991, Warwick, UK
58. Fawell JK (1991) In: Pesticides in soils and water: Current perspectives. BCPC Monograph No 47 p 205, 25–27 Mar 1991, Warwick, UK
59. Hoag DL, Hornsby AG (1992) J Environ Qual 21: 579
60. Crowe AS, Mutch JP (1992) Environ Monitor Assess 23: 19
61. Cohen SZ, Creeger SM, Carsel RF, Enfield CG (1984) In: Krueger RF, Seiber JN (eds), Treatment and disposal of pesticide wastes. ACS Symposium Series No 259. Washington DC
62. Jury WA, Focht DD, Farmer WJ (1987) J Environ Qual 16: 422
63. Aller L, Bennett T, Lehr JH, Petty R (1986) In: Garner WY, Honeycutt RC, Nigg HB (eds), Evaluation of pesticides in ground water: ACS Symposium Series No 315. Washington DC
64. Aharonson N (1987) Pure & Appl Chem 59: 1419
65. Jackson J, Peterson PJ (1989) In: Evaluation of selection schemes for identifying priority aquatic pollutants. Study Contract No B 6612/290/89, Commission of the European Communities, Directorate-General for the Environment Consumer Protection and Nuclear Safety, Brussels
66. Gustafson DI (1989) Environ Tox Chem 8: 339
67. Weber JB (1991) Proc South Weed Sci Soc 44: 45
68. Shiu WY, Ma KC, Mackay D, Seiber JN, Wauchope RD (1990) Rev Environ Cont Tox 116: 1
69. Shiu WY, Ma KC, Mackay D, Seiber JN, Wauchope RD (1990) Rev Environ Cont Tox 116: 15
70. Wauchope RD, Buttler TM, Hornsby AG, Augustijn-Beckers PWM, Burt JP (1992) Rev Environ Cont Tox 123: 5
71. Herner AE (1992) Weed Technol 6: 749
72. Bottoni P, Funari E (1992) The Science of Total Environ 123/124: 581
73. Jones RL (1990) In: Hazardous waste containment and treatment encyclopedia of environmental technology vol 4, Cheremisinoff PN Ed, Gulf Publishing, Houston TX
74. Michenfelder A, Schramm M (1992) Wasserwirt Wassertech 4: 206
75. GIFAP (1992) In: The regulatory use of lysimeter studies and computer modeling for the evaluation of pesticide mobility in soil. Technical Monograph (in preparation)
76. Jones RL, Hanks RJ (1988) In: De Coursey DG (eds), International symposium on water quality modeling of agricultural non-point sources, 19–23 Jun 1988, Logan UT
77. Russell MH, Layton RJ (1992) Weed Technol 6: 673
78. Kollig HP (1988) Tox. Environ Chem 17: 287
79. Martin E (1991) Pestic Sci 31: 399
80. Close ME, Canter LW (1990) The Environ Profess 12: 278

81. Goss DW (1992) Weed Technol 6: 701
82. Leonard RA, Knisel WG, Still DA (1987) Trans ASAE 30: 1403
83. Hollis JM (1991) In: Pesticides in soils and water: Current perspectives. BCPC Monograph No 47 p 165, 25–27 Mar 1991, Warwick, UK
84. Hornsby AG (1992) Weed Technol 6: 736
85. Swartz MS, Bird GW, Richtie JT (1989) J Nematol 21: 590
86. Thurman EM, Goolsby DA, Meyer MT, Kolpin DW (1991) Environ Sci Technol 25: 1794
87. Croll BT (1991) J Inst Water Environ Manage 5: 389
88. Cohen DB (1986) In: Garner WY, Honeycutt RC, Nigg HB (eds), Evaluation of pesticides in ground water: ACS Symposium Series No 315. Washington DC
89. Zaki MH, Moran D, Harris D (1982) Am J Public Health 72: 1391
90. Jones RL, Bostian AL (1989) Ground Water Monitor Rev 9: 75
91. GIFAP (1989) In: Prevention and reduction of pesticide residues in groundwater through good agricultural management practices. Guide C/98/TD/201, Mar 30 1989, Brussels
92. Jones RL, Rourke RV, Hansen JL (1986) Environ Toxicol Chem 5: 167
93. Jones RL, Black GW, Estes TL (1986) Environ Toxicol Chem 5: 1027
94. Pacenka S, Porter KS, Jones RL, Zecharias YB, Hughes HBF (1987) J of Contam Hydrol 2: 73
95. Jones RL (1992) Weed Technol 6: 683
96. Barrett MR (1992) Abstr Meet Weed Sci Soc Am 32: 77
97. Carter AD (1989) In: Proc Br Crop Prot Conf Weeds p 1157-1164
98. Woodburn KB, Knuteson JA (1991) Proc South Weed Sci Soc 44: 307
99. Preti F, Settesoldi D, Lubello C, Vazzana C, Lascialfaro P, Puglia S, Ghezzi G, Ghezzi R, Zanovello G (1992) Water Supply 10: 89
100. Vachaud G (1992) In: Measurement of water and solutes movement in soils: Methods and problems. Eurosol Conference, 6–12 Sep 1992, Maastricht
101. Knisel WG (1980) In: CREAMS: A field-scale model for chemicals, runoff, and erosion from agricultural management systems. U.S. Dept of Agriculture, Conservation Research Report No 26, Washington DC
102. Baker JL, Laflen JM, Johnson HP (1978) Trans ASAE 21: 886
103. Baker JL, Johnson HP (1979) Trans ASAE 22: 554
104. Isensee AR, Sadeghi AM (1992) Abstr Meet Weed Sci Soc Am 32: 75
105. Felsot AS, Mitchell JK, Kenimer AL (1990) J Environ Qual 19: 539
106. Buttle JM (1990) J Environ Qual 19: 531
107. Boiffin J, Papy F, Eimberck M (1988) Agronomie 8: 663
108. Papy F, Boiffin J (1988) Agronomie 8: 745
109. Duvoux B (1990) In: Protection rapprochée des cours d'eau contre les effets de l'érosion des terres agricoles. CEMAGREF ed p 86
110. Gril JJ, Lilin C, Paulet B (1986) In: Maîtrise de l'érosion des terres agricoles en Bavière (R.F.A.), CEMAGREF ed, Rapport de mission p 20
111. Gril JJ, Duvoux B (1991) In: Maîtrise du ruissellement et de l'érosion-Conditions d'adaptation des méthodes américaines. CEMAGREF DICOVA ed, p 157
112. Cooper JR, Gilliam JW, Daniels RB, Robarge WP (1987) Soil Sci Soc Am 51: 416
113. Paterson KG, Schnoor JL (1992) Water Environ Res 64: 274
114. Dillaha TA, Sherrard JH, Lee D, Mostaghimi S, Shanholtz VO (1988) Journal WPCF 60: 1231
115. Young RA, Huntrods T, Anderson W (1980) J Environ Qual 9: 483
116. Asmussen LE, White AW, Hauser EW, Sheridan JM (1977) J Environ Qual 6: 159
117. Magette WL, Brinsfield RB, Palmer RE, Wood JD (1989) Trans ASAE 32: 663
118. Nieswand GH, Hordon RM, Shelton TB, Chavooshian BB, Blarr S (1990) Water Res Bull 26: 959
119. Carlson CG, Dean R, Lemme G (1990) J Soil Water Conserv 17: 239

120. Clark L, Gomme J, Hennings S (1991) Pestic Sci 32: 15
121. Gomme J, Shurvell S, Hennings SM, Clark L (1992) J Inst Water Environ Manage 6: 172
122. Foster SSD, Chilton PJ, Stuart ME (1991) J Inst Water Environ Manage 5: 186
123. Caron G, Suffet I, Belton T (1985) Chemosphere 14: 993
124. Chiou CT, Malcolm RL, Brinton TI, Kile DE (1986) Environ Sci Technol 20: 502
125. Hassett JJ, Anderson MA (1982) Water Res 16: 681
126. Khan AT, Thompson MB (1990) Environ Tox Chem 9: 253
127. Lee DY, Farmer WJ (1989) J Environ Qual 18: 468
128. Lee DY, Farmer WJ, Aochi Y (1990) J Environ Qual 19: 567
129. Mc Gee (1991) Environ Sci Technol 25: 323
130. Carter ES (1991) In: Proc Br Crop Prot Conf Weeds p 503–512
131. Dizer H (1991) In: Examination of methods of treatment to remove pesticides from water intended for human consumption. Final Report, Institut für Wasser-, Boden- und Lufthygiene des Bundesgesundheitamtes, Apr 1991, Berlin
132. Bundesminister für Umwelt Naturschutz und Reaktorsicherheit (1990) In: Kenntnisstand der Entfernbarkeit von Pflanzenbehandlungs- und Schädlingsbekämpfungsmitteln (PBSM) bei Trinkwasseraufbereitungsverfahren. May 1990, Bonn
133. Miltner RJ, Baker DB, Speth TF, Fronk CA (1989) Journal AWWA 81: 43
134. GIFAP (1990) In: Removal of pesticides from drinking water. Technical Monograph No 15, Jul 1990
135. Chian ESK, Bruce WN, Fang HHP (1975) Environ Sci Technol 9: 52
136. Duranceau SJ, Taylor JS, Mulford LA (1992) Journal AWWA 84: 68
137. Price DRH (1991) In: Proc Br Crop Prot Conf Weeds p 1279–1284
138. Kern AD (1990) Proc South Weed Sci Soc 43: 14
139. Skinner AC (1992) In: Water management in agricultural catchments. Proceedings of the 3rd Silsoe Link Conference, 2 Sep 1992

Pesticide Metabolism in Aquatic Organisms

Margaret O. James

Department of Medicinal Chemistry, University of Florida,
Gainesville FL 32610-0485, USA

In recent years there have been a number of studies of the fate of pesticides and related model compounds in fish and shellfish. Some of these studies have been in vivo, and have addressed the questions of uptake and excretion of pesticides as well as biotransformation. These studies show that the ability of fish to biotransform pesticides is usually similar to that of other vertebrate species, although there is a tendency for phase 1 biotransformation to be slower in fish. The limited number of in vivo studies that have been conducted in aquatic invertebrates have revealed a number of significant differences between fish and shellfish in their ability to biotransform and excrete pesticides. In vitro studies of enzymes and transport systems involved in pesticide biotransformation, uptake and excretion have provided some explanation and understanding of the in vivo results, and are likely in the future to enable us to predict the likely biotransformation of a new pesticide molecule in a previously studied aquatic species. This information is important in the context of non-target species toxicity and body burdens of pesticides.

Chemistry of Plant Protection, Vol. 9

1 Introduction

Aquatic organisms may be exposed to pesticides from a variety of sources. These include agricultural run-off into drainage ditches and from there into larger bodies of water, direct application of pesticides to bodies of water to control unwanted insects and plants, run-off from domestic application and industrial effluent and atmospheric deposition [1, 2]. It is now clear that pesticides can be taken up by non-target aquatic species, resulting in potential adverse effects for the organism as well as for human and animal consumers of the pesticide-exposed fish and shellfish [3–5]. For some pesticides, fish and especially those shellfish that are closely biologically related to the insect targets, are more sensitive to pesticides than are mammals and birds. This may be due to differences in target organ or receptor sensitivity or to toxicokinetic factors such as more rapid uptake and slower detoxication resulting in a higher concentration of active compound at the site of toxicity. In recent years, advances have been made in studying the uptake, tissue distribution, biotransformation and excretion of pesticides, and model compounds containing groups commonly found in pesticides, in fish and aquatic invertebrates [6, 7]. The term "pesticide metabolism" is used in this context to include studies of pesticide uptake, tissue distribution, biotransformation and excretion, but with the focus on biotransformation.

2 Methodology Considerations

Most of the established methods for studying xenobiotic metabolism, including pesticide metabolism, were developed for studies in mammalian species. Studies of pesticide metabolism in aquatic species pose different methodological considerations in some cases than studies in mammalian species. The methodology used to study pesticide metabolism includes detailed whole animal studies with a variety of routes of administration of the pesticide, organ perfusion studies in situ or in an apparatus designed to accommodate the isolated organ, studies in organ slices, isolated cells, subcellular fractions and with purified enzymes. The method used will vary with the information expected from the study [8]. In vivo studies in fish are complicated by the need to keep the animal's gills (at least) under water for procedures of more than a minute or so in duration. Thus, although simple injections of pesticides under study can be done easily, obtaining sequential information on blood levels and urinary and fecal excretion is more difficult. The surgical procedures for placing indwelling catheters in fish veins and arteries, in ureters and at anal orifices have now been developed (see Ref. [8] for discussion), but in performing these procedures it is necessary to keep the anesthetized fish's gills perfused with water containing maintenance levels of anesthetic, usually tricaine methane sulfonate. As in studies with surgically modified rats, it is then desirable to allow recovery from surgery and anesthesia for at least a few days to ensure that any residual anesthetic has been eliminated. Studies of isolated perfused organs in situ can give valuable information on the contribution of that organ to overall biotransformation, but are often technically difficult [8, 9]. In vivo studies are sometimes complicated by the anatomy of the animal. For example, in vivo studies with crustacea and other shellfish are technically more difficult than with fish, due to anatomical differences and the difficulty in placing indwelling catheters in blood vessels and urinary and fecal orifices. Crustacean hemolymph clots rapidly on exposure to air, and their vasculature is not sufficiently organized to allow reproducible placement of catheters. The generally large blood volume of crustacea, typically 20–30% body weight, does permit frequent sampling without undue stress. Methodological advances in the area of cannulation of urinary and fecal orifices of crustacea have been made by Snyder and Chang [10] who placed external cannulae on lobsters in their study of ecdysone excretion and metabolism. Although the methodology for in vitro studies of enzyme activity in organ slices, isolated cells and subcellular fractions are generally similar for preparations from aquatic animals as from more commonly used laboratory animals, consideration of appropriate assay conditions is important. Usually preparations from poikilothermic aquatic animals will require a lower incubation temperature than preparations from small homeothermic mammals. Difficulties are also encountered in conducting in vitro studies in cases where the function of the organ is such that disruption of the cellular structure leads to protease release, and subsequent auto-digestion of biotransformation enzymes. This phenomenon has been observed in studying monooxygenase activity in crustacean hepatopancreas, where P450 reductase is destroyed during subcellular fractionation [11].

3 Uptake of Pesticides from the Water and Sediments by Fish and Shellfish

The extent to which pesticides may be taken up from contaminated water and sediments depends on several variables, including the physico-chemical properties of the pesticide and the preferred habitat, feeding behaviour and biological characteristics of the fish or shellfish. Compounds that are freely water soluble, but also somewhat lipid soluble, can be taken up across the gills of most fish and shellfish, but may also be readily excreted across the gills [12]. Water soluble compounds are rarely accumulated in tissues to any appreciable extent, unless they are anions or cations whose pKa is such that they are trapped by intra- and extra-cellular pH differences (see section on excretion). Compounds that are lipophilic and are mainly associated with sediments and particles in the water

Table 1. Routes of uptake of pesticides by aquatic animals

Form of pesticide	Route of uptake	Animals affected
Solution in water-column	Gill, skin	All animals
Particles in the water-column	Gill Gastrointestinal tract	All animals Filter-feeders
Particles on the surface of plants and zooplankton	Gastrointestinal tract	Filter feeders, plant eating animals
Particles in the sediment	Skin, carapace Gastrointestinal tract	Burrowing animals Bottom-feeding animals
Taken up by small animals	Gastrointestinal tract	Carnivorous animals

column will be taken up most readily by filter feeders such as molluscs. Thus, molluscs inhabiting polluted environments can accumulate high concentrations of lipophilic chemicals [13]. Pesticides such as DDT, Lindane and Kepone are known to bioaccumulate in molluscs. Fish and shellfish that consume filter feeders will be exposed to lipophilic pesticides both through direct contact with the dissolved and suspended chemical, and by consumption of the filter feeders. Furthermore, metabolites of the lipophilic pesticides that still possess lipophilic characteristics also have the potential to be taken up from the water or in the diet by fish and shellfish. Depending on the structure of the molecule, fish generally have a higher ability to metabolize pesticides than molluscs and shellfish (see below) and therefore generally do not bioaccumulate pesticides that can be metabolized and excreted. Exceptions are pesticides such as DDT, which is very slowly metabolized in all species [14], and other highly chlorinated pesticides such as lindane. Table 1 summarizes the most common routes of uptake of pesticides by various aquatic animals.

4 Biotransformation by Phase 1 Pathways

Phase 1 pathways are important in introducing or uncovering functional groups in pesticide molecules, thereby increasing their hydrophilicity [15]. For the majority of lipophilic pesticides, the most important phase 1 enzymes are the P450 monooxygenases, the FAD-monooxygenases (FMO) and the hydrolytic enzymes. Evidence for the occurrence and activities of these pathways in aquatic species has been gained from in vivo and in vitro studies. The P450 monooxygenases are a family of membrane-bound heme-thiolate proteins of molecular weight 48 000 to 60 000. In the past P450 has been called cytochrome P450, but recent nomenclature recommendations suggest the use of P450 or heme-thiolate protein P450 [16]. The P450s function to bind lipophilic substrates and,

Table 2. Phase 1 enzymes important in pesticide metabolism

Enzyme	Reaction catalysed	Representative substrate examples	Ref.
P450	Hydroxylation	Cypermethrin	22
	Epoxidation	Chlordene	24
	Sulfoxidation	Molinate	23
	N-hydroxylation	2-Acetylaminofluorene	154
	N-demethylation	Aminocarb	155
	P=S → P=O	Fenitrothion	156
FAD-monooxygenase	Sulfoxidation	Eptam, Aldicarb	64, 66
	N-oxidation	Dimethylaniline	65
Esterase	Hydrolysis	Chlorpyrifos	74
		Cypermethrin	22
		Fenitrothion	30
Epoxide hydrolase	Hydration	Styrene oxide	84
		Dieldrin	157

Structures and positions of biotransformation of these substrates are shown in Fig. 1.

in the presence of molecular oxygen, NADPH and P450 reductase, catalyse the insertion or addition of an atom of oxygen into the substrate [17]. In the absence of oxygen, P450 can also function as a reductase. P450 can catalyze dehalogenation of halogenated organics by either oxidative or reductive pathways [18]. The P450 enzymes involved in pesticide metabolism are expressed to the greatest extent in the liver, but are also found, usually in smaller amounts, in other tissues. Some forms of P450, especially members of the 2 and 3 families [16] are inducible by pesticides in mammalian species. In aquatic species, fish can be induced by some but not all of these pesticides, whereas chemical induction of P450 has not yet been conclusively demonstrated in invertebrates (see below). The FMOs are a family of membrane-bound enzymes that are important in mammalian species for the oxygenation of pesticides containing nucleophilic S, N and P [19]. Hydrolytic enzymes are found in liver and extrahepatic organs in mammalian species and fish, and are important in the phase 1 metabolism of esters and amides [20] and epoxides [21]. Characteristic reactions catalysed by these phase 1 enzymes and some representative pesticide substrates of each are shown in Table 2 and Fig. 1.

4.1 P450

Insight into the role of P450-dependent monooxygenation in the biotransformation of pesticides in fish and shellfish has been gained from in vivo and in vitro studies. In vivo studies are of importance in examining the overall fate of a pesticide, and are necessary to validate findings from in vitro studies.

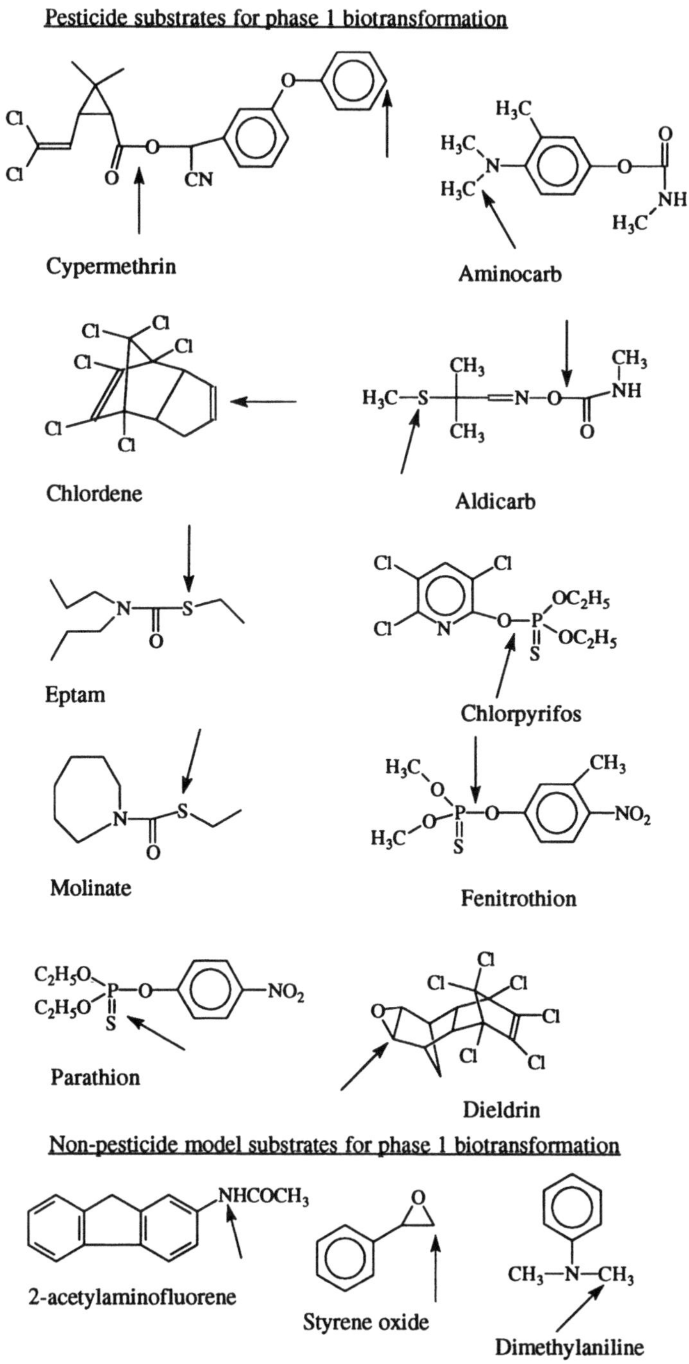

Fig. 1. Pesticide and model substrates for phase 1 biotransformation. The *arrow* indicates the point of metabolism

4.1.1 In Vivo Studies of Pesticide Monooxygenation

There are relatively fewer in vivo than in vitro studies of pesticide metabolism in aquatic species, due in part to the greater expense and technical difficulty of conducting in vivo studies. A discussion of the methodology used in conducting in vivo studies is contained in Ref. [8]. There is now abundent evidence that fish can monooxygenate pesticides. Studies of the fate of *cis*- and *trans*- cypermethrin in rainbow trout, *Oncorhyncus mykiss*, showed that 4′-hydroxylation, a P450-mediated reaction, was an important pathway for both isomers [22]. Unchanged cypermethrins were also excreted [22]. Concentrations of radioactivity derived from cypermethrins in the tissues of dosed trout were highest in intestine, bile and liver and were much lower in the carcass after either waterborne or oral administration [22]. The thiocarbamate herbicide molinate was oxidatively metabolized by both the white sturgeon *Acipenser transmontanus*, and the common carp, *Cyprinus carpio*, to 4-hydroxymolinate, molinate sulfoxide and other oxidative metabolites [23]. The enzymes catalyzing the oxidative reactions of molinate are not known, but the 4-hydroxylation at least is likely to be due to P450. The sulfoxide may be formed by either P450 or the FAD-monooxygenase. Another study showed that chlordene epoxide and chlordene were present in the edible portion of catfish, *Ictalurus punctatus*, environmentally exposed to chlordene [24] suggesting slow P450-dependent metabolism of chlordene to the epoxide and subsequent equilibration of the epoxide metabolite with all catfish tissues prior to the expected further metabolism and elimination. The uptake and elimination of hexachlorobenzene was studied in green sunfish, *Lepomis cyanellus* Raf., by feeding a diet containing [^{14}C]-hexachlorobenzene [25]. The fish tissues contained residues of radioactivity, with the lowest levels found in skeletal muscle and the highest in the alimentary canal. Analysis of selected tissues, including liver and muscle, showed that pentachlorophenol was a major metabolite [25]. While the route of metabolism of hexachlorobenzene to pentachlorophenol has not been established in fish, it is possible that P450 is involved. The bile of goldfish, *Carassius auratus*, treated with [^{14}C]-hexachlorocyclopentadiene contained more than eleven metabolites, which appeared to be dechlorination and oxidation products [26]. Complementary in vitro studies of hexachlorocyclopentadiene biotransformation in bluegill, *Lepomis machrochirus*, suggested that NADPH-dependent monooxygenation (probably catalysed by P450) and GSH-dependent conjugation were two pathways of metabolism [26]. The organophosphorous propesticide *O*-2,4-dichlorophenyl-*O*,*O*-diethylphosphorothionate was shown to form the oxon in exposed sheepshead minnows, *Cyprinidon variegatus* [27]. This reaction is normally catalysed by P450. Studies with non-pesticide substrates for cytochrome P450, such as polycyclic aromatic hydrocarbons and aflatoxin, also support the presence of several forms of active P450 in those fish species that have been examined [28, 29]. Although it is dangerous to generalize in this area, it can be stated that the rate of P450-dependent monooxygenation of most pesticides is likely to be slower in fish than in warm-blooded vertebrates.

Studies with invertebrates of pesticide P-450-dependent metabolism in vivo are very few. Johnson and Corbett showed that blue crabs, *Callinectes sapidus*, were able to metabolize fenitrothion to fenitrooxon, suggesting the presence of a P450 monooxygenase [30]. Freshwater crayfish, *Procambarus clarkii*, were able to convert methyl parathion to methyl paraoxon, whereas two prawn species, the Malaysian prawn, *Macrobrachium rosenbergii*, and the ridgeback prawn, *Sicyonia ingentis*, were unable to form methylparaoxon [31, 32]. Both crayfish and prawns were able to convert the model compound 4-nitroanisole to 4-nitrophenol, a demethylation reaction that is normally catalysed by P450. The model compound benzo(a)pyrene was metabolized rapidly by P450-dependent monooxygenation by the spiny lobster, *Panulirus argus* [33] but was metabolized only very slowly by the American lobster, *Homarus americanus* [34]. In the oyster, *Crassostrea virginica*, oxidative metabolism of the pesticide *O*-2,4-dichlorophenyl-*O*,*O*-diethylphosphorothioate was much slower than in the sheepshead minnow [27]. These limited studies suggest that among the invertebrates there is considerable variability in the capability of the P450 system to carry out pesticide monooxygenation. Several studies have indicated that among molluscs particularly, the capacity for P450-dependent monooxygenation is so slow that with continued exposure to a lipophilic pesticide, high concentrations of the pesticide will accumulate in the mollusc's body [13, 35].

4.1.2 In Vitro Studies of Pesticide Monooxygenation

In part because of the relative ease of in vitro experiments, there are more in vitro than in vivo studies of the monooxygenation of pesticides and model compounds. P450s have been studied in several freshwater and marine fish species in terms of their catalytic activities with various substrates, kinetic properties, regulation and, more recently, their molecular biology [36, 37]. As in other vertebrates, the liver contains the highest concentration of P450, and in the species studied to date multiple forms of P450 are present in liver [36–38]. Several pesticides and model compounds have been shown to be substrates for P450 in fish species. In addition to those listed in Table 2 these include parathion [39], aldrin [40], fenvalerate [41], ethoxycoumarin [42], ethoxyresorufin, benzphetamine and benzo(a)pyrene [42]. In the study of parathion activation, it was shown that products that inactivated acetylcholinesterase were formed in the presence of NADPH or NADH and that formation of the active metabolites was partially inhibited by the classic P450 inhibitors, CO and SKF-525A [39]. The major active metabolite was paraoxon, as expected, but other active products were also found. The fact that CO only partially inhibited parathion activation suggests that possibly the activation of parathion to paraoxon could be catalysed by other microsomal enzymes such as the FAD-monooxygenase.

Studies of pesticide monooxygenation in vitro in invertebrates have produced conflicting results. While Elmamlouk and Gessner [43] were unable to demonstrate paraoxon production from parathion in lobster hepatopancreas microsomes, Johnston and Corbett [44] did demonstrate production of fenitrooxon from fenitrothion in blue crab hepatopancreas microsomes. The discrepancy may be due to the fact that hepatopancreas microsomes from certain decapod crustacea contain inhibitors of monooxygenase activity that are released during the preparation of subcellular fractions [11, 45]. Spiny lobster hepatopancreas washed microsomal preparations had low NADPH-dependent benzphetamine *N*-demethylase activity, but if exogenous reductase were added, NADPH-dependent monooxygenation of several substrates was high (11,45 and unpublished observations). Similar preparations from American lobster hepatopancreas had much lower monooxygenase activity, even with added reductase [45]. In other invertebrates, such as bivalves, there is some evidence that monooxygenation can occur, but usually at a very much lower rate than in fish [35].

4.1.3 Regulation of P450

The effect of exposure to pesticides and other xenobiotics has been studied extensively in several fish species, both freshwater and marine. P450 1A1, the form that is induced by exposure of the animal to polycyclic aromatic hydrocarbons and related planar polycyclic compounds [46], is highly inducible in most fish species at exposure to very low concentrations of polyaromatic molecules such as 3-methylcholanthrene and β-naphthoflavone [47]. Most pesticides are not substrates for this form of P450. Kepone, DDT and related molecules can induce P450 in a manner similar to phenobarbital in the rat, but studies of the effects of these pesticides, and of phenobarbital itself on P450 in fish have shown weak or no inducing effect on P450 content or monooxygenase activity (reviewed in Ref. [48]). The resistance of fish to induction by phenobarbital and related compounds has been well established [49], and may contribute to the slow metabolism of certain pesticides. Studies in trout showed that the liver contained P450 isozymes similar to mammalian phenobarbital -inducible 2 family isozymes [16, 50], but neither mRNA synthesis nor P450 synthesis was increased by treatment with phenobarbital-like inducers [51].

Some of the imidazole- and triazole-based herbicides may be expected to inhibit or induce P450, based on studies in mammalian species. Those studies showed that imidazole and triazole derivatives acutely inhibit P450 monooxygenase activity, due to the unbonded electron pair on one of the nitrogens in the imidazole or triazole ring binding reversibly with the heme-bound iron of P450 at the oxygen binding site [52, 53]. When given chronically, some imidazoles and triazoles induce P450 content and monooxygenase activity in mammalian species [54–57]. Recent studies have shown that administration of prochloraz to trout affects hepatic P450 [58, 59]. Prochloraz inhibited aldrin epoxidase in vitro, and shortly after administration in vivo, but in vivo treatment caused P450 content of liver microsomes and ethoxyresorufin *O*-deethylase activity to increase [58, 59]. Thus, imidazole or triazole-based compounds are likely to affect pesticide P450-dependent metabolism in fish, but the effect observed will depend on how recently the fish was exposed to the imidazole or triazole-based pesticide. Recent exposure is likely to result in inhibition of monooxygenase activity, followed by increased monooxygenation as the P450 content increases in response to the inducing effects of the pesticide.

Studies of the effects of inducing agents on P450 content and monooxygenation in invertebrate species have generally shown little or no induction effect by either phenobarbital-like (e.g. DDT, lindane, kepone) or 3-methylcholanthrene-like agents [35, 45, 60]. Many invertebrates accumulate such compounds to high concentrations on continued exposure and may be a source of dietary transfer of such inducing agents to fish, mammalian predators and humans.

4.2 FAD-Monooxygenase

Many pesticides are potential substrates for the FMO as well as for P450. Such pesticides are those that contain nucleophilic S, N and P atoms in their structure, such as organophosphorothionates and thioether-containing pesticides [19, 61]. In vivo studies usually are not designed to determine if an observed monooxygenation is due to P450 or the FMO. Hodgson and coworkers have demonstrated by in vitro studies that in rodents numerous pesticides are metabolized by the FMO enzymes, although for some substrates there may also be some contribution of P450 in monooxygenation [61–63]. Recent in vitro studies have suggested that the FMO is also present in fish liver and possibly also in invertebrates. Figure 1 shows the structure of two postulated substrates for fish FMO, Eptam and Aldicarb. Cashman et al. [64] showed that the pesticide Eptam, which is widely used in California rice fields and leaches into rivers and lakes, may be metabolized to the S-oxide by an FMO in striped bass, *Morone saxatilis*, liver, although their studies did not rule out the possibility that P450 also catalysed this monooxygenation. Schlenk and Buhler also found evidence for FMO activity in the trout with the model substrate *N,N*-dimethylaniline and the pesticide aldicarb as substrates [65, 66] and in shellfish [67]. The fate of the pesticide aldicarb was studied in vivo in the rainbow trout following administration of the [^{14}C]-labeled pesticide at a dose of 48 µg/kg by gavage or of 1 mg/kg by intraperitoneal injection [68]. After the intraperitoneal injection of 1 mg/kg, 96.2% of the dose had been eliminated branchially after 24 hours and 3.6% was found in urine, with the remaining 0.2% dose in bile [68]. After oral administration of 48 µg aldicarb/kg, about 22% of the dose was still retained in the fish 24 hours later and the data indicated biliary excretion of aldicarb and metabolites [68]. Although there were differences in route of elimination of aldicarb-derived radioactivity between the oral and intraperitoneal doses, it was not clear if this was due to the differences in route of administration or the differences in dose used for each route. The aldicarb was extensively and rapidly absorbed from the oral route, and after both dosage regimes aldicarb was metabolized to aldicarb sulfoxide and aldicarb oxime (a hydrolysis product). Given the in vitro findings that aldicarb sulfoxide may be formed by the FMO [66], it is possible that this metabolite was formed in trout by the FMO as well as perhaps by the other major monooxygenase, P450. Schlenk and Buhler have shown that trout liver microsomes contain two proteins, mol. wts. 61 000 and 57 000, which cross-react with antibodies against two pig and rabbit FMO enzymes [69]. The content of anti-FMO-reactive protein in trout liver microsomes was less than that found in rat liver microsomes. Attempts to demonstrate FMO activity with *N*,*N*-dimethylaniline and methimazole in catfish liver microsomes were unsuccessful, and unlike trout, there was no cross-reactivity between catfish liver microsomes and the antibodies to pig and rabbit FMO enzymes [70]. Thus more studies are needed to determine the expression and properties of FMO in fish or shellfish and to ascertain how important this enzyme is in pesticide monooxygenation in fish and shellfish. Since sulfoxide metabolites of carbamate pesticides may be more toxic to fish than the parent compounds [64, 68] it is important to know the mechanism of formation of the sulfoxide metabolites and the likelihood of species differences in formation of the sulfoxide metabolites. This information will be of value in predicting which species may be most susceptible to toxicity and death following exposure to certain carbamate pesticides.

4.3 Hydrolytic Enzymes

4.3.1 Ester and Amide Hydrolases

Many pesticides contain ester or amide linkages in their structure. Ester linkages are most common and are found in pesticides that act in the central nervous system, including organophosphorothionates, carbamates and pyrethroids. Examples are chlorpyrifos, aldicarb, cypermethrin and related compounds (see Fig. 1). Hydrolysis of the ester linkage usually results in inactivation of the pesticide. The biotransformation of several ester-containing pesticides and model compounds has been studied in fish. It has been suggested that the rates of hydrolysis of pyrethroids in different species are inversely proportional to the toxicity of the pyrethroid, and that low esterase activity may explain, at least in part, the high sensitivity of many fish species to pyrethroids [7]. In rainbow trout, differences in the extent of ester hydrolysis of cypermethrin in vivo were observed between the *cis* and *trans* isomers, with more hydrolysis of the trans isomer and more oxidation of the cis isomer [22]. Permethrin was more slowly hydrolysed in vivo than it was oxidized, and the major metabolite was 4-hydroxypermethrin and its glucuronide in rainbow trout [71]. It was thought that the slow hydrolysis contributed to the higher toxicity of permethrin to trout than to mammals such as mice [71]. Later studies, however, showed that although biotransformation of permethrin to inactive metabolites was slower in trout than mice, there were also intrinsic target organ differences in sensitivity between trout and mice [72, 73]. Fenvalerate, a pyrethroid related to permethrin and cypermethrin was reported to be hydrolysed by carp as well as ring hydroxylated, but relative rates were not reported [41]. Rainbow trout hydrolysed both aldicarb and its metabolite aldicarb sulfoxide to the oxime metabolites following in vivo injection or gavage of aldicarb, but the rates of hydrolysis appeared to be low at the relatively high doses of 1 mg/kg ip or 48 μg/kg orally that were studied [68]. The major phase 1 metabolite of oral chlorpyrifos in catfish was the hydrolysis product, trichlorpyridinol, which was found in bile and urine as the glucuronide [74]. Following exposure of killifish (*Oryzias latipes*) and mullet (*Mugil cephalus*) to 0.1 ppm fenitrothion in running water, the pesticide was bioaccumulated by both fish [75]. The major route of metabolism was by ester hydrolysis in the killifish, but not the mullet, followed by glucuronidation of the liberated phenol [75]. When rainbow trout were exposed to waterborne diethylhexylphthalate at a level of 30 μg/L, there was little systemic accumulation of the diethylhexylphthalate due to rapid branchial hydrolysis of the diester to the monoethylhexylphthalate which was released to the water [76]. Thus, although the fish studied to date do have the capacity to hydrolyse various esters, there is wide variability in the extent of hydrolysis depending on the structure of the ester, the dose, the route of exposure and the species of fish.

Organophosphorothionate hydrolysis has been demonstrated in a few shellfish species. Fenitrothion was hydrolysed in vivo by crabs [30] and parathion and methyl parathion were hydrolysed by several shrimp and crayfish species [31, 32].

In vitro studies have also shown that ester hydrolysis occurs in fish, and the in vitro studies have confirmed that some fish species, including the widely used rainbow trout, have relatively low esterase activity in plasma and liver microsomes [72, 77]. In studies with another ester, Barron et al. [76] showed that gill as well as liver homogenates of the trout could metabolize diethylhexylphthalate to monoethylhexylphthalate, but the reported in vitro rates were very low and may not have been optimal. Among invertebrates, α-naphthyl acetate esterase activity has been found in the 27 000 × *g* supernatant from several organs of the mussel, *Mytilus galloprovincialis*, with highest activity in hepatopancreas [78]. Several electrophoretically separable protein bands with differing extents of esterase activity towards the model substrates α- and β-naphthyl acetate, α-naphthyl propionate and α-naphthyl butyrate were identified in hepatopancreas of the mussel [78]. These different mussel esterase protein bands showed differential inhibition to organophosphorothionate and carbamate insecticides, but the ability of the mussel esterases to hydrolyse the pesticides was not investigated.

Overall, it appears that many aquatic animals have lower esterase activity than mammalian species, and this may cause greater susceptibility to intoxication by pesticides that are inactivated by hydrolysis.

4.3.2 Epoxide Hydrolases

These enzymes catalyse the addition of water to epoxide rings giving rise to the *trans*-dihydrodiol [79]. Several pesticides are potential substrates for epoxide hydrolase, as they contain epoxide moieties in their structure. These include chlorinated pesticides such as dieldrin, endrin, HEOM and heptachlor epoxide, and juvenile hormone ester homologues. Other pesticides, such as naphthalene and chlordene may be metabolized by cytochrome P450 to epoxides or arene oxides. It has been shown by work in mammals that there are two major epoxide hydrolases with differing substrate selectivities and cellular locations. Epoxides with *trans* substituents adjacent to the oxirane ring, such as juvenile hormone ester, are potential substrates for the cytosolic epoxide hydrolase [80], while epoxide with *cis* substituents adjacent to the epoxide, such as cyclohexene oxide, naphthalene oxide and HEOM are substrates for the microsomal epoxide hydrolase [81]. Studies with purified microsomal epoxide hydrolase from mammals and an insect have shown that in both the mammals (rat and rabbit) and the insect (armyworm) dieldrin is a very poor substrate for epoxide hydrolase, while chlordane epoxide and HEOM are converted to diols somewhat more rapidly, and model compounds such as styrene oxide and 1,2-epoxyoctane are rapidly metabolized [82, 83]. Since microsomal epoxide hydrolase enzymes from such diverse species as rat and armyworm show similar substrate specificities, it is reasonable to assume that if microsomal epoxide hydrolase activity with a model substrate is found in another species, pesticides with epoxides in their structure will have the same potential for conversion to diols in the new species. Studies have shown that the microsomal form of epoxide hydrolase is present in liver or hepatopancreas of a variety of fish and shellfish species, and is particularly active in the hepatopancreas of decapod crustaceans such as the lobster, blue crab and spiny lobster [84–86]. These studies used the model epoxide substrates styrene oxide, epoxyoctane, benzo(a)pyrene-4,5-oxide and *cis*-stilbene oxide as substrates [84–86]. While little attention has been paid to the epoxide hydrolase activity of fish cytosol with trans epoxides, a recent study has shown that trout cytosol has good activity with the model substrate *trans*-stilbene oxide [86]. Clearly, more research is needed in this area for a better understanding of the role of epoxide hydrolase in the metabolism of epoxide-containing pesticides.

5 Biotransformation by Phase 2 Pathways

Phase 2 pathways, also known as conjugation pathways, are important in the biotransformation of molecules that already have certain functional groups, but are still relatively lipophilic. The phase 2 enzymes catalyse the conjugation of the functional group in the pesticide molecule with a small, usually polar endogenous molecule to give a metabolite that is readily excreted. Functional groups such as hydroxyl (aromatic or aliphatic), α-β-unsaturated carbonyl, carboxylate, epoxide, halogen, amine and thiol groups are some of the common sites of phase 2 biotransformation of pesticides. These groups may have been introduced by phase 1 metabolism, or may be present in the original pesticide. The most common phase 2 pathways which pesticides and their phase 1 metabolites undergo in aquatic species are glutathione conjugation, sulfation, carbohydrate conjugation, acetylation and taurine conjugation [87, 88]. Table 3 lists these pathways and some known pesticide substrates for each, which are shown in Fig. 2.

5.1 Glutathione Conjugation

This pathway is important in the biotransformation of pesticides with electrophilic centers in their molecular structure. The electrophilic atom in the pesticide substrate forms a thioether bond with the –SH group of the endogenous tripeptide, glutathione (GSH), in a reaction that can occur non-enzymatically as well as being catalysed by GSH-S-transferases [89]. The GSH conjugate, once formed, is then further metabolized in liver and kidney by peptidases to give firstly the cysteinylglycine conjugate and secondly the cysteine conjugate: the

Table 3. Phase 2 pathways important in pesticide metabolism

Enzyme	Endogenous cosubstrate	Representative substrate	Reference
GSH-S-transferase	Glutathione	Dibromoethane	158
		Chlorodinitrobenzene	84
		Molinate sulfoxide	91
Glucuronosyl- and glucosyl-transferases	UDP-glucuronic acid or UDP-glucose	β-Naphthol	151
		Pentachlorophenol	114
		4′-Hydroxypermethrin	71
		3-Trifluoromethyl-4-nitrophenol	30
Sulfotransferase	PAP-sulfate	Pentachlorophenol	114
Acetyltransferase	acetylCoA	Sulfadimethoxine	132
Taurine *N*-acyl-transferase	taurine[1]	Triclopyr	147
		2,4-Dichlorophenoxy-acetic acid	146
		2,4,5-Trichlorophenoxy-acetic acid	146

[1] the pesticide substrate must be activated to the coenzyme A derivative in order to form a peptide bond with taurine. The structures and positions of biotransformation of these substrates are shown in Fig. 2

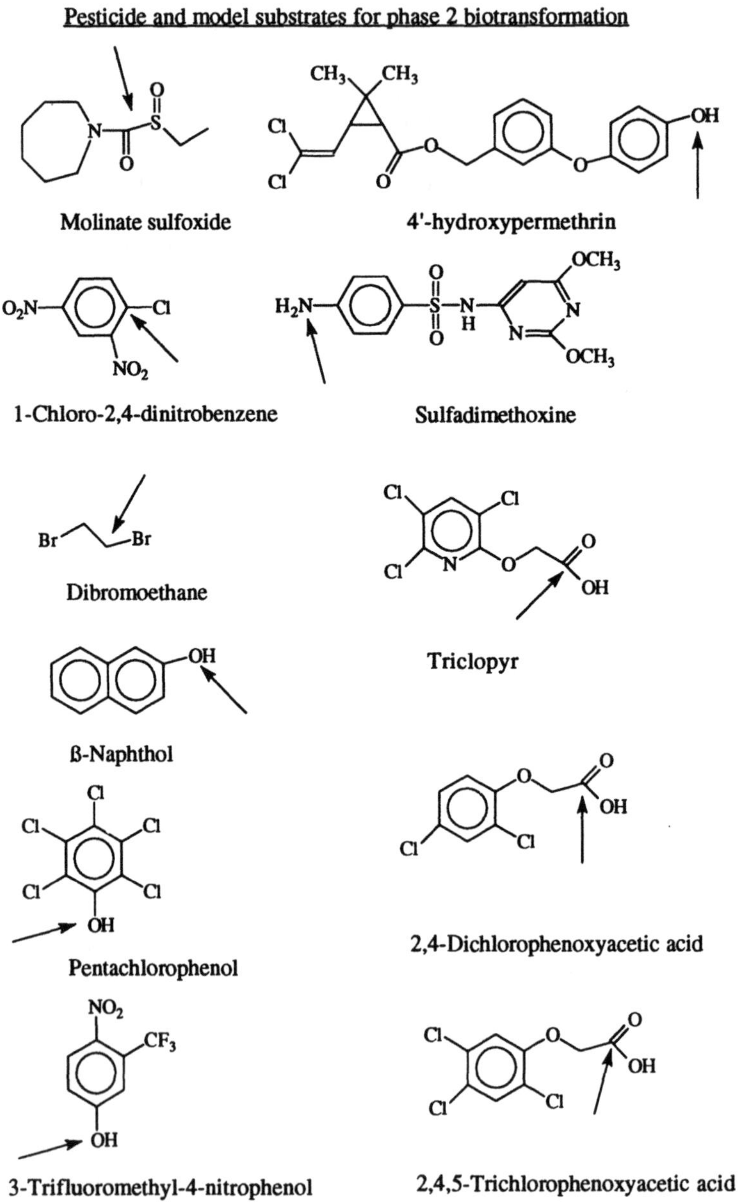

Fig. 2. Pesticide and model substrates for phase 2 biotransformation. The *arrow* indicates the point of metabolism

cysteine conjugate is then metabolized by acetyl transferase to give the *N*-acetylcysteine conjugate also known as the mercapturic acid. The mercapturic acid is the metabolite that is normally excreted from the animal, usually in urine, although the unacetylated cysteine conjugate may also be excreted in urine.

Examples of pesticides that form GSH conjugates are halogenated compounds such as dibromoethane and related chlorinated alkanes and alkenes that are used as nematocides, thiocarbamates such as molinate, and epoxides such as those discussed above in the section on epoxide hydrolase. GSH conjugation of most substrates gives a conjugate that is less toxic and reactive than the parent compound, but in the case of halogenated alkanes and alkenes with more than one halogen in the molecule, GSH conjugation can give rise to a metabolite that is more reactive than the parent compound. This is the case with dibromoethane, which is activated by GSH conjugation to an episulfonium ion that can attack DNA bases [90].

5.1.1 In Vivo Studies

In vivo studies have shown that GSH conjugates and their products of further metabolism, cysteine and *N*-acetylcysteine (mercapturic acid) conjugates are excreted in urine and bile of fish exposed to substrates for GSH conjugation. A well-documented pathway for biotransformation of molinate in carp [23, 91], sturgeon [22] and bass [92] was phase 1 metabolism to molinate sulfoxide followed by GSH conjugation and excretion of the mercapturic acid. Molinate was more toxic to carp than sturgeon, and it was shown that more molinate metabolites were excreted by sturgeon than carp 24 hours after exposure to 100 μg molinate/L water, and more molinate mercapturic acid was excreted by sturgeon than carp [23]. These results suggest that slow overall biotransformation or less efficient GSH conjugation of molinate or molinate sulfone in the carp may contribute to the higher toxicity of molinate to carp than sturgeon. There are several examples in the literature showing that differences between species and individuals in the efficiency of GSH conjugation of reactive metabolites can influence toxicity. For example, the reactive metabolite of acetaminophen, *N*-acetyl-benzoquinone imine, will cause liver necrosis only after GSH has been depleted from liver, as it may be detoxified by conjugation with GSH [93]. The reactive metabolite of aflatoxin, *exo*-aflatoxin B_1-8,9-oxide can be detoxified by glutathione conjugation and it has been shown that among mammals, species differences in GSH conjugation efficiency correlated with the likelihood of tumors formation after exposure to aflatoxin [94]. It should be noted for this example, however, that in trout and salmon no enzyme-catalyzed aflatoxin B_1-8,9-oxide glutathione conjugation occurred, perhaps due to a lack of the specific alpha class GST enzymes which catalyse aflatoxin B_1-8,9-oxide glutathione conjugation [95]. Species differences in aflatoxin carcinogenicity in fish appears to relate to differences in the efficiency of formation of aflatoxin B_1-8,9-oxide by P450 [96].

More detailed studies of the pathways for mercapturic acid formation in fish were carried out by Yagen et al. [97]. These investigators showed that when the GSH-conjugate of styrene oxide was injected into flounder, *Pseudopleuronectes americanus*, the urine contained not only the expected mercapturic acid conjugate but also substantial amounts of unacetylated cysteine conjugate [97]. This suggests that the enzyme catalyzing acetylation of cysteine conjugates of xenobiotics may have low activity in the flounder.

5.1.2 In Vitro Studies

Although there are few in vivo studies showing glutathione conjugation, there have been a number of reports showing the presence of GSH S-transferase (GST) activity with model substrates in vitro in tissues of fish and other aquatic species, and several reports of the isolation of purified GST enzymes from aquatic species. In most of these studies the model compound substrate 1-chloro-2,4-dinitrobenzene (CDNB) was used to probe activity. Although most of the available data is for this model compound, rather than specific pesticides, these studies are relevant to an understanding of the expected ability of different aquatic species to form GSH conjugates of pesticides because with few exceptions GST enzymes have overlapping substrate specificity and most GST studied to date have activity with CDNB. Species with high GST activity may be expected to be less sensitive to the toxicity of pesticides that are detoxified by GSH conjugation, and more sensitive to the toxicity of pesticides, such as dibromoethane, that are activated by GSH conjugation. Also of interest in this context are studies of the enzymes that further metabolize GSH conjugates to mercapturic acids, namely γ-glutamyl transpeptidase, dipeptidase, and in particular cysteine conjugate-*N*-acetyl-transferase, since this will give an indication of the ability of the species in question to form the readily excreted mercapturic acids [98]. Table 4 lists several aquatic species from which GST enzymes have been isolated, and the number of isozymes obtained. Many more aquatic species have been shown to possess GST activity with one or more substrates [87, 99]. As in mammals, the GST enzymes isolated from aquatic species consist of 2 subunits of molecular weight 23–28 kDa. Attempts have been made to determine if the GST isozymes from aquatic species are similar to mammalian GSTs, in terms of sequence similarity or immunochemical relatedness. These studies have shown similarities between some fish GST and mammalian GSTs in the pi

Table 4. Glutathione S-transferase isozymes purified from fish and shellfish

Species	Number of isozymes	Reference
Squid	1	100
Blue crab	2	159
Little skate	5	160
Thorny skate	2	161
Rainbow trout	7	162
Plaice	2	99
Atlantic salmon	6	163

class, but, perhaps not surprisingly, several of the fish GST isozymes have not shown extensive similarity to mammalian GSTs based on immunoreactivity or sequence homology [99, 100]. A more thorough discussion of GST in fish is found in the recent review by George [99].

The inducibility of GST in aquatic species is a question of considerable interest. In mammalian species, GST is readily induced by exposure to phenobarbital-type inducing agents, methylcholanthrene-type inducing agents, antioxidants and other compounds [95]. There is little evidence for induction of GST by phenobarbital-like inducing agents, as is the case for P450 in fish [48–51]. Following treatment of fish with 3-methylcholanthrene-type inducing agents, GST in fish organs are less likely to be markedly induced than would be the case with rat [99]. Indeed, some fish exhibit no induction of GST after treatment with 3-methylcholanthrene [101, 102]. The compound *trans*-stilbene oxide, which causes marked induction of epoxide hydrolase and GST in the rat, caused a 1.5 to 2-fold induction of GST in flounder, *Platichthys flesus* and plaice, *Pleuronectes platessa* [102], but had no effect on GST activity in English sole, *Parophrys vetulus* [103]. There have been some recent studies of the effects of pesticide exposure on GST activity in fish. Bach and Snegaroff [57] showed that treatment of rainbow trout with the fungicide prochloraz by intraperitoneal injection or in the diet did not result in induction of GST in liver, although following dietary treatment there was a small increase in the GST activity of the pyloric caecum. Treatment of fish with Aroclor 1254 or other polychlorinated biphenyls resulted in slight induction of GST activity in channel catfish liver [104]. In contrast, studies of the effect of dibromoethane on drug metabolizing enzymes in medaka showed that medaka, *Oryzias latipes*, exposed to 1.3 mg/L for 14 days had more than double the GST activity of control fish, and that hepatic cytosol from these medaka showed increases in a protein band at molecular weight 26 kDa (Fig. 3, Hawkins et al., in preparation). Overall, these induction studies showed that fish GST is not as sensitive to induction as GST in mammalian species and that of the inducing compounds studied to date, only *trans*-stilbene oxide (two out of three species studied) and dibromoethane (medaka) have been shown to induce GST in fish 2-fold or more. The induction of dibromoethane glutathione transferase is of toxicological importance, however, in that the episulfonium ion formed after GSH conjugation of dibromoethane is a highly reactive species implicated in mutagenesis [90].

More studies of the regulation, substrate selectivity and occurrence of GST in fish are needed to be able to understand and predict the likely toxicity of the many pesticides whose toxicity is altered by GSH conjugation.

5.2 Carbohydrate Conjugation and Sulfate Conjugation

A large number of pesticides contain nucleophilic groups that can be conjugated with glucuronic acid or glucose. Examples of such groups are aliphatic and aromatic hydroxyls, carboxylic acids, nucleophilic carbon atoms, amines and

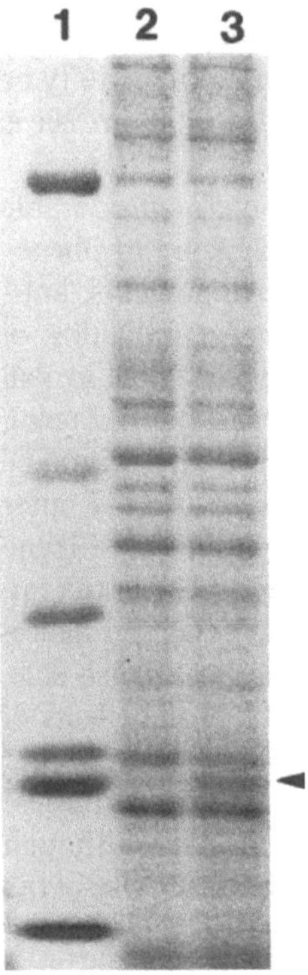

Fig. 3. Induction of glutathione S-transferase in hepatic cytosol from medaka treated with dibromoethane. *Lane 1*: Molecular weight standards of 66; 45; 36; 29; 26; and 20.1 kDa. *Lane 2*: Cytosol from the liver of control medaka. *Lane* 3: Cytosol from the liver of medaka that were exposed to dibromoethane, 1.3 ppm, for 14 days prior to sacrifice. The *arrow* indicates the protein at molecular weight 26 kDa which is more prominent in dibromoethane-exposed fish

thiols. Hydroxy groups (and less usually amine groups) are also substrates for sulfation, and for pesticides that contain hydroxy groups, carbohydrate conjugation and sulfation are competing reactions. For this reason carbohydrate conjugation and sulfation are discussed together in this review. If they are not present in the parent pesticide, groups that may be conjugated with carbohydrate (hydroxyls, carboxylates, amines) or sulfate (hydroxyl groups) may be introduced by phase 1 metabolism of the pesticide. In fish, the major carbohydrate conjugation pathway is glucuronidation, whereas in those invertebrates where carbohydrate conjugation has been studied the major carbohydrate conjugation pathway is glucosidation [88, 105–107]. Carbohydrate conjugation is catalysed by UDP-glucuronosyl- or UDP-glucosyl-transferases, with UDP-glucuronic acid and UDP-glucose respectively as cosubstrate [105]. Studies in mammals have shown that UDP-glucuronosyl-transferases are a family of membrane-bound enzymes with differing substrate selectivities [108]. Sulfation is catalysed by one or more of a family of cytosolic sulfotransferase enzymes with PAP-sulfate as cosubstrate [109]. In considering whether or not sulfation or carbohydrate conjugation will predominate for a given substrate, important factors are the affinity of the pesticide substrate for the enzyme and the availablility of the cosubstrate. In many instances where this has been studied, sulfation has been shown to be a high affinity, low capacity pathway, while carbohydrate conjugation is often a low affinity, high capacity pathway [107, 110]. The cosubstrate PAP-sulfate is generally present in cells in small amounts, compared to UDP-glucose and UDP-glucuronic acid, so that even though many substrates show higher affinity to the sulfotransferase than the glycosyltransferase enzymes, the low concentration of PAP-sulfate can limit the extent of xenobiotic sulfation in the presence of large amounts of xenobiotic. It has been shown that the concentration of UDP-glucuronic acid (0.35 mM) in trout liver was similar to the UDP-glucuronic acid concentration found in livers of several mammalian species [111]. Since aquatic species are generally exposed to pesticides at relatively low concentrations, it may be expected that if the same factors apply to carbohydrate conjugation versus sulfation in fish and shellfish as apply in mammals, sulfation should predominate. To date, however, there have been very few systematic studies of the dose dependency of conjugation of pesticides or model compounds in fish or shellfish. Another factor that affects the nature of the final excreted metabolite is the site of formation of the conjugate, and its route of excretion. If the initial exposure to the pesticide is at a high enough concentration so that glucuronidation predominates in the fish liver, the pesticide glucuronides may be excreted into bile. Once the bile is secreted into the intestine the pesticide glucuronide may be hydrolysed by intestinal β-glucuronidases and the parent pesticide re-absorbed and re-conjugated in a process known as enterohepatic cycling [112]. In the second conjugation cycle, the concentration of pesticide will be lower so that sulfate as well as glucuronide conjugates may be formed, or sulfation may predominate, and the ultimate urinary metabolites may be predominantly sulfate conjugates with smaller amounts of glucuronides. This scenario has been observed or deduced in fish for some xenobiotics, as described below.

5.2.1 In Vivo Studies

There have been several in vivo studies of pesticide fate in fish in which glucuronides have been shown to be important intermediate or ultimate metabolites (see Table 5 for selected examples). The fate of pentachlorophenol, a biocide that already has a functional group suitable for glucuronidation, has been studied in several fish species, including goldfish [113, 114], zebra fish, *Brachydanio rerio* [115], fathead minnow, *Pimephalis promelas*, sheepshead minnow, firemouth, *Cichlasoma meeki*, and rainbow trout [114]. In most of these fish species, the major metabolite excreted into the tankwater was pentachlorophenylsulfate [113, 114], although in the zebra fish, a species commonly used in toxicity testing, roughly equal amounts of glucuronide and sulfate were excreted into tank water at the exposure concentration (0.01 ppm) used [115]. Kobayashi [113] found that in the bile of goldfish exposed to 0.1 ppm of pentachlorophenol, only the glucuronide conjugate was present. Although different doses of pentachlorophenol were also studied, there was no information concerning the dose-dependency of metabolite excretion [113]. In goldfish, there was evidence that the pentachlorophenylglucuronide was hydrolysed in intestine and when re-absorbed was preferentially conjugated with sulfate and excreted across the gills and in urine as the sulfate conjugate [113]. Stehly and Hayton [114] found that the bile of goldfish and other fish exposed to 5 μg/L pentachlorophenol (i.e. 0.005 ppm) contained both glucuronide and sulfate conjugates, as well as in some cases parent compound, with the relative proportion of each conjugate varying with species. The discrepancy between the two studies in goldfish may perhaps be reconciled by considering the effect of the difference in exposure concentration. Since sulfation is usually a high affinity, low capacity pathway, it is possible that the reason no sulfate conjugate was found in bile when goldfish were exposed to 0.1 ppm was because at that concentration glucuronidation predominated, whereas at the lower concentration sulfation and glucuronidation occurred. Studies of phenol conjugation in goldfish, have also shown that although the glucuronide conjugate was excreted into bile, the major metabolite excreted in urine or into tank water was phenylsulfate [116–117]. Most workers think the frequently observed discrepancy in metabolite composition between bile and tankwater can be explained by assuming the general importance of enterohepatorenal cycling in the metabolism of hydroxylated compounds in fish [7, 114], although this is not universally accepted [115]. These studies illustrate the point that although fish bile is a rich source of pesticide metabolites in pesticide-exposed fish [118], in some cases examination of the bile for metabolites gives an incomplete picture of the ultimate fate of a pesticide.

Pesticides that form glucuronide or sulfate conjugates in fish after phase 1 metabolism include synthetic pyrethroids and thiocarbamates [7]. For example, 4′-hydroxypermethrin and 4′-hydroxycypermethrin were present in the bile of permethrin- or cypermethrin-treated rainbow trout as glucuronide conjugates [22, 71]. After phase 1 hydroxylation of the thiocarbamate herbicide, molinate, in the hexahydroazepine ring to 4-hydroxymolinate, the glucuronide metabolite was formed and excreted into tankwater by sturgeon and carp [23]. Recently it was shown that trichloropyridinol, a phase 1 metabolite of the organophosphorothionate pesticide chlorpyrifos, was excreted in catfish urine and bile mainly as the trichloropyridinol-glucuronide [74]. In this study other polar metabolites (one of which could have been the sulfate conjugate) were also found.

Thus, glucuronidation and sulfation are competing processes that are usually efficient in fish for conjugating hydroxy groups in pesticides and pesticide metabolites, so that very little unconjugated pesticide is excreted in urine.

There have been few studies of the sulfate and carbohydrate conjugation of pesticides and related compounds in shellfish. Results from some of these studies are presented in Table 5. The fate of the model compound, phenol, in the lobster was dose-dependent: at low doses, the predominant route of excretion was urinary, with phenylsulfate as the major metabolite, whereas at high doses unchanged phenol was predominantly excreted very rapidly across the gills and a lower percentage was excreted in urine as phenylsulfate [119]. Very little

Table 5. Conjugation of pesticides and phenolic pesticide metabolites in different aquatic species

Pesticide or model compound	Species	Parent compound	Phenolic metabolite	Sulfate	Glucoside	Glucuronide	Unknown	Reference
Phenol	Lobster	+ + +	–	+ +	trace	–	–	119
β-Naphthol	Lobster	+ +	–	trace	+ +	–	trace	
Methylparathion[1]	Ridgeback prawn	+	+ +	+ +	+	–	+ + +	32
4-Nitroanisole[2]	Ridgeback prawn	+ + +	+	+	trace	–	+	32
Methylparathion[1]	Crayfish	+	+ +	+	+ +	–	–	32
4-Nitroanisole[2]	Crayfish	+	+ +	+	+ + +	–	–	32
Fenitrothion[3]	Blue crab	+ + +	+ +	+	+	–	–	30
Pentachlorophenol	Zebrafish	+ +	–	+ +	–	+ +	–	115
Phenol	Zebrafish	+	–	+ +	–	+ + +	–	115
4-Nitrophenol	Zebrafish	+ + +	–	+	–	+ +	–	115
2-Cresol	Zebrafish	+ +	–	+ + +	–	+	–	115
4-Chlorophenol	Zebrafish	+ +	–	+ + +	–	+ +	–	115
1-Naphthol	Catfish	+	–	+ + +	–	+ +	–	151
Chlorpyrifos[4]	Catfish	+	+	–	–	+ + +	+	74, 164

The phenolic metabolites of the pesticides are: [1] 3-methyl-4-nitrophenol; [2] 4-nitrophenol and 4-aminophenol; [3] 3-methyl-4-nitrophenol and 3-methyl-4-aminophenol; [4] trichloropyridinol

Occurrence is as follows: + + +, > 50%; + +, 10–49%; +, 1–9%; trace, < 1%; –, not found phenylglucoside was excreted in urine [119]. A somewhat more lipophilic compound, β-naphthol, had a quite different fate in lobsters: β-naphthol was excreted in urine of dosed lobsters much more slowly than phenol and predominantly as the glucoside conjugate (Li and James, unpublished observations). The fate of methylparathion was investigated in crayfish, Malysian prawns and ridgeback prawns, and it was found that at non-toxic concentrations of this pesticide, the glucoside and sulfate conjugates of 4-nitrophenol, a phase 1 metabolite of methylparathion, were excreted into the tankwater [32]. Small amounts of galactoside and glucuronide conjugates were also reported as possible (based on HPLC retention time) metabolites of 4-nitrophenol in ridgeback prawns and freshwater crayfish (putative glucuronide only), but while the galactoside was confirmed by independent analysis, the authors could not confirm the identification of the putative glucuronide conjugate by other analytical methods. The relative amounts of sulfate and glucoside varied with species when all were exposed to the same concentration of methylparathion, 10 μg/L. In the ridgeback prawn additional, unidentified conjugates of 4-nitrophenol and 4-nitro-*m*-cresol were found after exposure to methylparathion, 4-nitroanisole or 4-nitro-*m*-cresol, suggesting that novel pathways for xenobiotic conjugation may remain to be discovered in invertebrates [32].

5.2.2 In Vitro Studies

In vitro studies of UDP-glucuronosyltransferase, UDP-glucosyltransferase and PAP-sulfotransferase have been carried out in several aquatic species. The substrates used in these in vitro studies are usually model compounds, rather than pesticides or pesticide metabolites, but the findings are likely to be applicable to a general understanding of carbohydrate and sulfate conjugation in the studied species.

5.2.2.1 UDP-Glucuronosyltransferase

Studies with fish have confirmed that UDP-glucuronosyltransferase is localized in the endoplasmic reticulum and shown that, as in mammals, the enzyme activity can be increased by adding detergent or otherwise disrupting the microsomal membrane. Recently, UDP-glucuronosyltransferases have been partially purified from plaice liver microsomes and shown to have molecular weights (52 to 57 kDa) similar to those of mammalian UDP-glucuronosyltransferases [120]. Several studies of microsomal UDP-glucuronosyltransferase activity have used *p*-nitrophenol as a substrate [99, 111, 121–123], and this is relevant to pesticides in that it is a metabolite of parathion and methylparathion. In other studies, 1- or 2-naphthol (also potential pesticide metabolites), 4-methylumbelliferone and 3-, 7- and 9-hydroxybenzo(a)pyrene have been shown to be substrates for fish UDP-glucuronosyltransferase [58, 99, 104, 123, 124].

The inducibility of fish UDP-glucuronosyltransferase has been investigated. In one set of experiments, rainbow trout were treated with prochloraz under several different protocols, which were: i.p. injection of doses ranging from 50 to 300 mg/kg and sacrifice 24 hours later; i.p. injection of 150 mg/kg and sacrifice at 1, 2, 4 or 8 days after the dose; and dietary administration of 60 mg prochloraz/kg for 7 days with sacrifice 24 hours later. Following i.p. injection, 1-naphthol-UGT was unchanged or inhibited in liver, and was unchanged in intestine, stomach and pyloric caeca, except for a less than 2-fold induction in the pyloric caecum at the 100 mg/kg dose [58]. Dietary treatment with prochloraz-sprayed pellets resulted in marked inhibition of 1-naphthol-UGT in liver, intestine, stomach and pyloric caeca. It was not clear why the prochloraz should inhibit UGT, but accidental treatment of trout with this fungicide could slow down glucuronidation and therefore excretion of prochloraz metabolites or other environmentally encountered xenobiotics. Rainbow trout treated with β-naphthoflavone, 50 mg/kg i.p., had induced *p*-nitrophenol-UGT in liver and kidney microsomes at 14 days after the dose, and hepatic *p*-nitrophenol-UGT was also induced at 3 and 7 days after the dose, although treatment with lower doses of 5 and 0.5 mg β-naphthoflavone/kg did not alter activity [125]. Aroclor 1254 and 3-methylcholanthrene were shown to induce hepatic and renal (Aroclor only), but not intestinal or branchial 1-naphthol-UGT in the marine fish, plaice, whereas no induction of testosterone or bilirubin-UGT was observed after either treatment [120]. Even in liver, the induction response of UGT to planar polycyclic-type inducing agents in fish was weak, requiring high doses of inducing agent, of low magnitude, typically less than 2-fold induction, and took longer to achieve, compared with the induction of P450 1A1 [104, 120, 125, 126].

5.2.2.2 UDP-Glucosyltransferase

Carbohydrate conjugation has been studied in vitro in tissues from several invertebrates. UDP-glucosyltransferase activity has been demonstrated in crustacean and molluscan species. Microsomal fractions of hepatopancreas, antennal gland and other organs of the lobster, spiny lobster, crayfish and gumboot chiton, *Cryptochiton stelleri*, possessed UDP-glucosyltransferase activity with a variety of substrates including phenol, *p*-nitrophenol, 4-methylumbelliferone, 1-naphthol, 2-naphthol and 3-hydroxybenzo(a)pyrene [106, 107, 127]. No UDP-glucuronosyltransferase activity with several of these substrates was found in the microsomes from either lobster species, although this was specifically examined [106, 107], whereas UDP-glucuronosyltransferase activity was reported for the chiton [128]. The question of whether or not invertebrates are capable of glucuronidation of pesticides and other xenobiotics is not definitively answered at this time. Although a few papers suggest that glucuronides are excreted by invertebrates [32] or formed in in vitro incubations [128–130], the methods used in these studies were either not rigorous in demonstrating product glucuronide formation [128–130] or did not give unequivocal identification of the glucuronide [32]. Since glucuronides are more water soluble than glucosides, and are likely to exist as the anion at physiologically important pH, glucuronides are excreted better than glucosides, which are not be anions at physiologically important pH, by vertebrate species that possess transport systems for organic anions (see below). Invertebrates often have inefficient excretion systems, so that there is little advantage and possibly a disadvantage for invertebrates to have an anionic metabolite, which may in some cases be ion-trapped in cells of invertebrates and less efficiently excreted than a polar, but unionized metabolite such as a glucoside. More studies are needed to determine if glucuronidation is an exclusively vertebrate phase 2 pathway, and to address the point of excretion characteristics of glucosides versus glucuronides in vertebrates and invertebrates.

5.2.2.3 Sulfotransferase

Sulfotransferase is a cytosolic enzyme of which there are relatively few studies in fish and shellfish. Gregus et al. [111] showed that trout had low hepatic 2-naphthol sulfotransferase activity compared with several mammalian species. Recent studies have, however, shown that catfish liver and intestinal mucosa possess efficient sulfotransferase activity with 3-, 7- or 9-hydroxybenzo(a)pyrene as substrate (124 and unpublished studies). In view of the in vivo findings that sulfate conjugates are frequently excreted into tank water (see above), studies of sulfotransferase activity in kidney and gill should be undertaken to determine if the excreted sulfate conjugates are formed in liver or at the organ of excretion, as is suggested by some studies. To date the sulfotransferase enzyme has not been purified from any fish species. Among invertebrates, sulfotransferase actvity was found in antennal gland cytosol of two crustacean species, the lobster and the spiny lobster, where good activity was found with phenol, 4-nitrophenol, 2-naphthol, 4-methylumbelliferone and 3-hydroxybenzo(a)pyrene [106, 107]. No activity was found in hepatopancreas cytosol in these studies although some sulfatase activity, which is normally microsomal, was found in cytosol suggesting that in these studies sulfotransferase may have been masked by solubilized sulfatase [106, 107]. In the intact cell, sulfatase and sulfotransferase would be separated. Others have found sulfotransferase activity with 4-nitrophenol in lobster hepatopancreas cytosol [43] and evidence for sulfate conjugates of 3-methyl-4-nitrophenol (a fenitrothion metabolite) in blue crab hepatopancreas [30]. There was some suggestion that the properties (Km values) of antennal gland β-naphthol sulfotransferase altered depending on the molting stage of the lobster [107]. Since variations in pesticide sensitivity are reported for crustacea in different molting stages, further study of this aspect is warranted. The in vitro studies of phenol sulfotransferase and glucosyltransferase in the lobster [107] provided an explanation for the low amount of phenylglucoside formed in vivo by lobsters [119], in that the Km for phenol glucosyltransferase was in the mM

Table 6. Properties of PAPS-sulfotransferase and UDP-glucosyltransferase in two lobster species (data from references 106 and 107)

Substrate	Spiny lobster				American lobster					
	Sulfotransferase[1]		Glucosyltransferase[2]		Sulfotransferase[1]		Glucosyltransferase[1]		Glucosyltransferase[2]	
	K_m[3]	V_{max}[4]	K_m[3]	V_{max}[4]	K_m[3]	V_{max}[4]	K_m[3]	V_{max}[4]	K_m[3]	V_{max}[4]
Phenol	–[5]	–	–	–	22	0.02	2110	0.48	1600	1.35
2-Naphthol	36	2.14	190	0.12	112	0.36	9	0.59	90	1.04
4-Nitrophenol	728	1.48	240	0.58	–	–	–	–	–	–

[1] antennal gland activity
[2] hepatopancreas activity
[3] μM
[4] nmole/min/mg protein
[5] not measured

range, whereas that for sulfotransferase was in the μM range (Table 6). This was not, however, a general rule, as some of the substrates examined had similar Km values for both sulfotransferase and glucosyltransferase (Table 6).

5.3 Acetylation

Pesticides containing amino groups are potential substrates for acetylation. This pathway has been demonstrated in trout as a major pathway for biotransformation of amino group containing environmental chemicals, such as 2-naphthylamine and 2-aminofluorene [109], and drugs such as isoniazid, benzocaine and its metabolite p-aminobenzoic acid, and sulfadimethoxine [111, 131, 132]. It may therefore be expected that pesticides containing amino groups would be acetylated by trout. Carp have been shown to form *N*-acetylsulfadimidine as a minor metabolite of sulfadimidine [133], but other species have not yet been examined and little is known about acetyltransferase activity in fish.

There have been few studies of the fate of the amino group in invertebrates. Studies have shown that *p*-nitroanisole was reduced to *p*-aminoanisole then acetylated by the sea urchin, *Strongylocentrotus purpuratus*, and the gumboot chiton, while the starfish, *Pisaster ochraceus*, and sea cucumber, *Cucumaria miniata*, did not acetylate the *p*-aminoanisole [134]. In an unusual metabolic pathway related to acetylation, mussels and oysters produced the *N*-formyl metabolite of *o*-toluidine [135]. After administration of sulfadimethoxine to lobsters, only very small amounts of *N*-acetylsulfadimethoxine were found in hemolymph, hepatopancreas and urine, and these may have been an artifact of the analysis method [136].

5.4 Taurine Conjugation

Pesticides that contain carboxylic acid groups are potential substrates for conjugation with amino acids [137]. In the aquatic species in which carboxylic acid conjugation has been studied, the only positively identified amino acid to form conjugates with xenobiotic carboxylic acids is taurine. In several species of marine fish, the flounder, the shark, *Squalus acanthias* and the southern flounder, *Paralichthyes lethostigma* [138] 2,4-dichlorophenoxyacetic acid (2,4-D) was eliminated to varying extents as the taurine conjugate, with the balance excreted as the unchanged 2,4 D. Other pesticide and non-pesticide carboxylic acids have also been eliminated from fish as taurine conjugates, including benzoic acid [139–141], 2,4,5-trichlorophenoxyacetic acid (2,4,5-T) [88, 142], phenylacetic acid [143–144] and the DDT metabolite, bis(4-chlorophenyl)acetic acid (DDA) [138, 145].

Taurine conjugation has also been identified as a major pathway of metabolism of carboxylic acid groups in the small number of invertebrates (both crustacea) that have been studied to date. It was shown that injected 2,4-D, 2,4,5-T, DDA and phenylacetic acid were conjugated to various extents with taurine prior to excretion from the spiny lobster [146]. The unconjugated balance of the pesticide was excreted unchanged in the spiny lobster urine [146]. For these carboxylic acids, taurine conjugation did not facilitate excretion, as the unconjugated acid was more rapidly eliminated than the conjugate [146]. In studies with the freshwater crayfish, Barron et al. [147] exposed crayfish to the herbicide triclopyr for 11 days, and followed the elimination for 36 days. The hepatopancreas and the hemolymph contained higher concentrations of triclopyr residues than the tail muscle, and the principle component of both hepatopancreas and hemolymph was parent triclopyr. The hepatopancreas contained one major and several minor metabolites, and the major metabolite was confirmed to be the taurine conjugate [147].

Although taurine conjugation appears to be an important pathway for conjugation of carboxylic acids in aquatic animals, there have been very few studies of the enzymology of taurine conjugation. By analogy with glycine conjugation in mammals, the taurine *N*-acyltransferase enzyme was found in the mitochondrial matrix of kidney and liver of several fish species, and required the coenzyme A derivative of the xenobiotic carboxylic acid as substrate with taurine as cosubstrate [138]. Recently, the pathway for amino acid conjugation of bile acids was examined in fish [148] and taurine conjugation was found to predominate, although glycine was an alternate, high Km substrate for forming bile acid conjugates.

As more pesticides with carboxylic acids in their structure are studied it is to be hoped that the factors governing the formation of amino acid conjugates will be elucidated.

6 Elimination of Pesticides and Their Biotransformation Products by Fish and Shellfish

Pesticides, and their more water-soluble biotransformation products, may be excreted from the fish or shellfish body by several routes (Table 7). The actual route of excretion taken by a given pesticide depends on several factors, including dose, route of exposure and species, and most pesticides and their metabolites are excreted by more than one route. Excretion across the gill was shown a number of years ago [12] to be an important pathway for small molecules that are freely water soluble but that have some lipophilic character, characteristics that pertain to a number of pesticides. Subsequent studies have for the most part supported this general rule for branchial excretion, although there have been few systematic studies to examine, for example, species differences and dose dependancy in the extent of branchial excretion of pesticides and

Table 7. Major routes of elimination of pesticides and pesticide metabolites by aquatic species

Route	Physico-chemical property of excreted compound	Example	Species	References
Gills	Small molecule with hydrophilic and lipophilic properties	Phenol	Lobster	119
		Phenol	Zebrafish	115
		Monoethyl-hexylphthalate	Trout	76
Urine	Small water-soluble molecules, preferably weak acids with pKa < 6	Phenylsulfate	Lobster	119
		Naphthylglucuronide	Catfish	151
		2,4-Dichlorophenoxyacetic acid	Lobster	146
		2,4,5-Trichlorophenoxyacetyl-taurine	Shark	142
Bile	Anions and cations larger than about 300 molecular weight	Trichloropyridinol glucuronide	Catfish	74, 164
		Pentachlorophenyl sulfate	Goldfish	114
		Molinate-glutathione	Carp	91
		2,4-dinitrophenylglutathione	Skate	98

other xenobiotics. The model compound, phenol, was eliminated across the gills of lobster despite being highly bound to hemolymph proteins, and the extent of branchial excretion of phenol was markedly dose-dependent, being higher at higher doses [119]. Phenol was also excreted across the gills of goldfish following a dose of 10 mg/kg [115]. Diethylhexylphthalate was taken up across the gills of rainbow trout and metabolized to monoethylhexylphthalate, apparently in gills, and the monoethylhexylphthalate was efficiently excreted across the gills without reaching the systemic circulation [76].

Following systemic uptake and biotransformation of pesticides to more water soluble metabolites, such as sulfate conjugates, glucuronides, taurine conjugates and mercapturic acid conjugates, urinary excretion is often the major route of elimination, as these metabolites are all substrates or potential substrates for the renal organic anion transport system which in vertebrates, including all teleost fish studied to date, facilitates excretion of anions into urine [149]. It has been shown that sulfate conjugates of 3-hydroxybenzo(a)pyrene are more readily excreted than glucuronide conjugates by the flounder renal organic anion transport system, leading to a predominance of sulfate conjugates in urine in the first few hours after administration [150]. The same trend was seen for 1-naphthol conjugates excreted by catfish. While the sulfate conjugate of 1-naphthol was the major urinary metabolite at all times in the urine of catfish given 1-naphthol intravascularly or orally, the proportion of glucuronide conjugate excreted in urine went up with time [151]. Studies of the fate of several carboxylic acids in the southern flounder showed that carboxylic acids that were extensively metabolized to taurine conjugates were more slowly excreted than carboxylic acids that were poorly metabolized [88]. Comparison of the extents of metabolism and the rates of excretion of benzoic acid and bis-(4-chlorophenyl)acetic acid (DDA, a metabolite of DDT) in different fish species, showed that those species that extensively metabolized these acids to the taurine conjugates also excreted them more slowly [139–141, 145]. These results suggested that taurine conjugates may be poorer substrates for the organic anion transport system than the parent carboxylic acid. Examination of the actual transport of benzoic acid, benzoyltaurine and benzoylglycine (the mammalian metabolite of benzoic acid) in southern flounder renal tubules showed that benzoyltaurine was more efficiently transported than benzoic acid, but that benzoylglycine (which was not produced by the fish) was better excreted than either [145]. Thus differences between species in the rates of excretion of carboxylic acids are more likely to be due to species differences in the efficiency of the organic anion transport system for particular substrates than to differences in biotransformation. Several crustacean species have been shown to have organic anion transporters in bladder or antennal gland, but in some species the net result is reabsorption of the anion into blood rather than excretion [152]. Examination of organic anion transport in spiny lobster hepatopancreas brush border vesicles showed that uptake of the model anion, *p*-aminohippurate, was dependent only on concentration and pH gradients (greater uptake at high concentration and when pH in the vesicle was higher than pH in the medium) and was not energy dependent or susceptible to inhibition by competitive substrates [153]. In other words, the uptake was by diffusion of the unionized portion and higher pH inside the vesicle led to trapping of the ionized molecule. This suggests that excretion of anionic metabolites from the hepatopancreas of the spiny lobster into the intestine for excretion in feces will be slow, as only the unionized portion will be excreted by diffusion, and this finding may partially explain the slow excretion of phenylacetyltaurine and the taurine conjugates of DDA, 2,4-D and 2,4,5-T from spiny lobster hepatopancreas [146]. It is not known how widespread this deficiency in hepatopancreas organic anion transport is among crustaceans.

In fish, many pesticides and pesticide metabolites are excreted into bile, which often contains high concentrations of the metabolites of xenobiotics to which the fish has been exposed [118]. The major types of metabolites found in bile are glucuronide and glutathione conjugates, and the further metabolites of glutathione conjugates. These metabolites are of higher molecular weight than metabolites such as sulfate, acetyl and taurine conjugates, and although the relationship between physico-chemical properties of xenobiotic metabolites and biliary excretion has not been systematically investigated in fish it does appear that, as in mammals, very small molecules are not excreted into bile [118]. It has recently been shown in the thorny back skate that further metabolites of glutathione conjugates may be formed right in the liver and secreted into bile, thereby facilitating excretion [98]. Clearly, efficient excretion of pesticide metabolites will minimize the potential for toxic effects of pesticides in fish and shellfish.

7 Status of the Field

Considerable progress in understanding the likely pathways of pesticide biotransformation in fish and shellfish has been made in the past ten years. While it is not yet possible to state the fate of a new pesticide in a given fish species with certainty, likely pathways of biotransformation and excretion of typical small organic molecule pesticides can be predicted with some degree of confidence, at least for the more commonly used species, such as trout and catfish. Less is known about the likely long term effects of pesticide exposure on pesticide metabolism. It is now widely recognized that for most biotransformation pathways, fish possess qualitatively the same capabilities as other vertebrates, although there are considerable quantitative differences in pesticide biotransformation by a particular pathway between fish species, as there are between mammalian species. An exception is that fish appear to be deficient in glycine conjugation and form only taurine conjugates of carboxylic acids. The species differences reflect differences in the protein structure and active site configuration of the enzymes, and tools are now available for the study of these questions. The fish that have been studied possess multiple forms of P450, UDP-glucuronosyltransferase and glutathione S-transferase, some of which are orthologous to mammalian forms. There are, however, major differences between fish and mammalian species with respect to the regulation of expression of some of these enzymes. It has been established that fish are resistant to induction by phenobarbital-like inducing agents, and fish may be a valuable tool in understanding the mechanism of induction by this class of compounds. More research should be undertaken in order to understand and predict the responses of fish to pesticides that contain imidazole and other heterocyclic rings and are therefore potential inducers and inhibitors of P450 and other xenobiotic-metabolizing enzymes. A better understanding of the biochemical and physiological basis of the processes of pesticide uptake, biotransformation and excretion is needed in order to fully understand species differences in pesticide metabolism. The use of the tools of molecular biology in obtaining further knowledge about the enzymes and transport proteins important in pesticide metabolism should provide further advances in this field.

There is a much smaller base of knowledge concerning pesticide biotransformation pathways in shellfish, and their regulation, and it is consequently difficult to accurately predict pesticide metabolism in shellfish. Some shellfish have a very low capacity for monooxygenation, while others possess efficient monooxygenation pathways. Very little is known about the inducibility of xenobiotic-metabolizing enzymes in shellfish. Recent studies suggest that crustacea at least have efficient pathways for phase 2 biotransformation, particularly sulfation. The major pathway of carbohydrate conjugation appears to be different between fish and shellfish, in that most of the available evidence shows that glucosides, not glucuronides, are the major metabolites formed in shellfish. Although careful studies in a few crustacean species have shown no evidence for glucuronyltransferase, but good glucosyltransferase activity, there remains some confusion in the literature about the ability of shellfish to form glucuronide conjugates, and this deserves further investigation.

8 Conclusions

Examination of the literature pertaining to pesticide metabolism in aquatic species has shown that these studies are recognized as important for predicting pesticide toxicity in non-target species, and for predicting the likelihood that pesticide residues will remain in the edible parts of food-producing fish and shellfish. While considerable progress in our understanding of the fate of pesticides in different species has been made in the recent past, more studies of the basic processes of biotransformation, including structure-activity relationships at the enzymatic level, are needed before it will be possible to accurately predict the fate of a new pesticide in any given aquatic species.

9 References

1. Kraybill HF (1977) Ann NY Acad Sci (USA) 298: 80
2. Atlas E, Giam CS (1981) Science 211: 163
3. Hargrave BT, Harding GC, Vass WP, Erikson PE, Fowler BR, Scott V (1992). Arch Environ Contam Toxicol 22: 41
4. Ahmed MT, Ismail SM (1991) J Egypt Public Health Assoc 66: 557
5. Van der Oost R, Heida H, Opperhuizen A, Vermeulen NP (1991). Comp Biochem Physiol C 100: 43
6. Khan MAQ, Lech JJ, Menn JJ (1979) Pesticide and xenobiotic metabolism in aquatic organisms. American Chemical Society Symposium Series 99, Washington
7. Huckle KR, Millburn P (1990) In: Hutson DR, Roberts TR (eds) Environmental fate of pesticides. Wiley, London, p 175
8. Kleinow KM, James MO, Lech JJ (1992) In: Hutson DH, Hawkins DR, Struble CB, Paulson G (eds) The fate of xenobiotics in food-producing animals. American Chemical Society, Washington DC. p 98 (ACS Symposium Series No 503)
9. James MO, Kleinow KM (1994) In: Ostrander G and Malins DR (eds) Aquatic Toxicology. CRC, Boca Raton, FL (In press).
10. Snyder MJ, Chang ES (1991) Biol Bull 180: 475
11. James MO (1990) Arch Biochem Biophys 282: 8
12. Maren TH, Embry R, Broder LE (1968) Comp Biochem Physiol 26: 853
13. Mix MC (1984) In: Hodgson E (ed) Reviews in environmental toxicology, vol 5. Elsevier, North Holland
14. Addison RF, Willis DE (1978) Toxicol Appl Pharmacol 43: 303
15. Gibson CG and Skett P (1986) Introduction to drug metabolism. Chapman and Hall, London

16. Nelson DR, Kamataki T, Waxman DJ, Guengerich FP, Estabrook RW, Feyereisen R, Gonzalez FJ, Coon MJ, Gunsalus IC, Gotoh O, Okuda K, Nebert DW (1993) DNA and Cell Biol 12: 1
17. Ortiz de Montellano P (1986) Cytochrome P450. Plenum, New York
18. Anders MW (1982) In: Jakoby WB, Bend JR, Caldwell J (eds) Metabolic basis of detoxication: Metabolism of functional groups. Academic New York, p 29
19. Hodgson E, Levi PE (1988) Drug Metabol Drug Interact 6: 219
20. Heymann E (1980) In: Jakoby WB (ed) Enzymatic basis of detoxication. Academic, New York, vol II p 291
21. Wixtrom RN, Hammock BD (1985). Biochem Pharmacol Toxicol 1: 1
22. Edwards R, Millburn P and Hutson DH (1989) Xenobiotica 17: 1175
23. Tjeerdema RS, Crosby DG (1988) Xenobiotica 18: 831
24. Castillo GD, Jeffus MT and Kenner CT (1978). J Assoc Off Anal Chem 61: 1
25. Sanborn JR, Childers WF and Hansen LG (1977) Ag and Food Chem 25: 551
26. Podowski AA, Sclove SL, Pilipowicz A and Khan MAQ (1991) Arch Environ Contam Toxicol 20: 488
27. Hale R (1989) Mar Env Res 28: 67
28. Varanasi U and Stein JE (1991) Environ Health Persp 90: 93
29. Loveland PM, Wilcox JS, Pawloski NE, Bailey GS (1987) Carcinogenesis 8: 1065
30. Johnston JJ and Corbett MD (1986) Toxicol Appl Pharmacol 85: 181
31. Foster GD and Crosby DG (1986) Environ Toxicol Chem 15: 1059
32. Foster GD and Crosby DG (1987) Xenobiotica 17: 1393
33. Little PJ, James MO, Pritchard JB, Bend JR (1985) Toxicol Appl Pharmacol 77: 325
34. James MO, Schell JD, Magee V (1989) Bull MDIBL 28: 119
35. Livingstone DR (1991) In: Gilles R (ed) Advances in comparative and environmental physiology, vol 7 Springer, Berlin Heidelberg New York
36. Goksøyr A, Förlin L (1992) Aquatic Toxicol 22: 287
37. Stegeman, JJ (1993) In: JB Schenkman (ed) Handbook of experimental pharmacology. (in press)
38. Miranda CL, Wang JL, Henderson MC, Buhler DR (1990) Biochim Biophys Acta 1037: 155
39. Cunha Bastos J, Cunha Bastos VLF, Rossini A, Fortini H, Castro Faria MV (1992) Comp Biochem Physiol 102C: 561
40. Khan MAQ, Korte F, Payne JF (1977) In: Khan MAQ (ed) Pesticides in aquatic environments. Plenum, New York, p 191
41. Miyamoto J (1989) In: Kato R, Estabrook RW, Cayen MN (eds) Xenobiotic metabolism and disposition. Taylor and Francis, London, p 257
42. James MO, Khan MAQ, Bend JR (1979) Comp Biochem Pharmacol 62C: 155
43. Elmamlouk TH, Gessner T (1976) Comp Biochem Physiol 53C: 19
44. Johnston JJ, Corbett MD (1986) Pesticide Biochem Physiol 26: 193
45. James MO (1989) Xenobiotica 19: 1063
46. Nebert DW, Gonzalez FJ (1987) Ann Rev Biochem 56: 945
47. Stegeman JJ, Lech JJ (1991) Environ Health Persp 90: 101
48. Kleinow KM, Melancon MJ, Lech JJ (1987) Environ Health Persp 71: 105
49. Kleinow KM, Haasch ML, Williams DE, Lech JJ (1990) Comp Biochem Physiol C 96: 259
50. Miranda CL, Wang J-L, Henderson MC, Buhler DR (1989) Arch Biochem Biophys 268: 227
51. Haasch ML, Kleinow KM, Lech JJ (1988) Toxicol Appl Pharmacol 94: 246
52. Wilkinson CF, Hetnarski K, Hick LJ (1974) Pesticide Biochem Physiol 4: 299
53. Kahl R, Friederici DE, Kahl GF, Ritter W, Krevs R (1980) Drug Metab Disp 8: 191
54. Riviere JL (1983) Pesticide Biochem Physiol 19: 44
55. Lavrijsen K, Van Houdt J, Thijs D, Meuldermans W, Heykants J (1986) Biochem Pharmacol 35: 1867

56. Hostetler KA, Wrighton SA, Molowa DT, Thomas PE, Levin W, Guzelian PS (1989) Mol Pharmacol 35: 279
57. Manning BW, Franklin MR (1992) Xenobiotica 22: 1395
58. Bach J, Snegaroff J (1989) Xenobiotica 19: 1
59. Snegaroff J, Bach J (1989) Xenobiotica 19: 255
60. James MO (1989) In: Varanasi U (ed) Metabolism of polycyclic aromatic hydrocarbons in the aquatic environment. CRC, Boca Raton, p 69
61. Levi PE, Hodgson E (1988) Xenobiotica 18: 29
62. Tynes RE, Hodson E (1985) Arch Biochem Biophys 240: 77
63. Smyser BP, Hodgson E (1985) Biochem Pharmacol 34: 1145
64. Cashman JR, Olsen LD, Young G, Bern H (1989) Chem Res Toxicol 2: 392
65. Schlenk D, Buhler DR (1991) Aq Tox 20: 13
66. Schlenk D, Buhler DR (1991) Xenobiotica 21: 1583
67. Schlenk D, Buhler DR (1990) Mar Biol 97: 47
68. Schlenk D, Erickson DA, Lech JJ, Buhler DR (1992) Fundam Appl Toxicol 18: 131
69. Schlenk D, Buhler DR (1993) Biochim Biophys Acta 1156: 103
70. Schlenk D, Ronis MJ, Miranda CL, Buhler DR (1993) Biochem Pharmacol 45: 217
71. Glickman AH, Hamid AA, Rickert DE, Lech JJ (1981) Toxicol Appl Pharmacol 57: 88
72. Glickman AH, Weitman SD, Lech JJ (1982) Toxicol Appl Pharmacol 66: 153
73. Glickman AH, Lech JJ (1982) Toxicol Appl Pharmacol 66: 162
74. Barron MG, Plakas SM, Wilga PC (1991) Toxicol Appl Pharmacol 108: 474
75. Takimoto Y, Ohshima M, Miyamoto J (1987) Ecotoxicol Environ Safety 13: 104
76. Barron, MG, Schultz IR, Hayton WL (1989) Toxicol Appl Pharmacol 98: 49
77. Glickman AH, Lech JJ (1981) Toxicol Appl Pharmacol 60: 186
78. Ozretic B, Krajnovic-Ozretic M (1992) Comp Biochem Physiol 103C: 221
79. Armstrong RN (1987) CRC Critical Reviews in Biochemistry 22: 39
80. Casas J, Harshman LG, Messeguer A, Kuwano E, Hammock BD (1991) Arch Biochem Biophys 286: 153
81. Guenthner TM (1990) In: Mulder GJ (ed) Conjugation reactions in drug metabolism. Taylor and Francis, London p 365
82. Walker CH, Timms CW, Wolf CR, Oesch F (1986) Biochem Pharmacol 35: 499
83. Mullin CA, Wilkinson CF (1980) Pesticide Biochem Physiol 14: 192
84. James MO, Bowen ER, Dansette PM, Bend JR (1979) Chem Biol Int 25: 321
85. James MO, Heard CS, Hawkins WE (1988) Aquatic Toxicol 12: 1
86. Parker LM, Laurén DJ, Hammock BD, Winder B, Hinton DE (1993) Carcinogenesis 14: 211
87. James MO (1987) Env Hlth Persp 71: 97
88. James MO (1986) In: Menn JJ, Caldwell J, Hutson D, Paulsen G (eds) Xenobiotic conjugation chemistry. American Chemical Society Symposium Series. No. 299, p 29
89. Ketterer B, Coles B, Meyer DJ (1989) In: Kato R, Estabrook RW, Cayen MN (eds) Xenobiotic metabolism and disposition. Taylor and Francis, London, p 55
90. Guengerich FP, Peterson LA, Cmarik JL, Koga N, Inskeep PB (1987) Environ Health Persp 76: 15
91. Lay MM, Menn JJ (1979) Xenobiotica 9: 669
92. Tjeerdema RS, Crosby DG (1987) Aq Toxicol 9: 305
93. Mitchell JR, Jollow DJ, Potter WZ, Gillette JR, Brodie BB (1973) J Pharmacol Exp Therap 187: 211
94. Raj HG, Clearfield MS, Lotlikar PD (1984) Carcinogenesis 5: 879
95. Ketterer B, Mulder GJ (1990) In: Mulder GJ (ed) Conjugation reactions in drug metabolism. Taylor and Francis, London, p 307
96. Goeger DE, Shelton DW, Hendricks JD, Pereira C, Bailey GS (1988) Carcinogenesis 9: 1793

97. Yagen B, Foureman GL, Ben-Zvi Z, Ryan AJ, Hernandez O, Cox RH, Bend JR (1984) Drug Metab Disp 12: 389
98. Simmons TW, Hinchman CA, Ballatori N (1991) Biochem Pharmacol 42: 2221
99. George SG (1994) In: Ostrander GK, Malins DR (eds) Aquatic Toxicology: molecular, biochemical and cellular perspectives. CRC, Boca Raton
100. Harris T, Coles B, Meyer DJ, Ketterer B (1991) Comp Biochem Physiol 98B: 511
101. James MO, Bend JR (1980) Toxicol Appl Pharmacol 54: 117
102. Leaver MJ, Clarke DJ, George SG (1992) Aquatic Toxicol 22: 265
103. Collier TK, Varanasi U (1991) Arch Environ Contam Toxicol 20: 462
104. Ankely GT, Blazer VS (1988) Can J Fish Aquat Sc 45: 132
105. Dutton GJ (1980) Glucuronidation of drugs and other compounds. CRC, Raton
106. Schell JD, James MO (1989) J Biochem Toxicol 4: 133
107. Li C-LJ, James MO (1993) Aq Toxicol 26: 57
108. Burchell B (1993) In: Tavoloni N, Berk PD (eds) Hepatic transport and bile secretion: physiology and pathophysiology. Raven, New York, p 489
109. Mulder GJ (1981) Sulfation of drugs and related compounds. CRC, Boca Raton
110. Mulder GJ (1990) In: Mulder GJ (ed) Conjugation reactions in drug metabolism. Taylor and Francis, London, p 41
111. Gregus Z, Watkins JB, Thompson TN, Harvey MJ, Rozman K, Klaassen CD (1983) Toxicol Appl Pharmacol 67: 430
112. Smith RL (1973) The excretory function of bile. Chapman and Hall, London
113. Kobayashi K (1979) In: Khan MAQ, Lech JJ, Menn JJ (eds) Pesticide and xenobiotic metabolism in aquatic organisms. American Chemical Society Symposium Series 99, Washington, p 131
114. Stehly GR, Hayton WL (1989) Xenobiotica 19: 75
115. Kasokat T, Nagel R, Urich K (1987) Xenobiotica 17: 1215
116. Layiwola PJ, Linnecar DFC, Knights B (1983) Xenobiotica 13: 107
117. Layiwola PJ, Linnecar DFC (1981) Xenobiotica 11: 167
118. Guarino AM (1991) Vet Hum Toxicol 33 Suppl 1: 54
119. James MO, Schell JD, Barron MG, Li C-LJ (1991) Drug Metab Disp 19: 536
120. Clarke DJ, Burchell B, George SG (1992) Toxicol Appl Pharmacol 115: 130
121. Lindström-Seppä P, Koivusaari U, Hänninen O (1981) Comp Biochem Physiol 69C: 259
122. Short CR, Flory W, Flynn M (1988) Comp Biochem Physiol 89C: 153
123. Devaux A, Rivière JL (1988) In: Siest G, Magdalou J, Burchell B (eds) Cellular and molecular aspects of glucuronidation. John Libbey, London, p 293
124. James MO, Altman AH, Feistner H, Gahne Å (1993) The Toxicologist 13: 63
125. Pesonen M, Celander M, Förlin L, Andersson T (1987) Toxicol Appl Pharmacol 91: 75
126. Andersson T, Pesonen M, Johansson C (1985) Biochem Pharmacol 34: 3309
127. Hänninen O, Lindström-Seppä P, Koivusaari U, Väisänen M, Julkunen A, Juvonen R (1984) Biochem Soc Trans 12: 13
128. Schlenk D, Buhler DR (1988) Aq Toxicol 13: 167
129. Kurelec B, Britvic S, Krca S, Zahn RK (1986) Mar Biol 91: 523
130. Kurelec B, Britvic S, Krca S, Müller WEG, Zahn RK (1987) Comp Biochem Physiol 86C: 17
131. Meinertz JR, Gingerich WH, Allen JL (1991) Xenobiotica 21: 525
132. Kleinow KM, Beilfuss WL, Jarboe HH, Droy BF, Lech JJ (1992) Can J Fish Aq Sci 49: 1070
133. Grondel JL, Nouws JFM, Haenen OLM (1986) Vet Immunol Immunopathol 12: 281
134. Landrum PF, Crosby DG (1981) Xenobiotica 11: 351
135. Knezovich JP, Crosby DG (1985) Environ Toxicol Chem 4: 435

136. James MO and Barron MG. Disposition of sulfadimethoxine in the lobster. Vet Human Toxicology, 30: Supplement 1, 36–40, 1988
137. Hutt AJ, Caldwell J (1990) In: Mulder GJ (ed) Conjugation reactions in drug metabolism. Taylor and Francis, London, p 273
138. James MO (1979) In: Aitio A (ed), Conjugation reactions in drug biotransformations. Elsevier/North Holland, Amsterdam. p 121
139. James MO, Pritchard JB (1987) Drug Metab Disp 15: 665
140. Burke AB, Millburn P, Huckle KR, Hutson DH (1987) Drug Metab Disp 15: 581
141. Plakas SM, James MO (1990) Drug Metabolism Disp 18: 552
142. Guarino AM, James MO, Bend JR (1977) Xenobiotica 7: 623
143. James MO, Bend JR (1976) Xenobiotica 6: 393
144. Cravedi JP, Tulliez J (1987) Xenobiotica 17: 1103
145. Pritchard JB, James MO (1979) J Pharmacol Exp Ther 208: 280
146. James MO (1982) Drug Metab Disp 10: 516
147. Barron MG, Hansen SC, Ball T (1991) Drug Metab Disp 19: 163
148. Vessey DA, Benfatto AM, Zerweck E, Vestweber C (1990) Comp Biochem Physiol B 95: 647
149. Pritchard JB, Miller DS (1991) Am J Physiol 261: R1329
150. Pritchard JB, Bend JR (1991) Environ Health Persp 90: 85
151. Stehly GR, Plakas SM (1992) Drug Metab Disp 20: 70
152. Miller DS, Smith PM, Pritchard JB (1989) Am J Physiol 257: R501
153. James MO, Pritchard JB (1991) In: Frehse H (ed) Pesticide chemistry, Advances in international research, development and legislation. VCH, Weinheim, p 277
154. James MO, Hawkins WE, Walker WW (1993) Aq Toxicol (in press)
155. Szeto SY, Holmes SB (1982) J Environ Sci Hlth B17: 51
156. Miyamoto J, Takimoto Y, Mihara K (1979) In: Khan MAQ, Lech JJ, Menn JJ (eds) Pesticide and xenobiotic metabolism in aquatic organisms. American Chemical Society Symposium Series 99, Washington, p 3
157. Sudershan P, Khan MAQ (1981) Pestic Biochem Physiol 15: 192
158. Hawkins WE, Walder WW, James MO (in preparation)
159. Keeran WS, Lee RF (1987) Arch Biochem Biophys 255: 233
160. Foureman GL, Bend JR (1984) Chem-Biol Interact 49: 89
161. Sugiyama Y, Yamada T, Kaplowitz N (1981) Biochem J 199: 749
162. Ramage PIN, Nimmo IA (1984) Comp Biochem Physiol 78B: 189
163. Ramage PIN, Rae GH, Nimmo IA (1986) Comp Biochem Physiol 83B: 23
164. Barron MG, Plakas SM, Wilga PC, Ball T (1993) Environ Toxicol Chem 12: 1469

Fate of Pesticides in Aquatic Environments

Georg Matthess

Universität Kiel, Geologisch-Paläontologisches Institut und Museum,
Olshausenstr. 40/60, D-24098 Kiel, FRG

Drinking water supplies are based in Germany on subsurface and to a minor extent on surface water. Both types of water resources may be contaminated by pesticides. These include among others triazines, urea derivatives, carbamates sulfonylureas and phenoxy alkane carboxylic acids, halogenated hydrocarbons and organic phosphorus derivatives. For a risk assessment of surface and ground water contamination, the occurrence and fate of these chemicals in aquatic environments is to be considered. It requires detailed knowledge of the flow regime and of the geochemical behaviour of the pesticides in water and soil with respect to the physical, chemical and microbial processes controlling their persistence and transport in the different aquatic systems. These processes can be summarized as dissolution/precipitation, complexation, adsorption/desorption, filtration, degradation (oxidation/reduction, photolysis (in surface water), hydrolysis, metabolism), transport (including spray drift, volatilisation and gas exchange) processes. The consequences for pesticide use and for future research needs are discussed.

Chemistry of Plant Protection, Vol. 9

1 Introduction

This chapter deals with the pesticide transport within the terrestrial part of the hydrologic cycle with emphasis on the subsurface water. The occurrence and fate of pesticides in surface and subsurface aquatic environments are of interest with respect to drinking water supplies. In West-Germany about 72% of drinking water is derived from ground water and about 17% is bank filtered river water and artificially recharged ground water, whereas river water, lake water and dam reservoir water contribute only 0.5, 3.4, and 6.8%, respectively [1]. All of these water supplies may be contaminated by pesticides, as was shown by Zullei-Seibert [2] for German water resources, by RIVM and RIZA [3] for European ground waters and by Cohen et al. [4, 5] for US water resources. For the protection of public water supply in Germany and other countries, protection areas with zones of specified restrictions are established.

Pesticides are mostly synthetic organic chemicals used for plant protection against predators and competing herbs. According to their radius of action they can be separted into herbicides, insecticides, nematicides, acaricides, molluscicides, rodenticides, fungicides, and bactericides. Because of their widespread use and their persistence, ground water contamination by pesticides is most probable by herbicides, insecticides and nematicides:

- herbicides, with triazines, urea derivatives, sulfonylureas and phenoxy alkane carboxylic acids as the most important chemical groups,
- insecticides, with chlorinated hydrocarbons, carbamates and organic phosphorus derivatives as the most important representatives,
- nematicides, with the main groups chlorinated respectively brominated hydrocarbons, carbamates and organic phosphorus derivatives.

For a risks assessment of surface and ground water contamination, detailed knowledge of the flow regime and of the geochemical behaviour of the pesticides in water and soil with respect to the physical, chemical and microbial processes controlling their persistence and transport in the different aquatic systems is necessary. These processes can be summarized as dissolution/precipitation, complexation, adsorption/desorption, filtration, degradation (oxidation/reduction, photolysis, hydrolysis, metabolism), transport (including spray drift), volatilisation and gas exchange processes. Agricultural and – to a lesser degree – forest areas often cover the flood plains composed of Quaternary sand and gravel deposits containing important ground water bodies. Thus competing interests arise between plant producers, who use pesticides, and ground water protectors who want them banned. Pesticides are normally spread over agricultural or forestal areas where they come into contact with rain, surface water, and – by accidental spills – directly with ground water. The type of plant production being established in the respective area has to be considered. In the case of crop rotation the applied pesticides are changed every year and the risk of surface and ground water contamination is relatively low because of retardation and degradation processes, and the use of different pesticides. In soils used for special-purpose crops, an accumulation of pesticide residues is possible, because the same substances are applied every year. Especially on sandy soils the leaching of residues can lead to severe ground water contamination [6]. Other sources for ground water contamination by pesticides are waste disposal sites, where pesticides are transported in the leachate to the underlying ground water, e.g. at Gloucester landfill site, Canada, the pesticides DDT, lindane and pentachlorophenol [7].

Surface water in rivers and in natural and artificial lakes may be contaminated by pesticides following different pathways, such as air transport followed by wet or dry deposition, from erosion of pesticide-containing soils, from industrial and municipal sewage systems or from accidental spills directly into surface water. Their transport behaviour is controlled by similar processes as in the subsurface environment. Major differences are the significantly higher flow velocities in the surface stream channels and the higher efficiency of photolysis and volatilisation processes. Due to artificial recharge or river bank filtration, pesticides may be transported from the surface water into the ground water [8].

2 Hydrologic Background

Ground water originates from the infiltration of atmospheric precipitation and, where a landward gradient of the water table occurs, of surface water, too. After flowing in the subsurface environment for varying distances it feeds the surface water systems, rivers and lakes. Only a small portion of ground water flows directly into the sea or evaporates where the water table comes close to the ground surface. During its movement from the surface of the ground through the soil, the water-unsaturated and the water-saturated zones to the discharge areas where it re-emerges, water is subject to numerous interactions between the aqueous and the solid phases through the above-mentioned physical, chemical and microbial processes, which lend it its typical properties [9–11].

The subsurface is hydrologically separated to water-unsaturated and -saturated zones (Fig. 1). The water-unsaturated zone includes the soil sensŭ stricto and, in most cases, the subsoil (intermediate layers) lying below down to the ground water table. In this zone solid materials coexist with air (called ground air) and water in the form of adhesion, capillary, and percolating water, which are referred to here as pore water [10]. The boundary between water-unsaturated and -saturated zones is approximately described by the ground water table, which in a strict sense is defined as the surface of unconfined water along which the hydrostatic pressure is equal to the atmospheric pressure. The zone below the water table is generally called the saturated or ground water zone, neglecting the fact that the lower part of the capillary fringe is also

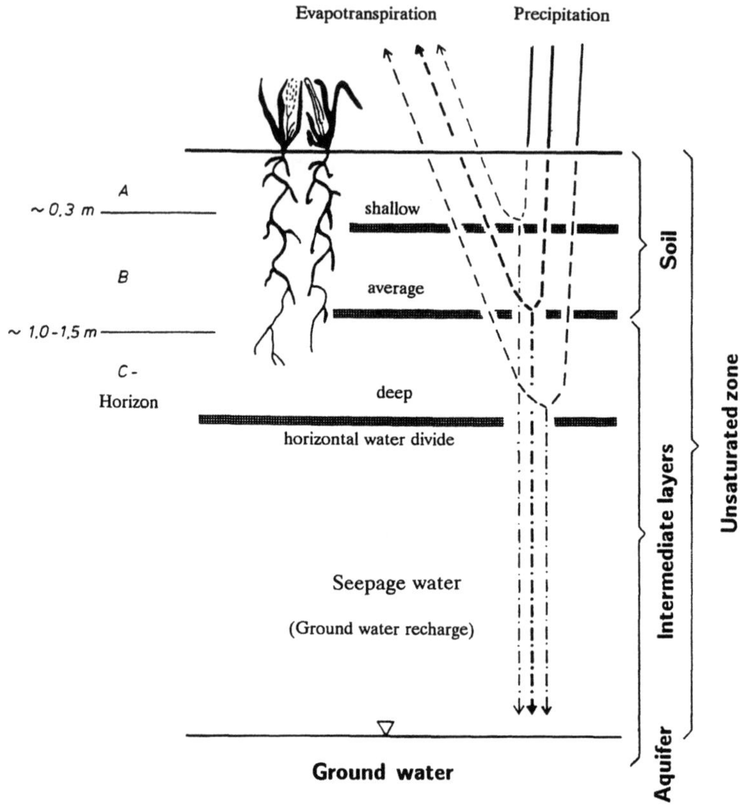

Fig. 1. Hydrological subsurface categories

water-saturated and participates in the ground water flow with almost the same velocity as the ground water just below the water table [1]. This simplification is acceptable in this context in as far as no detailed knowledge is available with respect to the processes in the capillary fringe.

In the water-saturated zone below the water table, the voids of the earth's crust are filled by ground water, the behaviour of which is controlled only by gravity. In the water-unsaturated zone the voids are filled partially by ground air and partially by water. As mentioned above a part of this water is bound by adhesive forces to the surface of the solid material, and a part is held by surface tension and moves by capillary action. Percolation occurs only within the intergranular spaces when the water content – for example, by infiltrated water after heavy rainfalls or after snow melt – exceeds a certain limiting value, the field capacity. The field capacity in coarse underground material is significantly smaller than in finer material. The excess water percolates down due to gravity and may reach the water table. The thickness of the water-unsaturated zone depends on the climate and the surface relief. In lowlands and in humid climates, the thickness of the unsaturated zone may be zero or only a few centimeters; in arid zones and mountain areas, its thickness may range up to hundreds or even thousands of meters.

Ground water, which is in contact with the atmosphere or the ground air through open spaces in the unsaturated zone, is called unconfined water. Ground water enclosed by impermeable layers is referred to as confined water or artesian water [1, 9].

Ground water flows in permeable geologic formations known as aquifers, which may be subdivided into porous aquifers, fractured aquifers, and karstic aquifers. In porous aquifers – i.e. sands and gravels – ground water flows in pores that form a dense network of interstices. The flow velocities range from fractions of 1 m/d up to few m/d. Velocities of greater than 10 m/d are rare. In fractured rocks, ground water movement is restricted to fractures, fissures, and cracks produced by mechanical stresses in any hard rock or by contraction of cooling magmas in magmatites. The fractures are enlarged by dissolution processes in karstic rocks, such as limestones and dolomites, where dissolution voids with lengths of several kilometers may occur. In fractured and karstic rocks, ground water flow velocities range from less than 1 m/day up to tens or hundreds of m/d. Extreme velocities of 8000 m/d and 26 000 m/d have been reported. The velocity decreases generally with increasing depth.

Analogous relations hold for the water flow under water-unsaturated and saturated conditions. In the water-saturated zone, the flow velocity (filter velocity) v_f is described by the Darcy equation (1)

$$v_f = -k_f \cdot \operatorname{grad} h \tag{1}$$

with the hydraulic conductivity k_f (in m/s) and the hydraulic gradient grad h (dimensionless). For water-unsaturated conditions, both the hydraulic conductivity and the hydraulic gradient, depend on the water saturation w. The hydraulic conductivity in the unsaturated zone k_u decreases with increasing water tension, i.e. with decreasing water saturation. The seepage velocity v_u can be described by the Richards equation (2), derived from the Darcy equation (1)

$$v_u = -k_u \cdot \operatorname{grad} \psi \tag{2}$$

with the hydraulic conductivity in the unsaturated zone k_u, for which holds for $w < 100\%$, $k_u < k_f$ and for $w = 100\%$, $k_u = k_f$. The hydraulic potential in the unsaturated zone ψ is mainly composed of the gravitational potential ψ_z and – with a change of sign – the matrix potential ψ_m. The flow direction of the water in the unsaturated zone is controlled by the direction of the hydraulic gradient, which results from the absolute values of the gravitational potential ψ_z and the matrix potential ψ_m, the latter depending on the nature and granulometry of the solid material and on the water saturation w. The water loss by evapotranspiration at the soil surface is compensated by an upward water flow. The lower boundary of this flow region is a horizontal water divide, which develops in the vegetation period (Fig. 1). In the humid climate of Central Europe, however, the downward movement predominates, thus causing an annual net transport of water and its dissolved and suspended contents toward the ground water [1].

Because the field capacity is significantly smaller and the hydraulic conductivity k_u is significantly larger in coarse than in fine grained subsurface materials, seepage velocity in rocks with low field capacity and high hydraulic conductivity k_u is relatively high (in the order of magnitude of some meters or tens of meters per day), e.g. in materials rich in coarse rock debris and in gravel. However, in subsurface materials with high field capacities the seepage water will percolate downward very slowly. At seepage velocities of about 1 m/a, which may be assumed as a representative value for sandy-loamy subsurface materials, the residence time in the unsaturated zone will be some years depending on the local thickness of this zone. Under such conditions the seepage water develops an "age stratification", as has been shown by isotope-hydrologic investigations. Recently infiltrated precipitation water pushes the lower, older, water layers downward, so that the percolate that enters the water table in the depth a few hours or days after a rain event, is "old" water from the water layer directly above the water table. Besides the dominant intergranular percolation water movement, certain amounts of local water can move rapidly through macropore, cleavage and fracture systems. This is especially important for low permeability clay profiles in which shrinkage cracks occur as a result of summer droughts [1, 10].

Besides this perferential flow due to structural inhomogeneities such as macropores and fracture systems another hydraulic mechanism, namely capillary barriers, strongly affects the transport of contaminants in naturally stratified water-unsaturated zones. On the boundary of fine-grained materials overlying coarser-grained layers seepage water is partially retained by capillary forces up to a certain limiting pressure head. This capillary water body is drained laterally following the gradient of its upper surface (Fig. 2). The levels of the natural strata boundaries vary locally. Their depressions form local sinks for the lateral water flow. Here, after surpassing a critical capillary head a breakthrough occurs through the capillary barrier (Fig. 3) [12]. This process explains the multifold observations of locally fixed downward fingering wetting and transport fronts.

The distribution of a dissolved substance and the flow velocity are controlled by the hydraulic conductivity and porosity of the rocks and by the local hydraulic gradient of the water table. Besides the aquifers that generally have appreciable permeabilities and porosities, formations occur that are characterized by low permeabilities. These formations are practically incapable of transmitting ground water. They may be subdivided according to their porosity into aquicludes, containing water in very minute pores (e.g., clays), and aquifuges, which are practically free of voids (e.g., solid granite).

The pathways of seepage and ground water may be visualized in a generalized way as streamline patterns (Fig. 4). As a consequence in a vertical profile the age of ground water in most cases increases with depth and simultaneously it comes from infiltration points increasingly upstream of the sampling point.

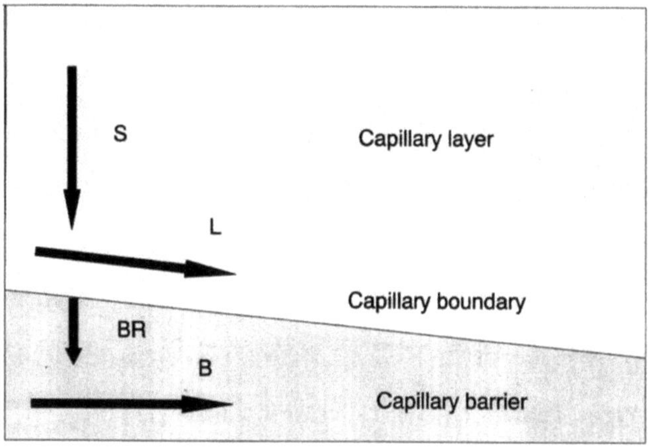

Fig. 2. Flow conditions in a capillary barrier system (L = lateral flow, B = soil run off, BR = breakthrough through the capillary barrier, S = seepage water) [12] (reprinted with permission of the Chair of Applied Geology, Karlsruhe University)

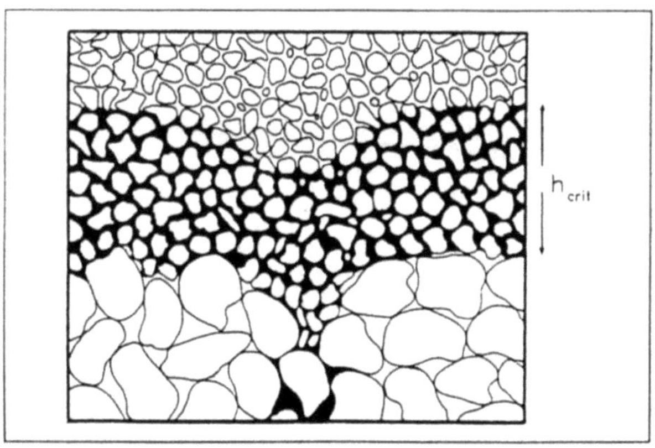

Fig. 3. Breakthrough site in a capillary barrier [12] (reprinted with permission of the Chair of Applied Geology, Karlsruhe University)

A given ground water unit will spend a specific residence time along its subsurface pathway. Surface water samples and ground water samples taken at a spring or a well represent a mixture of water of different age, which – considering all possible pathways – are described by the cumulative distributions of residence times (Fig. 5). In this graph three different residence time distributions are shown. Curve A represents ground water flow from a basin with a shallow cover of porous material on impermeable bedrock: 50% of the water has a subsurface residence time of less than 2 years, less than 10% spends up to 25 years. Curve B represents a system with a relatively thin layer of porous aquifer

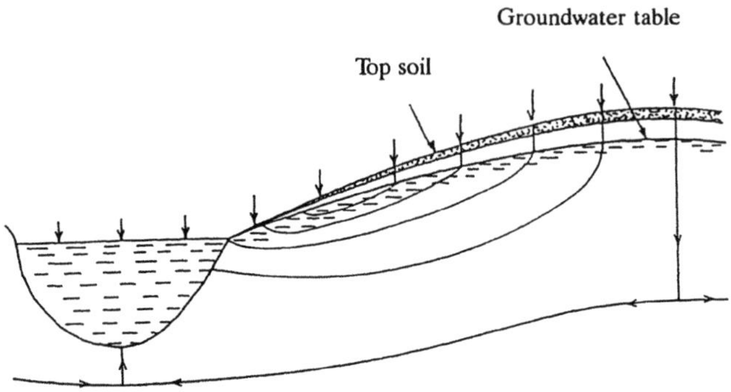

Fig. 4. Patterns of water pathways in a basin [13]

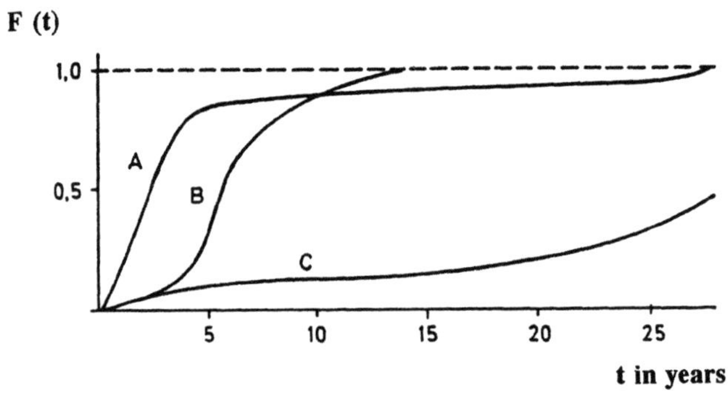

Fig. 5. Frequence distribution of residence times for water in a basin [13]. $F(t)$ is the fraction of the flow (or recharge), for which water spends a time equal to or less than given on the time axis

material underlain by a highly fractured, permeable bedrock. In this case, only about 20% of the recharge spends less than 5 years in the basin, the major part has a subsurface residence time between 4 to 10 years. Curve C represents a fairly deep sand and gravel aquifer, in which ground water spends tens to hundreds of years, in very thick sedimentary basins eventually even thousands of years.

The residence time of ground water ranges from a few days to tens or thousands of years depending on the flow velocity and the dimensions of the aquifer system. The distance between recharge and discharge areas and aquifer thickness are also important parameters. Nace [14] quotes a mean residence time of 5000 years for ground water between 0 to 4000 m depth. Some deep-lying bodies of water are cut off from the hydrological cycle by practically impermeable rocks. These stagnant waters are called fossil waters if they originate in pre-Holocene times [10].

The presence of an often thick unsaturated zone, combined with large saturated zone storage and relatively slow ground water flow velocities, causes a very late appearance of the effects of contamination, but also a very long residence time of contaminants in the aquifer system [15].

The possibility of entry and propagation of contaminants through the ground water depends on the subsurface conditions. Fissured rocks and karst aquifers, and very coarse sandy-gravelly outwash deposits with high ground water flow velocities are particularly susceptible to contamination, especially when an extensive soil or subsoil cover is lacking.

As flowing surface waters, all those water bodies with flow velocities >0.1 km/h are included, even backwaters of dams without thermostratification. The pesticides entrained in the river channel undergo permanent changes in their concentration depending on the flow time between their entrance into the river, on dispersion, on dilution by additional clean water inflows, on sorption and desorption processes on sediments of the channel bottom, on suspended solids and on aquatic organisms, on volatilisation and on photolytic, non-biotic or biotic degradation.

The flow velocities of the surface and ground water can be deduced from their very different mean residence times in the respective water compartments of the global water cycle. For deep (inactive) and for shallow (active) ground water a mean residence time of 1400 years and 300 years respectively are quoted, whereas for lakes 17 years and for rivers 16 days are listed [16].

The flow time of a contaminated water unit depends on the flow velocity which itself is controlled by the hydraulic gradient and the water mass. The flow times in German navigable rivers vary at mean flow between 2 and 5 km/h with higher and lower values in flood and draught periods. Thus according to Schmitz [17] there are flow times in the order of

	mean flow	mean high flow	mean low flow
Rhine river Basle – Emmerich	140 h	116 h	260 h
Weser river Hann Münden – Bremerhaven	132 h	68 h	168 h
Danube river Kelheim – Passau	60 h	30 h	84 h
Neckar river Plochingen – Mannheim	80 h	26 h	84 h

It is obvious that low flow in rivers with weirs cause appreciably retarded flow times, e.g. in the Neckar river, whereas these obstacles do not affect the flow times at high flow to such extent. However, even in the Neckar, the flow velocity does not decrease below 0.2 km/h [17].

In estuaries, the flow velocities are affected by the tidal movements, which are connected with periodical reversals in water flow. The resulting mean flow velocity is e.g. in the Elbe river about 0.06 m/s, thus, for the distance Geesthacht–Cuxhaven (150 km) a flow time of 740 h can be calculated [17].

3 Transport of Pesticides

3.1 Dissolution and Precipitation

Substances can be transported in water as ions, as complexes but also in particulate forms such as colloids. For a great number of organic substances their solubility in water is the controlling factor for the concentration which may appear in seepage and ground water. Organic acids, bases and salts are highly soluble in water and there are various pesticides belonging to these groups, e.g. the phenoxyalkane carboxylic acids like 2,4-D or 2,4,5-T, which are used as herbicides in cereal production. Polar organic derivatives, for example the nitrophenols like the herbicide DNOC, are easily dissolved in water as molecules. Non-polar substances, e.g. the chlorinated hydrocarbons, normally have a very low solubility in water, e.g. aldrin with 0.01 mg/l. They form separate phases as liquids immiscible in water if their solubility is exceeded.

The solubility of ions in water, which is dependent on pressure, temperature and other ions in solution, can be calculated with the help of modal relations from the chemical equilibrium [19]. For the dissolution of a solid [AB] the solubility product can be written:

$$K_{SP} = [A]^{\nu_A} \cdot [B]^{\nu_B} \tag{3}$$

where ν_A and ν_B are the stochiometric coefficients of the chemical species A and B. Tables of solubility products of numerous substances exist and can be used to calculate the solubility of a substance in a given water.

If the solubility product is exceeded as a result of changes in chemical composition of solutes, changes of pH-values or of redox potentials the dissolved substance may be precipitated at the same point along its surface or subsurface flow path.

The solubility of non-polar organic substances S can be derived from empirical regression equations between solubility and the octanol/water distribution coefficient K_{OW}. With the help of this constant the solubility (in mmol/l) may be estimated, using empirical equations, e.g. Eq. (4) [20]

$$\ln S = 7.494 - \ln K_{OW} + 6.8(1 - T_m/T) \tag{4}$$

which considers the melting temperature $t_m > t$ (specific temperature), or Eq. (5) [6]

$$\log S = -1.45 \log K_{OW} + 4.42 \qquad (n = 69; r^2 = 0.91) \tag{5}$$

Table 1. Molar mass, water solubility, vapour pressure and octanol/water partition coefficient of pesticides [18] (at 20 °C, *25 °C, **26 °C, ***22 °C)

	Molar mass g	Solubility mg/kg	Vapour pressure mPa	$\log K_{OW}$
Halogenated hydrocarbons				
Lindane	290.85	7	1.2×10^{-5}	3.76
Triazines				
Anilazine	275.54	8	8.2×10^{-9}	3.02
Atrazine	230.24	33	4.0×10^{-7}	2.34
Simazine	201.66	5	8.1×10^{-9}	1.96
Terbuthylazine	229.72	8.5	1.5×10^{-6}	3.04
Urea derivatives (Carbamates etc.)				
Aldicarb	190.27	9000	10^{-4}*	1.14
Chlortoluron	212.69	700	1.7×10^{-7}	2.29
Isoproturon	206.32	65	$< 1 \times 10^{-5}$	≈ 2.5***
Methabenzthiazuron	221.29	59	5.9×10^{-8}	2.64
Phenoxy fatty acids				
Dichlorprop	235.06	620	8.8×10^{-7}	1.74***
Mecoprop	214.65	620	$< 3.1 \times 10^{-6}$	0.09*
MCPA	200.62	300	2.7×10^{-6}	2.86
Phosphoric acid esters				
Methidathion	302.33	250	1.9×10^{-6}	2.52
Parathion-ethyl	291.27	11	8.9×10^{-6}	3.83
Aniline derivatives				
Dichlofluanid	333.23	130	1.4×10^{-7}	3.7
Pendimethalin	281.31	0.3**	$\approx 1.3 \times 10^{-5}$*	5.18*
Triazol derivatives				
Bitertanol	337.4			
A-Isomer		2.9	2.2×10^{-12}	4.1
B-Isomer		1.6	2.5×10^{-11}	4.4
Penconazol	284.2	80	2.1×10^{-6}	3.70
Triadimefon	293.76	64	$< 10^{-5}$	3.18
Triazinon derivative				
Metamitron	202.22	1700	8.6×10^{-9}	0.83

with the number of data couples n and the regression coefficient r. The octanol/water distribution coefficients of many substances exist; those of new substances can easily be measured or estimated from the chemical molecule structure.

The solubility of non-polar organic substances can be significantly higher in natural water than in pure water due to their interaction with dissolved organic compounds, especially with humic substances.

3.2 Complexation

The pesticides may interact with natural dissolved organic carbon (DOC) producing associates or complexes with physical and chemical properties different from the pure agents. These interactions are favoured in the top soils where reaction time and natural organic substances with their hydrophobic and hydrophylic properties are available. The DOC of seepage water ranges between about 5 and 20 mg/l, decreasing rapidly with depth, where in the ground water DOC varies between about 0.5 and 3 mg/l [21, 22]. Thus, the formation of pesticide/DOC-associates are often postulated and sometimes quantitatively proved. There are examples where the mobility and the degradability of a hydrophobic trace component are appreciably influenced by DOC, however, there are contradictory findings as well. These phenomena may be addressed using either kinetically and thermodynamically based sorption experiments, or a chromatographic approach [23].

An experiment with terbuthylazine and the Bs-horizon of the podzol-site Segeberger Forst using a kinetic approach discriminated the most mobile "colloid" fraction $< 12\ \mu m > 0.45\ \mu m$, the "predominantly free" (i.e. dissolved) fraction $< 0.45\ \mu m$ and the particulate fraction $> 12\ \mu m$. The agent distributed itself among the three fractions in the ratio 30%–30%–40%, reaching equilibrium after 50 h. Desorption experiments found a 40% remobilisation of the sorbed agent, an indication that no dramatic increase of "free" terbuthylazine may occur and that about 25% of the agent was irreversibly bound to the solid fraction [23].

Gelchromatographic studies using an organic polymer (TSK) and a porous glass (CPG) in low pressure columns (LC) (particle size 40–80 μm) showed that the polymer interacted appreciably with the DOC of the water separating the substances according to their molar masses and different hydrophobicity. However, this procedure may destroy labile complexes. Porous glass, on the other hand, hardly interacts with the DOC, thus enabling careful chromatographic separation conditions, but only a small separation effect. This made it possible to study the colloid fraction chromatographically. Using the Karlsruhe local ground water as DOC source it was found that a DOC concentration of about 0.5 mg/l can bind terbuthylazine in concentrations of 30 up to about 100 μg/l in an apparent "high molecular" fraction. For this effect DOC structure units of about 50 C-atoms are necessary to fix one molecule of the agent terbuthylazine. The observation, that the gels are able to adsorb terbuthylazine without breaking labile associates indicates that the bonds are relatively stable. From the clearly differentiated fractions in the UV-detection curves it was deduced that the interaction mechanisms between the pesticide and the DOC are not uniform along the whole concentration range, but are different in specific fractions of DOC. The integral sum of UV-active fractions stayed nearly constant albeit with the addition of terbuthylazine. This indicates structural changes in the DOC-aggregates due to the addition of the pesticide as well as the limited power of the UV-absorption measurements for quantitative investigations of such interactions [23].

Chromatographic experiments using a C-14-labelled pendimethalin agent and porous glass or polymers as column material exhibit that minor quantities of DOC impede the sorption of the agent on both glass and polymers, the affinity of the agent to the polymer being greater than to glass. However, due to the character of solutions it is not clear whether it is still the original agent or a metabolite. In any case the elution in the high molecular range indicates an association with humic-like substances [23].

It could be concluded, that organic carbon in particulate, colloidal and dissolved form is able to bind pesticides, partially irreversibly. It can be assumed that the pesticide traces in ground water are present in such complexes. The chromatographic results indicate that the sorption equilibria of dissolved pesticides on polymer organic phases (TSK) and glass (CPG) in the presence of DOC are shifted almost completely towards the "dissolved phase" after a contact time of 24 h, which reduces drastically the retention time of the pesticides. It is reasonable to transfer these observations to natural conditions, because with respect to the type of materials and the sorption equilibria the TSK-gel may be considered as a model of an organic carbon-rich top soil and the porous glass-gel as that of a (tecto-)silicate mineral background. The DOC-content of about 1 mg/l in many ground waters exceeds by far the concentration necessary for the complexation of the studied pesticides to an extent beyond the maximum permissible concentration for pesticides in drinking water [23].

These observations have to be seen in the context of the mobility of humic substances, which may act as carriers for contaminants. The solubility of these substances decreases in the series fulvic acids – humic acids – humins, whereas their affinity to sorption on mineral particles increases [24–26]. In the seepage water the content of dissolved humic substances decreases rapidly with depth, possibly as a result of precipitation, flocculation, formation of unsoluble metalo-complexes, sorption on minerals or microbial degradation. It should be mentioned that with increasing depth the macromolecular humic acids are preferentially eliminated, mainly by sorption or precipitation, whereas the more mobile fulvic acids are degraded [11].

3.3 Sorption and Retardation

Risk assessment for chemical substances has also to take into account the mobility of the substance in the surface and subsurface environments. The mobility strongly depends on its adsorption/desorption properties. Many solid substances in the ground or suspended in surface and subsurface water tend to release certain constituents into solution and to remove dissociated and non-dissociated components from solution by binding them to the surface of solid particles due to intermolecular interchanges [11, 19, 27]. Binding of a compound to a solid material as it occurs in the unsaturated and saturated zones is generally referred to as sorption. Distinction is made between adsorption, or binding on a two dimensional interface or surface, and absorption, or partitioning into the three dimensional bulk of the sorbing phase, which in natural systems cannot be discriminated in all cases. Phase transfer reactions of this kind include transfer of neutral organic compounds from the aqueous phase to bulk non-aqueous phases, transfer of ion-pairs from aqueous phases to bulk non-aqueous phases, adsorption of organic ions from aqueous phase onto lipophilic surface with counter ions in electric double layer and ligand exchange of organic molecules with surface hydroxyl groups at inorganic oxide surfaces [28]. The equilibrium distribution of a non-polar solute between an aqueous phase and a natural sorbent can be likened to the distribution of the solute between an aqueous phase and a non-aqueous phase [29].

Adsorption is characterized as a surface-dependent, phase-overlapping relationship which results in an equilibrium between the concentration of ionic, polar or non-polar substances dissolved in water and that sorbed into or onto the solid material [30, 31]. This equilibrium is often described by the Freundlich or the Henry isotherms. Eq. (6) gives the Freundlich isotherm

$$c_a = k \cdot c_w^{1/n} \tag{6}$$

with the concentration at the sorbent c_a (in meq/g or mmol/g), the concentration in solution c_w (in meq/ml or mmol/ml) and the Freundlich constants k and $1/n$, specific for each species and solid matrix. For the Henry isotherm:

$$c_a = K_d \cdot c_w \tag{7}$$

with a distribution coefficient K_d (in ml/g), the exponent equals unity. However, the behaviour of many substances deviates more or less from the assumption $1/n = 1$ (Table 2).

The sorption constants k are horizon- and substance specific (Table 2, Figs 6–8), but a direct comparison is allowed only if the slopes of the isotherms, the

Table 2. Sorption characteristics of some pesticides for different horizons of the podzol-site Segeberger Forest [32, 33] (reprinted with permission of the authors)

Agent		Ah	Bs	Ss	Sk
Atrazine	k	14.9	1.51	0.08	0.1
	1/n	0.88	0.92	0.94	0.93
	r^2	0.99	0.97	0.94	0.95
Terbuthylazine	k	60.3	3.83	0.15	0.22
	1/n	0.88	0.92	0.93	0.98
	r^2	0.99	0.97	0.97	0.89
Simazine	k	17.9	4.32	–	–
	1/n	0.80	0.97	–	–
	r^2	0.99	0.98	–	–
Chlortoluron	k	29.7	3.28	–	–
	1/n	0.77	0.82	–	–
	r^2	1	1	–	–
Isoproturon	k	8.33	1.57	–	–
	1/n	0.80	0.83	–	–
	r^2	1	0.98	–	–

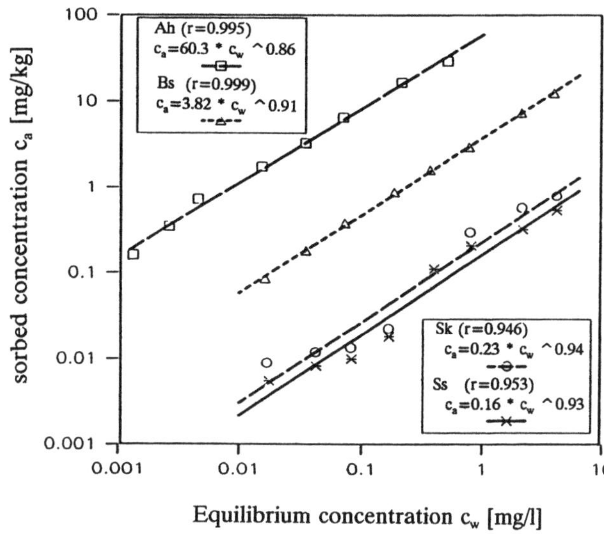

Fig. 6. Sorption isotherms for terbuthylazine in soil samples from different horizons of the podzol-site Segeberger Forest [33, 35] (reprinted with permission of the authors)

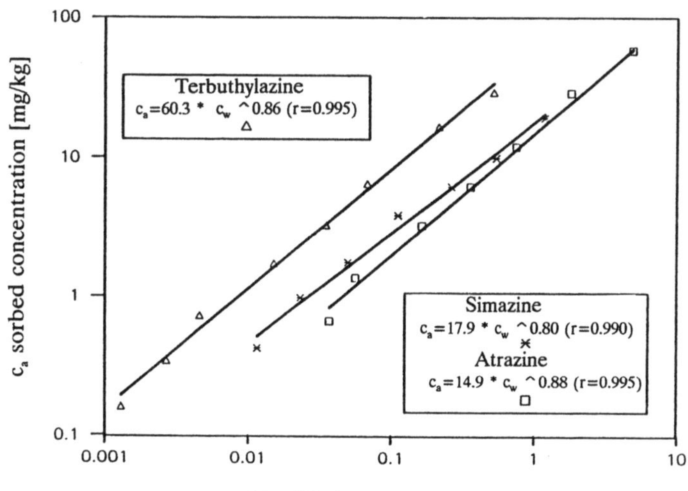

Fig. 7. Sorption isotherms of the *s*-triazines terbuthylazine, atrazine and simazine in soil samples of the Ah-horizon of the podzol-site Segeberger Forest [33] (reprinted with permission of the authors)

$1/n$ values, are equal [34]. This, however, does not hold in many cases. Nevertheless it is evident, that the mobility of terbuthylazine in the Ah-horizon at the podzol-site Segeberger Forest north of Hamburg is lowest and increases from the Bt-horizon to the lime-containing (Sk) and the lime-free sands (Ss). Furthermore it indicates, for the Ah-horizon, an increasing mobility following the

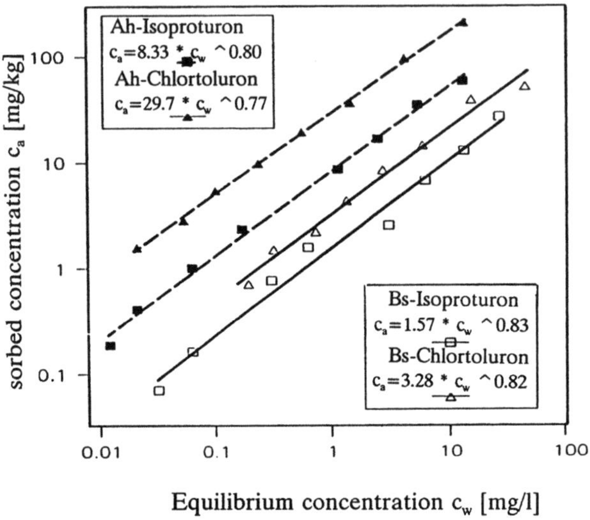

Fig. 8. Sorption isotherms for chlortoluron and isoproturon in soil samples of the Ah- and Bs-horizons of the podzol-site Segeberger Forest [33] (reprinted with permission of the authors)

sequence terbuthylazine – chlortoluron – simazine – atrazine – isoproturon, whereas the mobility in the Bt-horizon increases in the sequence simazine – terbuthylazine –chlortoluron – atrazine.

Both relationships show that an increase in the concentration of a solute will raise its adsorbed quantity and a decrease in its concentration will result in desorption. The empirical distribution coefficient K_d is a useful measure for the affinity of a specific natural solid material for a certain contaminant. It is defined as the ratio of the adsorbed substance in meq/100 g solid to the concentration of the same substance in solution in meq/l. In diluted solutions K_d equals the coefficient of the Freundlich isotherm.

The adsorbing particles in surface streams and in the subsurface are mainly organic substances, especially humic substances, clay minerals, as well as oxides and hydroxides of iron, manganese and aluminium. Under ground water conditions (pH ~7) these particles usually have a negative net charge, but also contain a few positive charge sites which are balanced by equivalent amounts of countercharged ions or polar molecules.

The aptitude for adsorption of non-polar organic chemicals, as for example chlorinated hydrocarbons, is correlated to their solubility in organic solvents. In a macroscopic view chlorinated hydrocarbons and other non-polar organic compounds are increasingly adsorbed the more lipophilic they are. As measure of lipophilicity the octanol/water partition coefficient K_{OW} is commonly used as reference solvent. Many data exist for the octanol-water system, and empirical rules have been developed for the calculation of octanol-water partition coefficients [29, 36]. In a first approach a corrected distribution coefficient K_{OC}, which refers to a 100% organic phase, which is assumed to be uniform, is derived usually from the octanol-water distribution coefficient K_{OW} (Table 3).

The distribution coefficients favourable to the non-aqueous phase are primarily the result of the large positive entropy change that occurs upon dehydration of the solute molecules ($\Delta G(W \rightarrow 0) = \mathrm{RT} \ln K_{OW}$). In the aqueous phase, the non-polar solute is surrounded by an envelope of highly structured water molecules. As the solute is transferred from the aqueous phase to the non-aqueous phase, the structural envelope of water molecules breaks down and the entropy of the system increases [28, 40].

Linear free-energy relationships allow us to estimate partition coefficients and to assess the lipophilicity of a solvent or sorbent [29]. However, one should be aware that these linear free- energy relationships are not strict thermodynamic functions [28].

The distribution coefficient K_d between water and solid is related to the corrected sorption constant K_{OC} and the concentration of organic carbon in the solid phase c_{OC} (in g/g) by Eq. (28) [41]:

$$K_d = c_{OC} \cdot K_{OC} \qquad (\mathrm{mg/g}) \tag{8}$$

Thus, the degree of sorption of a given non-polar organic compound, e.g., a halogenated hydrocarbon, by soil and aquifer materials can be estimated from the organic carbon content of the natural sorbent and from the octanol/water partition coefficient of the compound neglecting the effects of the other sorbents mentioned above.

As a result of continuous adsorption–desorption processes, the transport of contaminants is retarded with respect to water flow. This can be formulated in a one-dimensional retardation equation, Eq. (9):

$$R_d = \frac{v_w}{v_s} = 1 + \frac{\rho_b}{n} \cdot K_d \tag{9}$$

In this equation v_s is the mean transport velocity of the substance, v_w the mean flow velocity of the water, ρ_b the bulk density and n the porosity of the

Table 3. Regression equations for the relations between K_{OW} and K_{OC} (after Briggs [37], Karickhoff et al. [30], Giger et al. [38] and Matthess et al. [39].

Equation	n	r^2	Ref.
$\log K_{OC} = 0.524 \log K_{OW} + 0.618$	30	0.92	[37]
$\log K_{OC} = 1.00 \log K_{OW} - 0.21$	10	1.00	[30]
$\log K_{OC} = 0.72 \log K_{OW} + 0.5$*	–	0.97	[38]
$\log K_{OC} = 0.83 \log K_{OW} + 0.30$	30	0.93	[39]

n = numer of cases, r = correlation coefficient; * for $C_{org} > 0.1$ *mass-%*.

aquifer; K_d is the mean distribution coefficient between water and rock. For K_d greater than 1, v_s is lower than v_w, the movement of the substance is retarded with respect to the movement of water. The substance-specific K_d values and R_d factors are the reason for the different transport velocities of the dissolved substances (chromatography effect). The relative mobility of a substance with respect to the transport medium water can be predicted with the help of K_d values measured under controlled conditions in the laboratory or calculated with Eq. (8) and the relationships listed in Table 3. Thus, retardation factors can be used for an evaluation of the transport velocity of a contaminant in the subsurface water.

Under flow conditions at which sorption equilibrium can be assumed (<0.001 cm/s $\sim$ <1 m/d), the retardation factor of a given compound in a given aquifer can be estimated. Field and laboratory observations indicate that many non-polar organic substances are very mobile in the subsurface. Furthermore, at high flow velocities (>0.01 cm/s $\sim$ >1 m/d) these compounds may be transported faster than it can be predicted from their distribution coefficients due to slow sorption kinetics [42].

In many cases the sorption isotherms deviate from the desorption isotherms, because many sorption processes are not reversible or the sorption or desorption rate constants are different. Thus, for organic substances specific hysteresis effects are observed, which cause a retarded desorption [43]. The velocity of adsorption–desorption processes depends on the molecular diffusion rate of the substance which is adsorbed on the adsorbing particle as well as on the rate of adsorption at the surface of the adsorbing particle. It is proportionally related to the concentration in solution. As an example the relative amounts of adsorbed and desorbed pesticides are compared for two sandy soils with different organic content in Table 4.

More than 85% of the sorption by aquifer materials takes place at the grain size fraction <125 μm. Thus, this fraction can be taken as representative of the whole sample for sorptive properties [28].

Table 4. Adsorption/desorption studies of simazine and lindane with two soils [44]

Initial concentration (μg/g)		Adsorption		Desorption	
		% simazine	% lindane	% simazine	% lindane
HB soil	40	58.3	93.0	22.0	3.0
	20	59.2	93.0	15.2	3.1
	10	62.0	94.4	15.8	3.1
	5	62.8	94.2	13.7	2.5
BBA soil	40	18.1	55.2	48.8	33.6
	30	18.5	–	44.4	–
	20	18.7	55.9	41.4	33.8
	10	21.6	56.2	38.9	32.0
	5	–	56.0	–	33.6

For sorbents containing less than 0.1% organic carbon (c_{OC} <0.001 g/g), relatively small partition coefficients were found. In the absence of organic carbon, the specific surface area and the nature of the mineral surface have a greater impact on the degree of sorption [42].

To get a rough quantitative estimate of the retention behaviour of non-polar substances, e.g. a halogenated hydrocarbon in aquifers, one can approximate transport in the subsurface as a one-dimensional process at constant flow in a homogeneous porous medium. If the organic carbon content of the grain size fraction <125 μm of the aquifer material is known, retardation factors [38] can be calculated for any halogenated hydrocarbon, provided that its K_{OW} value is available:

$$R_d = 1 + f(s) \cdot K_d(s) \rho_b \left(\frac{1-n}{n} \right) \tag{10}$$

with the bulk density of the aquifer material ρ_b (g/cm^3), the total porosity n, the grain size fraction <125 μm $f(s)$ and the distribution coefficient of aquifer material <125 μm $K_d(s)$ (Assumption: homogeneous distribution of this grain size fraction).

Recent investigations show that a natural system has to be considered as a complex totality of many individual phases [45]. Therefore, the relatively simple two phase concept of the subsurface consisting of a solid phase and water used, e.g. by Karickhoff et al. [30] and Giger et al. [38], should be replaced by three- or multi-phase systems solid – colloids – water [46, 47]. Even the treatment of the mobility behaviour using a three-phase system indicates the predominant influence of colloidal phases. This can be elucidated by the thermodynamically inconsistent observation that a hydrophobic pesticide agent in water is the more soluble the higher the solid/water ratio is [48, 49]. This apparent contradiction is due to the colloid fraction in the previous studies being neglected. The amount of colloids necessarily grows with increasing quantities of solids, thus shifting as "quasi-dissolved solid phase" the equilibrium toward the "dissolved" phase [45].

The complexation of pesticides (Sect. 3.2) also influences the yield of solid phase extraction with C18-resin. Thus, Johnson et al. [50] found significantly lower recoveries at higher DOC-concentrations.

K_d- and R_d-values have been measured for numerous pesticides in the laboratory and in the field covering the spectrum of the most important porous aquifers in central Europe [39]. The retardation factors for various pesticides were found experimentally to be between 0.4 and 10.

The intermittent flow conditions in the unsaturated zone seem to retard pesticide transport. Changes in physical and chemical environmental conditions, especially the decrease of ionic strength of ground water, induce desorption. Nevertheless, the transport retardation of pesticides increases the residence time of the pesticides in the unsaturated and saturated zones. Thus, time-dependent degradation processes are favoured.

Table 5. Mobility classes of pesticides [51] (reprinted with permission of American Society of Agronomy, Madison, WI)

Class	R_f	Pesticides
1. Not mobile	0.00–0.09	Chloroxuron, Deiquat, Chlorphenamidine, Fluorodifen, DDT, Moresten, Dieldrin, Paraquat, Trifluralin, Endrin
2. Slightly mobile	0.10–0.34	Siduron, Prometryne, Propanil, Diuron, Dichlobenil, Chlorpropham, Azinphosmethyl
3. Intermediate mobile	0.35–0.64	Propachlor, Prometone, 2,4,5-T, Propham, Fluormeturon, Diphenamid, Monuron, Atrazine, Simazine, Ametryne, Propazine, Trietazine
4. Mobile	0.65–0.89	Picloram, Fenac, MCPA, Amitrole, 2,4-D, Bromacil, Fluoromidine
5. Very mobile	0.90–1.00	TCA, Dalapon, Dicamba, Chloramben

In Table 5 pesticides are classified in mobility classes based on R_f values from thin-layer chromatography on soil plates. The R_f value is derived from the distance which the substance has passed in a certain time on the soil plate relatively to the eluent (H_2O) and is closely related to the retardation factor. Helling [51] classified 40 test substances into 5 mobility classes: The most mobile pesticides are organic acids in classes 4 and 5; triazines and urea derivatives have a mean mobility (classes 2 and 3) whereas the mobility of chlorinated hydrocarbons and organic cations is very low (class 1).

The binding of organic compounds to solid material in an aquifer retards their migration and is also very often of indirect importance as the first step in their biological or chemical transformation. The agents may also be bound to suspended negatively charged colloid particles, which are rejected by the negatively charged mineral grains. This allows a preferential transport in the macropores with a higher velocity than the mean water flow. In this way, the mobility of hydrophobic organic compounds including many herbicides is increased by their sorption to preferentially transported macromolecules [52].

In surface water, the sorption occurs on surfaces of aquatic organisms and of suspended or sedimented solids. The pesticides adsorbed on suspended particles are transported following the same controls as the dissolved substances, but they may be sedimented when the flow velocity decreases. These fine-grained sediments may, however, be remobilized and transported further by floods. This transport of the sorbate is another poorly predictable and erratic process of pesticide transport in surface water, which in the similar case of radionuclide transport is not covered by the model of Shih & Gloyna [53]. Generally it can be stated, that if sorption and desorption are involved in pesticide transport in rivers, they cause a smearing of the descending branch of the breakthrough curve of a single event of contamination.

3.4 Filtration

Pesticides may be transported in surface and ground water adsorbed unto organic or inorganic particulate matter. This propagation may be limited if the particles are removed from water by filtration during infiltration of surface water or within the aquifer. The pore sizes of natural porous media are usually about in the same order of magnitude as the sizes of the particles considered here (Fig. 9). Thus, comparison of the sizes of particles with that of pores (Fig. 10) indicate that mechanical filtration can at best only very incompletely retain such particles. A closer view explains the filtration effect as the result of complex physical and chemical processes, which can be described in a macroscopic way by Eq. (11) with the filter efficiency (filter factor) λ_f of the porous medium and the decrease of the initial concentration of a suspension c_o along a flow length, s, to the observed concentration c_S:

$$\ln(c_s/c_0) = -\lambda_f \cdot s \tag{11}$$

The filter factor λ_f lumps together geometrical parameters such as the porosity, the grain sizes of filter material and of the suspended particles, and physical parameters as the Peclet number (the quotient of mean flow velocity and diffusion coefficient), the Boltzmann' constant, the temperature, the water viscosity, the Darcy velocity and the densities of suspended particles and of water, and finally the mechanisms affecting particle fixation on solid grains such as sedimentation, flow processes, diffusion and interception [57].

During contact with the grain surfaces the particles can be fixed by mass (Van der Waals) and electrical (Coulomb) forces. Under the chemical conditions

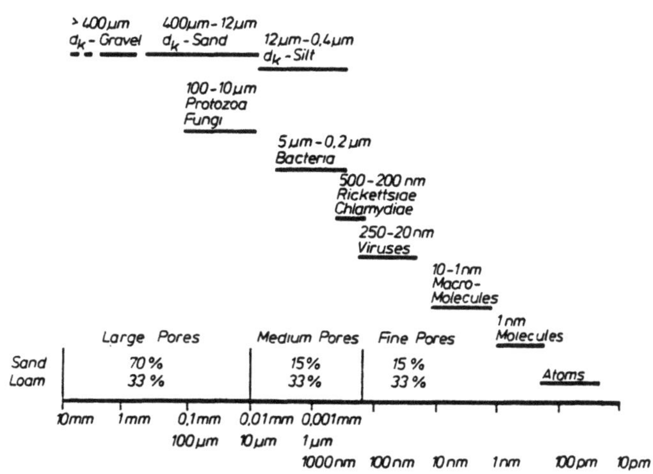

Fig. 9. Comparison of particle sizes with the hydraulic equivalent diameters of pore canals [54] (reprinted with permission of Elsevier Science Publishers BV, Amsterdam)

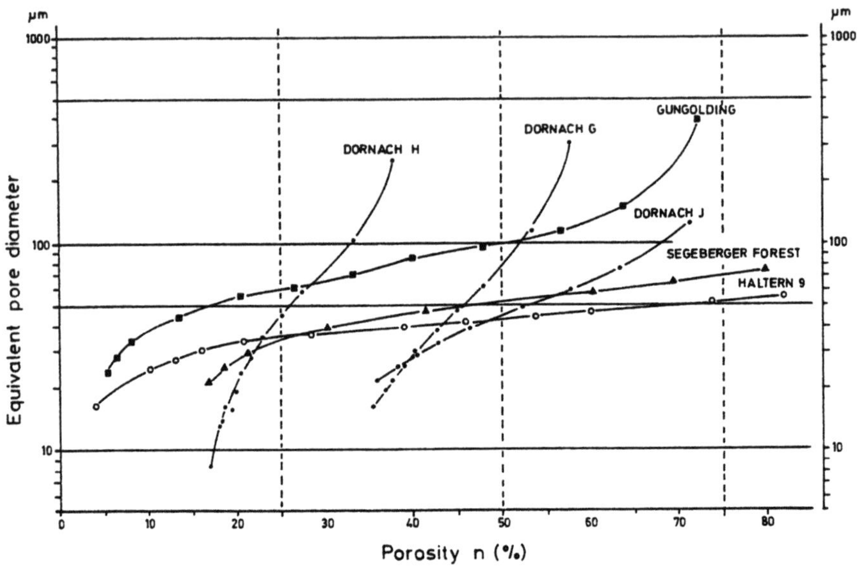

Fig. 10. Equivalent diameters of pores in non-consolidated aquifers from Dornach/Munich, Gungolding/Altmuehl valley, Haltern and Segeberger Forest [55] (reprinted with permission of Erich Schmidt Verlag Gmbtt & Co, Berlin)

of most ground waters, the net surface charges of the grains are negative. The adsorption behaviour is analogous to that of anions: it increases with increasing electrolyte content and decreasing pH of water. Decreasing grain sizes, i.e. increasing specific surfaces, and increasing content of clay and organic matter, i.e. increasing exchange capacity of the aquifer also increase particle adsorption. Therefore, the filter efficiency is dependent upon the chemical and physico-chemical properties of the ground water and filter medium.

The filter mechanisms depend on the hydraulic conditions (flow velocity, pore size distribution, interconnection of voids etc.). The filter efficiency (filter factor) of an aquifer due to the different filter mechanisms (transport to the grain surface: sedimentation, diffusion, pore flow and interception, fixation) may be used as a measure of the decrease in an initial concentration on a certain flow length. Laboratory experiments show that for sand the filter factor is highly correlated to the effective grain diameter, which is routinely determined in hydrogeology [58]. When these parameters change, the particles may by transported further. Thus, they may reach an observation point after a longer period. This may be formally treated as a 'retardation", because the transport velocity calculated from these observations is lower than that of the ground water.

During prolonged contamination of an aquifer, whether by river bank filtration, sewage injection or sewage land treatment, a filter layer develops, especially at the water/solid interface – similar to that on the surface of sand filters – which contains a biocoenosis (biofilm) with bacterial slimes and amorphous inorganic exudates [59] comparable to that on the sediment surface in rivers and lakes. Under continuous flow conditions, the hydraulic conductivity of such layers may be reduced to 5% of the original value [60].

Sand filters which have been used over very long periods show fatigue indications, which are mainly caused by the accumulation of toxic microbial metabolites [61]. However, by an intermittent flush with clean highly oxydized water the oxygen content of the sand filter can be increased and thus, the filter efficiency may be improved [62].

3.5 Degradation

Another very important factor for a risk assessment of ground water contamination by pesticides is their degradability. Whereas the user wants a substance which will persist in the soil over the whole protection period, a quick and quantitative degradation is necessarily demanded from the viewpoint of ground water protection.

Degradation in the subsurface is often described by a modified first order degradation equation, Eq. 12, [63], which indicates, that degradation very often begins after a time lag, in which no or only minor degradation occurs [64]:

$$c_t = c_0 \cdot e^{-\lambda_t(t - t_0)} \qquad (12)$$

with the initial concentration c_0, the concentration c_t at the time t, the length of the initial phase t_0, and the degradation constant λ_t, which is related to the half-life $t_{1/2}$ by $\lambda_t = \ln 2/t_{1/2}$. If – as at the degradation of atrazine and linuron [65] – the processes are influenced by adsorption and desorption, microbial activity, nutrient supply, competing reactions and the chemical or physical environment, the degradation follows a rate equation of higher order (nth. order [66].

3.5.1 Non-Biotic Degradation

In soil and aquifer, pesticides react chemically. They hydrolyze, oxidize or form isomers if water or oxygen is present. In surface water and in the upper centimeters of the soil, the degradation of substances can be accelerated by photolysis, an effect of light energy (UV-rotation). Quantitatively the photochemical degradation can reach the rates of microbial degradation for several groups of substances [67].

Pesticides in the atmosphere, in surface water or on surfaces such as leaves, vegetation, and soils can undergo either direct phototransformation through an excited state or indirect phototransformation by reacting with another chemical in an excited state. The indirect phototransformation may include not only the transfer of excitational energy from an excited donor molecule (sensitizer) to the acceptor molecule (quencher), which undergoes photoreactions (sensitized photodegradation), but also the degradation of the contaminants through their reaction with photochemically generated intermediate species (photoinduced degradation). Molecular oxygen may also be transfered to an acceptor (e.g. pesticides) as a result of light absorbed by a sensitizer in the system (photosensitized oxidation). The photoinduced degradation may also involve donors which are first excited and transfer electrons (not electric energy) to a substrate; the resulting excited-state charge transfer complex may degrade [68, 69]. Finally, a photochemically excited molecule can release its energy in other ways besides photochemical reactions [68]. The efficiency of these various processes depends upon the medium in which the molecule occurs (dissolved in a liquid, vapour phase, adsorbed on a solid) [70]:

The reaction quantum yield (φ) of a pesticide by direct phototransformation in water is defined by:

$$\varphi = \frac{R}{\int_{\lambda_1}^{\lambda_2} I_a} \tag{13}$$

with the number of molecules of a particular process R, the absorbed light quanta I_a and the range of wave length $\lambda_1 - \lambda_2$.

When a dilute solution of a contaminant of absorbance <0.02 at a certain wavelength λ is subjected to photolysis the following first-order kinetics equation can be used for the calculation of the rate constant ($k_{p,\lambda}$) (s^{-1}) for the photolysis (p) of pesticides [71]:

$$\ln(c_0/c_t) = k_{p,\lambda} \cdot t \tag{14}$$

In Eq. (14), c_0 and c_t are the concentrations of a pesticide (in mol/l) at times zero and t (expressed in s). Assuming first order kinetics in the substrate, the half-life for the sunlight induced degradation is defined by Eq. 15 [71]:

$$(t_{1/2})_{sp} = \frac{\ln 2}{k_{sp}} = \frac{0.693}{k_{sp}} \tag{15}$$

Published data of photodegradation of pesticides and other environmental chemicals in soils or on soil surfaces are rare [72, 73]. The photodecomposition of environmental contaminants adsorbed on solid surfaces is described usually with first-order rate kinetics, however the sunlight photolysis first-order rate constants (k_{sp}) of environmental contaminants adsorbed on solid surfaces and a theory for the solid phase photolysis are not yet available.

Indirect photolysis is based in natural water on the appreciable amounts of dissolved and suspended organic substances, which can alter the photolysis rates of environmental chemicals. For instance the well known triplet sensitizer acetone occurs in most natural aquatic environments [68]. It accelerates the photodegradation of several organochlorine xenobiotics and occasionally alters the photoproduct distribution arising from these contaminants and α-substituted *p*-chlorotoluenes, e.g. $p\text{-ClC}_6H_4CH_2-X(X=H, CN, OH, COOH)$ in CH_3OH at >290 nm [74], although these substances were non-reactive in several natural waters and CH_3OH, respectively [69].

The presence of light-absorbing materials in waters has a screening effect that makes the photolysis rates of pesticides slower than that in pure water solutions, except for those cases in which the presence of natural sensitizers accelerates the photoreaction [75].

Hydrogen peroxide (HOOH) is photochemically formed from humic acids, fulvic acids, and humins which are present in almost every aquatic and soil environment [76–78]. The concentration of hydrogen peroxide in surface freshwaters ranges between 100 to 7000 nM. Photo absorption results in the eventual generation of hydroxyl radicals (•OH) and organoperoxy radicals [77, 78]. The *s*-triazines atrazine and simazine are also degraded by H_2O_2 in presence of UV-light ($\lambda > 290$ nm) (Fig. 11), the photocatalytic process is apparently accelerated by the formation of •OH radicals from TiO_2 (Fig. 12). As an intermediate product desethyl atrazine was detected [79].

The oxidation of pesticides of •OH species is second-order [80]. However, first-order treatment can be applied to the oxidation of the environmental chemicals by •OH species for the determination of the pseudo-first-order rate constants [82].

Pesticides may be photochemically incorporated in water into humic macromolecules and their model monomers, (such as benzaldehyde, benzoic acid, vanillin, vanillic acid, syringaldehyde, and syringic acid). At >285 nm several polychlorobenzenes eventually form polychlorodibenzofurans containing 2 Cl atoms less than the initial chlorobenzene [83].

The most important chemical degradation reactions are acid and alkaline hydrolysis. The organic phosphorus derivatives, the chlorinated hydrocarbons, the carbamates and the triazines are degraded by these reaction types (Figs. 13, 14, 16). The hydrolysis can be accelerated by clay minerals or metal

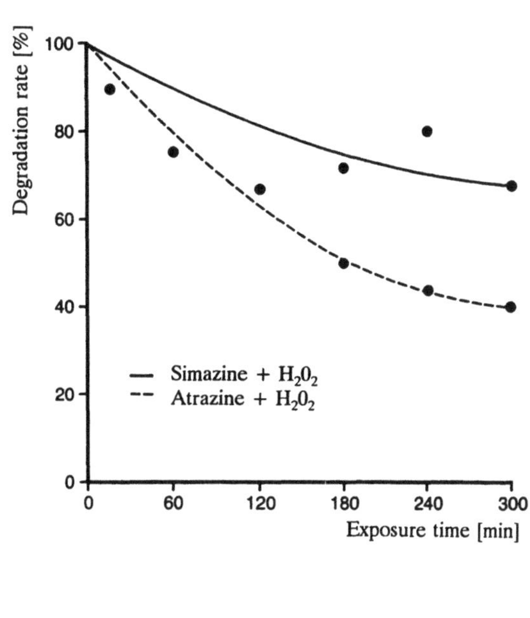

Fig. 11. Photolysis of atrazine and simazine (20 ppm each) in presence of H_2O_2 ($\lambda > 290$ nm; 55 ppm) [79] (reprinted with permission of VDI Verlag, Düsseldorf)

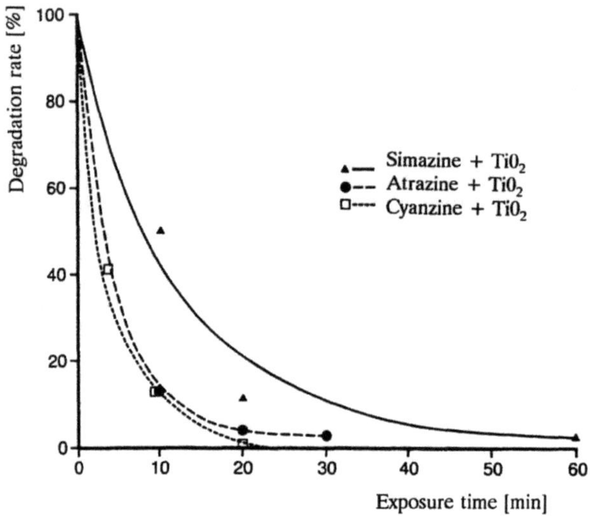

Fig. 12. Photocatalytic degradation of atrazine, simazine and cyanazine (20 ppm each) in presence of TiO_2 ($\lambda > 290$ nm; 250 mg/l) [79, 81] (reprinted with permission of VDI Verlag, Düsseldorf, and Gordon and Breach Science Publishers, New York)

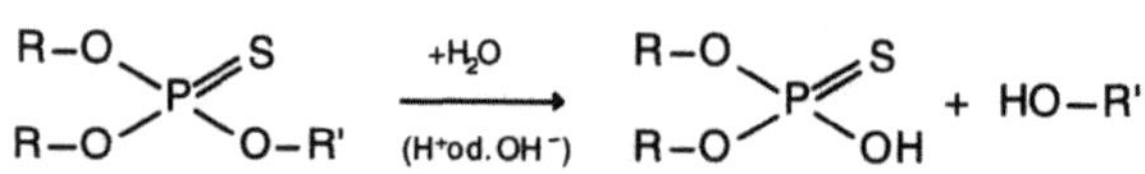

Fig. 13. Hydrolytic degradation of diazinon [84] (reprinted with permission of American Society of Agronomy, Madison WI)

oxides [85] (Fig. 14). The sorption on organic substances, however, impedes the degradation, e.g. that of chlorinated hydrocarbons as heptachlor and endrin [86]. In general, the hydrolysis rate constants for halogenated alkanes increase from primary to secondary to tertiary halides and are about one order of magnitude greater for alkylbromides than for the corresponding alkylchlorides [42].

The contribution of the different mechanisms can be shown for instance for metamitron in soil with the schematic degradation pathway (Fig. 15). The half-life of its hydrolytic degradation strongly depends on the pH, with a half-life of 10 days (pH 9), 4 weeks (pH 7) and 14 months (pH 4), measured in sterile solutions. However, these values are not relevant for natural conditions, where they are superimposed by photolytic effects and microbial degradation. The photolytic degradation uses the reaction quantum yield (Eq. 15).

In a degradation experiment with ^{14}C labelled metamitron in a sandy loam (82.7% sand, 13.3% silt, 4.0% clay; 2.15% C_{org}; pH(H_2O) 6.3) the balance found after 100 days 18% of the applied ^{14}C in CO_2, 44% in extractable form and

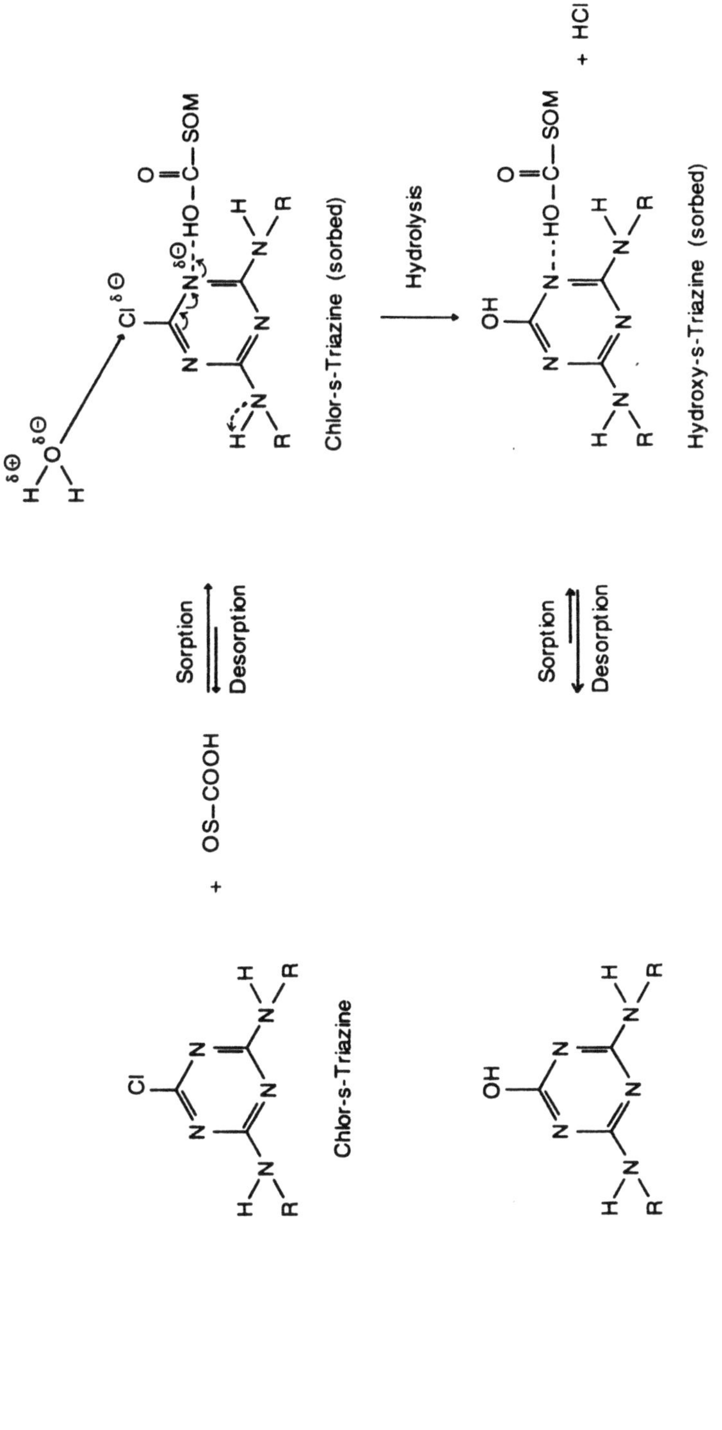

Fig. 14. Hydrolytic degradation of *s*-triazines (Simazine, atrazine) in an acid environment to hydroxy-triazines in presence of a catalytic active exchanger [84] (SOM = Soil Organic Matter) (reprinted with permission of American Society of Agronomy, Madison WI)

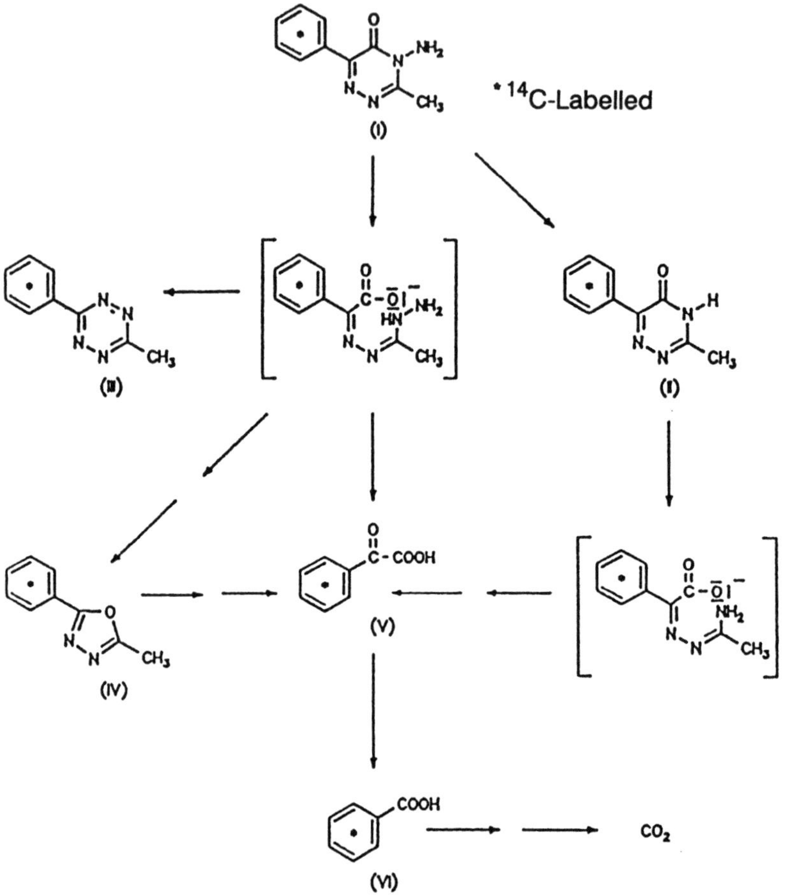

Fig. 15. Balance and schematic degradation pathway of metamitron in soil [87]

38% in bound residues [87]. If the reaction quantum yield measured for metamitron (= 0.0162) and the molar extinction coefficients between the wave lengths 297.5 (λ_1) and 370 nm (λ_2) are fed into the computer program GC-SOLAR [88], one gets the following data as environmental half-lives.

Season	Geographical latitude 30 th.	40 th.	50 th.	60 th.
	half-life (h)			
Spring	0.48	0.54	0.65	0.84
Summer	0.41	0.43	0.47	0.52
Autumm	0.69	0.96	1.56	3.29
Winter	1.01	1.74	3.98	14.16

Boundary conditions: Pure water; water surface; 30th., 40th., 50th. and 60th. latitude; 10th. longitude; half-life, integrated over the whole day.

The program allows one to simulate the photochemical degradation for different environments [87].

Pesticides containing sulfur are often oxidized, as e.g. the oxidation from parathion to paraoxon. In this case the sulfur in the molecule is exchanged by oxygen. These reduction–oxidation processes are usually microbially mediated (Fig. 16).

3.5.2 Biotic Degradation

Organic compounds are decomposed by microorganisms which obtain from these processes carbon and hydrogen for their cell synthesis. The energy necessary for their metabolism is supplied by the degradation of substances rich in energy into lower-energy metabolites (Fig. 17), and finally into CO_2 and H_2O. The enzymatic transformation of pesticides may involve a chemical change such that the inhibitory properties are lost. This detoxication or inactivation is performed through degradation, conjugation or complexing. Alternatively, non-toxic substances may be transformed to toxicants ("activation") or toxic parent substances may be converted to new substances with another, sometimes even

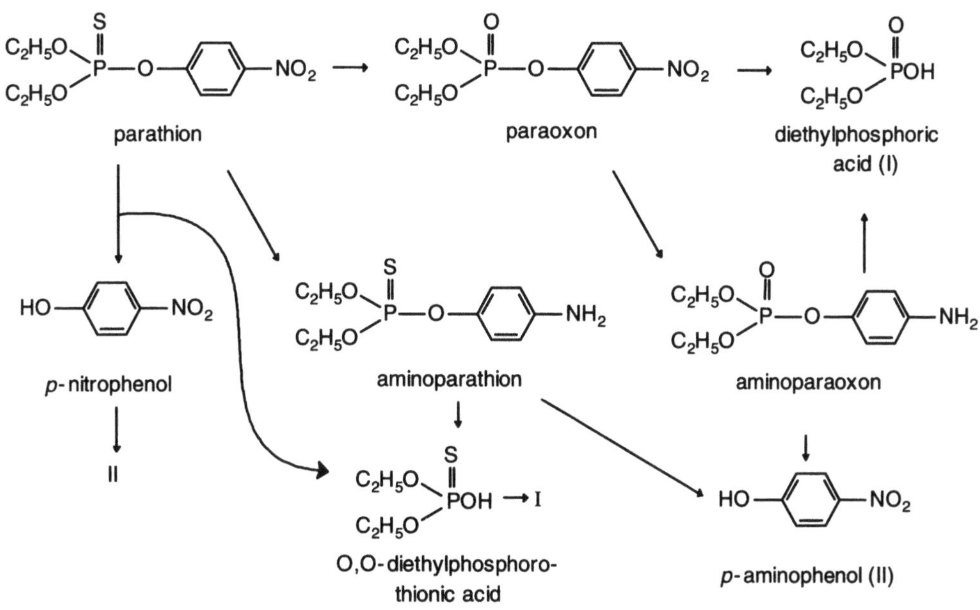

Fig. 16. Pathways for the metabolism of parathion and paraoxon [89, 90] (reprinted with permission of Academic Press, Orlando FL, and of UNESCO, Paris)

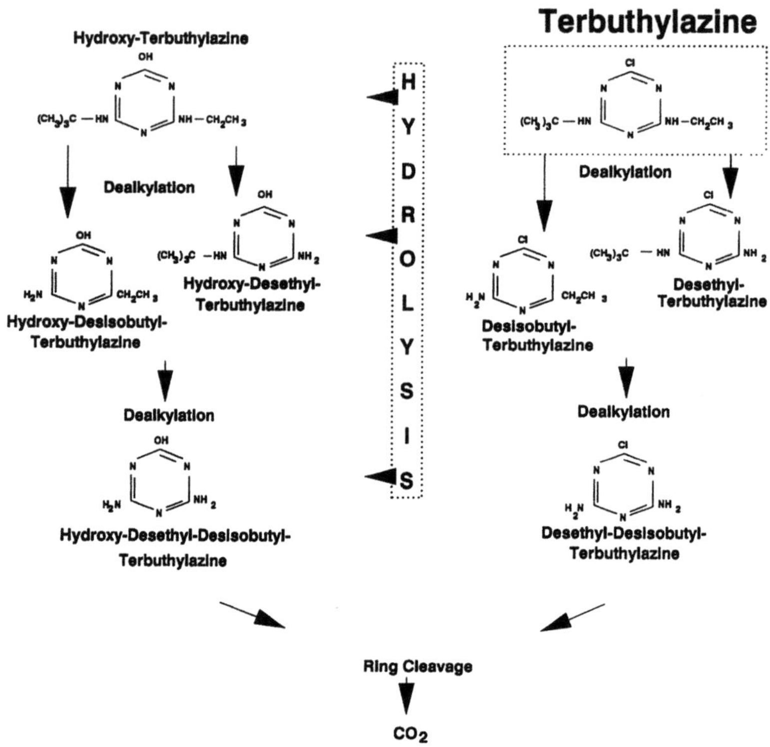

Fig. 17. Major metabolites of terbuthylazine [91] (reprinted with permission of Gustav Fischer Verlag, Stuttgart)

higher toxicity level or inhibitory spectrum. Finally the parent agents may be metabolized to new stimulatory substances [92, 93]. Under anaerobic conditions, the microorganisms receive the necessary oxygen by reducing oxygen-bearing compounds, particularly nitrates and sulfates. Substances not containing oxygen and with a low hydrolysis constant are more persistent in ground water than alcohols, ethers and esters. Oxygenated compounds which do not follow this hypothesis are generally alkyl-substituted or contain halogen atoms in the molecule, which results in a much lower degradability. As an example of a relatively persistent compound found in ground water Zoeteman et al. [94] mention bis(2-chloro-iso-propyl) ether. The elimination of organic compounds which is observed in the field does not make clear whether a secondary organic compound is formed or whether a total mineralization is reached [31]. ^{14}C-labelled pesticides proved to be a good tool to study the degradation pathways and rates and the fate of metabolites in the subsurface either in laboratory experiments [95, 96] or in experiments with monolyth lysimeters simulating close-to-realistic field conditions [97].

The directions of the microbial reactions are controlled by the thermodynamic relations of the respective system, but they proceed under favourable ecological conditions much faster than non-biotic physico-chemical reactions. Microbial reactions are caused by autochthonous microorganisms, which are adapted to local subsurface environments including the ground water milieu [10, 11, 98–101], which, until recently, was considered to be almost sterile. Microbial degradation is quantitatively more important for the decay of organic substances than pure chemical reactions. Bacteria, actinomycetes, fungi and – in soil – algae are the main groups of microflora involved in these processes. The composition of the microbial population depends on such ecological factors as pH, oxygen, water, temperature, nutrients and geochemical environment, and can vary greatly.

Subsurface microbial colonization is, as a general rule, greatest in the nutrient-rich humic upper soil layers and decreases generally with falling nutrient supply at greater depths. This led in the past to the false interpretation that no microorganisms can live at depth. In the presence of energy sources, particularly organic substances, which permit life, microbial metabolism can also take place below the humus layer. Whereas aerobic microorganisms are tied to free oxygen availability for their distribution, which generally decreases downward, anaerobes and facultative anaerobes are found at various depths. Besides the most important carbon compounds, nutrients necessary for microbial colonization are chiefly nitrogen compounds, such as ammonia or organic compounds of nitrogen, and phosphates and in the anaerobic zone oxygen compounds like nitrate and sulfate.

The metabolism of the nematicide fenamiphos follows under aerobic conditions the same degradation pathway in 16 soils from 11 countries with different climates, however with different output. After 90 days' incubation at 22 °C 1.1–39.0% of the applied ^{14}C-labelled agent was mineralized and between 18.5 and 61.8% was bound to the soils [102].

It was shown that microbial activity is to be expected in deeper layers of the substratum. Although the nutrient conditions are poor compared to those in the soil, there is a population density of adapted species in the unsaturated zone and in the aquifer in the order of 10^6 cells/g material [103–106]. However, only relatively few data are available on the contribution of microorganisms to degradation, the presence of microbial population indicates that microbial degradation of substances in the aquifer may occur in the same manner as it is known from soils. There are indications that the microbial degradation of organic trace contaminants needs threshold concentrations of other natural or man-made substrates, below which no further degradation occurs [107, 108].

Metabolism and velocity of degradation depend on the ecological conditions. Degradation under aerobic conditions is generally faster than under anaerobic conditions. Most pesticides can be degraded since the microbial population can adapt to specific substances. Therefore the degradation rates increase after several applications of the same pesticide, e.g. 2-4 D and dichlorprop [64]. However, pesticides can also have negative effects on the viability of microorganisms. The combination of several pesticides can change the degradation rates determined for single pesticides, e.g. cabaryl impedes the microbial degradation of chlorpropham [109]. However, other agents may cause an enhancement or have no effects on the degradation rates [64]. Finally, degradation is controlled by the amount of initial concentration of the pesticide as well as by many soil properties and climatic conditions [96, 102, 110–112]. Generally the absolute degradation rate decreases with increasing initial concentration [65, 113].

The adsorption of pesticides on organic substances or clay minerals inhibits their degradation in most cases, even if the microbial activity in these systems is high. The degradation is affected by the formation of bound (non-extractable) residues, which are defined by the International Union of Pure and Applied Chemistry (IUPAC) as non-extractable pesticide agents, metabolites or fragments in plant and soil, which result from a correct agricultural use of the pesticide and remain in the plant or soil after the application of extraction methods, which do not significantly change the structure of the agent. Not included are that non-extractable fragments, which stem from metabolism of natural constituents of plants and soils [114].

Most substances have the highest degradation rates at slightly acid to neutral pH conditions (range 5 to 7) [64]. Increasing temperature and water content in the unsaturated zone have an accelerating effect on microbial degradation. Thus, microbial degradation is highest during vegetation period and almost no degradation occurs in winter [115].

Table 6. Degradation of pesticides in soil and water depending on the pH value [39] (reprinted with permission of Erich Schmidt Verlag, Berlin)

Pesticide	$t_{1/2}$ (days)		pH-range Soil and Water
	Soil	Water	
Aldrin	194	26.6 (2)	7.4–8.0
Heptachlor	304–4,380	3.5 (2)	7.4–8.0
Chlordane	365–1,460	53.3 (2)	7.4–8.0
Telodrin		4.0 (2)	7.4–8.0
Lindane	591	>400 (1)	7.4–8.0
Dieldrin	365–2,560	>56 (2)	7.4–8.0
DDT	9–3,650	>56 (2)	7.4–8.0
Camphechlor	3,650		7.4–8.0
Endrin		>56 (2)	7.4–8.0
Diazinon	2.0	30.8–18.0 (3)	5.0–8.0
Malathion	0.8	26.9 (3)	9.0
Parathion	30.0	>4,250 (3)	5.0–6.0
Paraoxon	0.75	>4,250 (3)	5.0–6.0
Crotoxyphos	0.4	17.1 (3)	6.0
Atrazine	8.0	3.21 (3)	7.0
Triallate	56–77	680–1,170 (3)	6.0–8.0

(1) model aquifer; (2) river water; (3) groundwater

There are very few data on the degradation rates of pesticides in ground water. In this environment conditions are different from those in soils, because the main controlling parameters – temperature, pH and water content – are nearly constant and there is no free access of atmospheric oxygen. If there is no dissolved oxygen in the aquifer, anaerobic conditions occur, leading to a decrease in degradation. The first step towards predicting degradation in an aquifer may be the use of data known from river water (Table 6).

Eichelberger and Lichtenstein [116] have shown that organic phosphorus derivatives and carbamates in river water samples have very short half-lives, whereas the values for chlorinated hydrocarbons vary widely. However, these results cannot be transferred without further research into ground water conditions, because of the lack of photolysis effects present in river water.

Thus, for a risk assessment of ground water contamination by a certain pesticide, its geochemical properties due to solubility, degradation and adsorption–desorption behaviour must be investigated. These characteristics, which have different consequences in different locations, can be used for a classification of pesticides into classes of different risk.

In Table 7 the octanol/water distribution coefficients and the degradation constants of different pesticide groups are summarized. The data were taken from investigations with soil samples. The ranges of the values are very large, so no chemical group can be totally considered as extremely hazardeous.

Table 7. Adsorption- and degradation characteristics of different pesticide groups (octanol/water distribution coefficients (K_{OW}), half-lives ($t_{1/2}$), degradation rate constants (λ_t) [39] (reprinted with permission of Erich Schmidt Verlag, Berlin)

Chemical group	K_{OW}	$t_{1/2}(d)$	$\lambda_t(1/d)$
Not ionic:			
Halogenated hydrocarbons	250 –1,250,000	194 –4,380	0.0036–0.00016
Carbamates	6.8– 236	2 – 650	0.35 –0.0011
Anilines, anilides, amides	24 – 2,040	1.5– 321	0.073 –0.0022
Urea-derivatives	19.4– 2,040	22.3– 710	0.031 –0.00098
Org. phosphorous derivatives	1,458 – 91,400	6 – 602	0.12 –0.0012
Ionic:			
Kationic:			
Dipyridyles	94 – 253	almost no degradation	
Alkaline:			
Uracile	19 – 690	56 – 149	0.012 –0.0047
Triazines	19.4– 2,580	22 –2,460	0.032 –0.0018
Acid:			
Nitrophenols	488*	15.3	0.045
Phenoxy fatty acids	250 – 1,160	< 7 – 924	> 0.1 –0.00075

* calculated from: $\log K_{OC} = 0.6 \log S + 3.9$

Although there are complex relations between several factors controlling degradation of pesticides, several prognostic models for the estimation of degradation or accumulation of residues in soils have been developed since the 1950s. The empirical model of Walker [117, 118] gives a good estimation of the degradation of some herbicides in a certain location. As input data, soil properties and climatic conditions as well as the initial concentration of the respective herbicide are needed.

3.6 Advection and Dispersion

The extension of a contaminated ground water zone depends on its hydrogeological conditions which are defined by the position of the water table, the hydraulic gradient, the water balance, the distance to any wells or springs and the properties of the contaminant, such as solubility, sorption behaviour and persistence, and of the rocks, such as sorption capacity, porosity and hydraulic conductivity. The movement of a contaminant in the water-unsaturated zone is influenced by the moisture content, the water balance and the above mentioned contaminant and rock properties.

The spreading and dilution of a pesticide-contaminated ground water plume, initially well-defined in space are, during its transport with the flowing ground water (advection), due to hydrodynamic dispersion. This process includes the diffusion based on Brownian molecular movement and the hydromechanical dispersion due to the tortuosity of the pore channels. The hydromechanical dispersion depends on the textural and structural variety of aquifers mainly as a result of different flow velocities depending on the width and consistence of the aquifer voids. The relative importance of these two mechanisms depends on the flow velocity. The importance of diffusion can be assessed by the Peclet number Pe, the quotient of the flow velocity v_{w} and the coefficient of molecular diffusion D_{d} (Eq. 16).

$$Pe = v_{\mathrm{w}}/D_{\mathrm{d}} \tag{16}$$

For $Pe < 0.4$ the molecular diffusion predominates, for $0.4 < Pe < 5$ both the molecular diffusion and mechanical dispersion are in the same order of magnitude. At $Pe > 5$ the hydromechanical dispersion predominates [119]. The importance of hydromechanical dispersion will increase with increasing flow velocity relative to the other mechanisms. Both the molecular diffusion and the hydromechanical dispersion are in the unsaturated zone dependent on the water saturation. Furthermore, the retardation of interactive substances by adsorption–desorption and the elimination and – with respect to the propagation of particulate and colloidal substances – the filtration, has to be considered. Thus, the transport of pesticides in a ground water system may be treated using the general transport equation [120], including terms for retardation, for the biological elimination and – partially – for the filtration of these substances in the ground water milieu [6]:

$$\frac{\delta c}{\delta t} = \frac{D_l}{R_d} \cdot \frac{\delta^2 c}{\delta x^2} + \frac{D_t}{R_d} \cdot \frac{\delta^2 c}{\delta y^2} - \frac{v_w}{R_d} \cdot \frac{\delta c}{\delta x} - \lambda_t \cdot t - \lambda_f \cdot v_w \cdot t \tag{17}$$

with the coefficients of longitudinal and transversal hydrodynamic dispersion $D_l, D_t = D' + D_d$; D' = coefficient of hydromechanical dispersion, D_d = diffusion coefficient, the concentration gradient $\delta c/\delta x$, the average ground water flow velocity v_w, the retardation factor R_d (Eq. 9), the degradation constant λ_t (Eq. 12), and the filter factor λ_f (Eq. 11).

The ground water flow velocity, v_w, can be calculated either from the hydraulic conductivity, k_f, the hydraulic gradient, grad h and the effective porosity, n_e, as mean flow velocity (Eq. 18) or from tracer experiments, which determine the mean transport time, t_{50}, of an ideal tracer for flow length x (Eq. 19). The mean transport time, t_{50}, (Fig. 18) can be derived from the cumulative concentration time curve at 50% of recovered tracer quantity.

$$v_w = -\frac{k_f \operatorname{grad} h}{n_e} \tag{18}$$

$$v_w = \frac{x}{t} \tag{19}$$

In field and laboratory experiments a relatively smaller mean transport velocity of some pesticides could be observed for the solutes or for particles (v_s) than for water as indicated by a conservative tracer (v_w)($R_d = v_s/v_w$). With

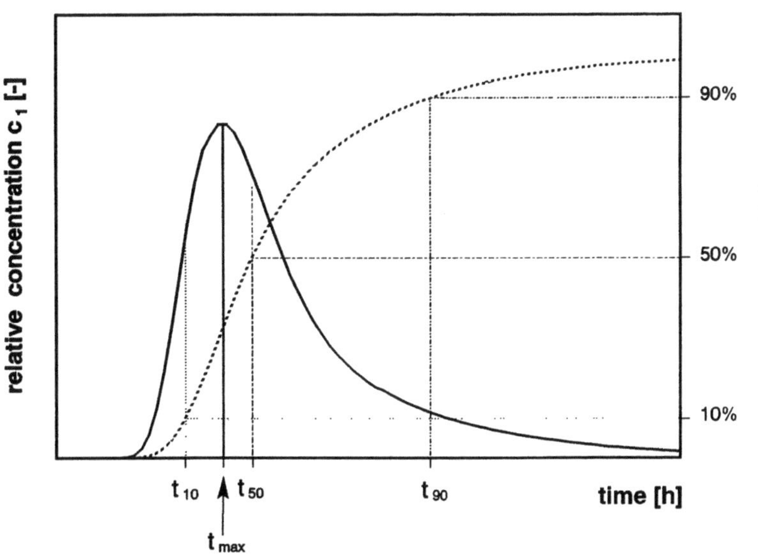

Fig. 18. Schematic breakthrough curve at an observation point

increasing dispersivity the maximum flow velocity will deviate more and more from the mean flow velocity.

The coefficient of longitudinal hydrodynamic dispersion D_l (in m^2/s) is a function of the mean flow velocity of ground water v_w (in m/s), thus the specific property of the solid subsurface material is the dispersivity α_l (in m):

$$D_l = \alpha_l \cdot v_w^b \tag{20}$$

with $0.9 < b < 1.2$.

In homogeneous aquifers the longitudinal dispersivity depends only on textural influences (grain size, unconformity, grain form, pore size distribution, pore connections). This so called microdispersivity deviates, however, from the so called macrodispersivity in most granular aquifers, in which structural features such as horizontal and sloping stratification predominate, which increases with increasing flow distance [121–126]. The transverse dispersion coefficient, D_t, is in homogeneous aquifers about one tenth of the longitudinal dispersion coefficient [127].

The pesticides in dissolved and suspended form spread also in surface water by molecular diffusion and turbulence contaminating adjacent water bodies. This dispersion process can be described by the same equations as those valid for ground water bodies. However, relatively little is known on the longitudinal dispersion coefficient in the rivers. Hull [128] suggests for natural surface water channels an empirical formula

$$D_l = 2.5 \cdot \sqrt{Q \cdot \bar{v}} \tag{21}$$

with Q = discharge (in m^3/s) and $\bar{v}$ = mean flow velocity (in m/s). For the Neckar river between Plochingen and Mannheim a dispersion coefficient between 1.75 and 3.75 can be derived using this formula [17]. The effect of dispersion in surface water is, as in ground water, a reduction of the concentrations and an extension of the duration of the contamination.

In porous media, the mobility of water depends on the pore sizes. In micro- and dead end pores water may be considered as immobile compared with that in macropores, which predominates in the fluxes of water and of the substances dissolved in it. This is considered for the transport of a reactive substance in the water-unsaturated zone in the concepts of the one-region and the two-region model of Van Genuchten and Wierenga [129].

The one-region flow model is described by:

$$R_d \frac{\delta c}{\delta t} = D \frac{\delta^2 c}{\delta z^2} - v_w \frac{\delta c}{\delta z} \tag{22}$$

with the retardation factor R_d (Eq. 9) and the dispersion coefficient D (Eq. 20). For the two-region model the following definitory equations hold

$$\beta R_d \frac{\delta c_m}{\delta T} + (1 - \beta) R_d \frac{\delta c_{im}}{\delta T} = \frac{1}{Pe_m} \frac{\delta^2 c_m}{\delta x^2} - \frac{\delta c_m}{\delta x} \tag{23}$$

$$(1 - \beta) R_d \frac{\delta c_{im}}{\delta T} = \omega (c_m - c_{im}) \tag{24}$$

with

$$\mathrm{Pe_m} = \frac{v_{\mathrm{wm}} L}{D_{\mathrm{m}}}; \quad \beta = \frac{\Theta_{\mathrm{m}} + \rho f K_{\mathrm{d}}}{\Theta + \rho K_{\mathrm{d}}}$$

$$\omega = \alpha \frac{L}{q}; \quad x = \frac{z}{L}; \quad T = \frac{v_{\mathrm{w}} t}{L} = \frac{qt}{\Theta L}$$

The latter model considers the diffusion-controlled substance exchange between both regions by a first order rate constant α, furthermore the retardation factor R_d, the rate constant ω and the part of the substance β, which is present in the mobile region in equilibrium with that in the immobile region. The 4-parameter model is written in a dimensionless form by the flow distance (column length) L, the dimensionless time (pore volumes) T, the Peclet number Pe, Eq. (16), and the Darcy-flow Q. The indices m and im mark the mobile or immobile water region. Thus, v_m is the pore water velocity in the mobile region, f is the number of binding positions in the mobile region and D_m the dispersion coefficient in the mobile region, whereas θ_m and θ_{im} are the volumetric water contents of the mobile and immobile water regions.

The analytical solutions of the model equations, (Eqs 22 and 23) [130] were used by Parker and Van Genuchten [131] in an non-linear curve fitting program on the basis of the least square method, which allows the assessment of unknown transport parameters by optimisation of curve fitting. Lennartz and Rambow [33] used these relations to evaluate the breakthrough curves of the

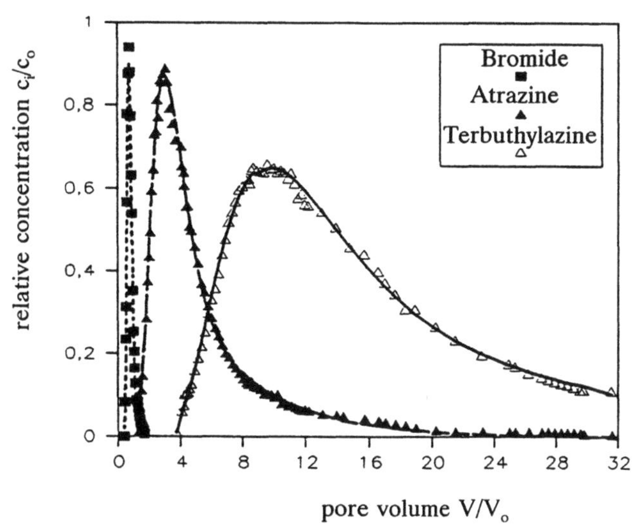

Fig. 19. Measured concentrations and fitted breakthrough curves of atrazine, terbuthylazine and bromide in the Bs-horizon of the podzol-site Segeberger Forest ($v = 10$ cm/d, $L = 14$ cm) [33] (reprinted with permission of the authors)

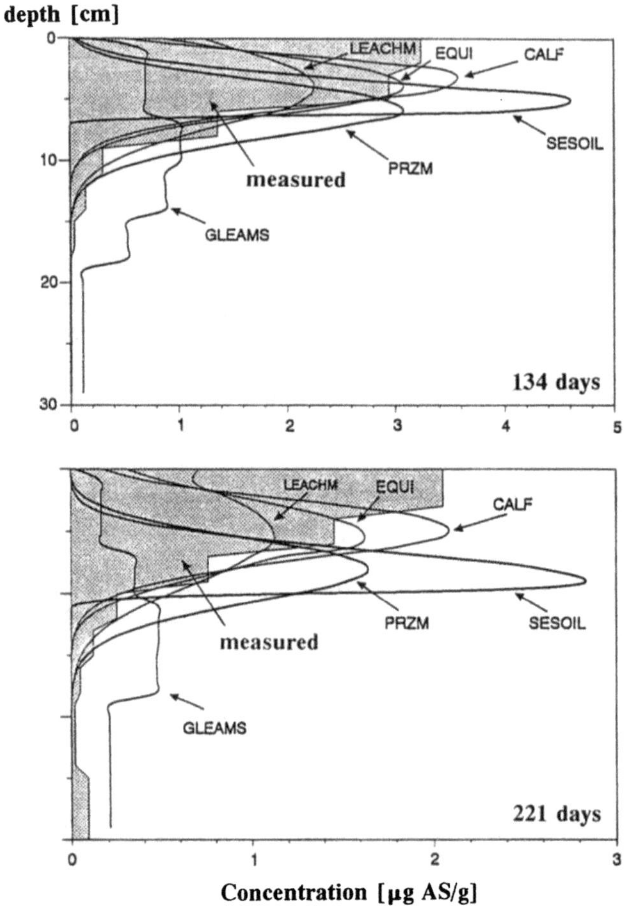

Fig. 20. Measured and calculated terbuthylazine concentrations of the field experiment in Ahlum 134 and 221 days after application [135] (reprinted with permission of the author)

pesticides atrazine and terbuthylazine through soil horizons of a podzol and to compare them with that of the conservative tracer bromide (Fig. 19).

A comparison of model calculations with CALF (FLOWCONC) [132], LEACHM [133] and PRZM [134] showed good agreement with the eluate concentrations of atrazine in column experiments, whereas for terbuthylazine poorer results were gained (Fig. 20). In the columns, the observed and the calculated residual contents of both pesticides do not match, mainly due to the general neglect of their sorption kinetics. Model calculations with CALF, EQUI [136], GLEAMS [137], LEACHM, PRZM and SESOIL [138] could not exactly describe the behaviour of the herbicides terbuthylazine, chlortoluron and isoproturon, observed in field experiments. Considering the sorption processes in the soil beyond the assumption of a constant sorption coefficient

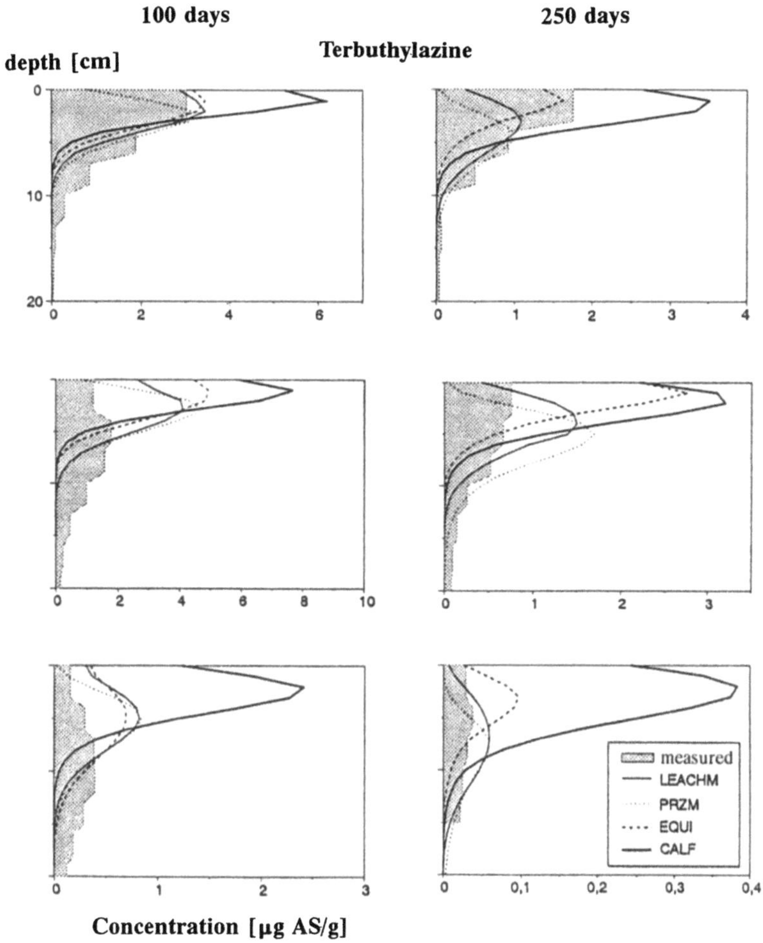

Fig. 21. Measured and calculated movement of three pesticides after 100 and 250 days at the test site Biologische Bundesanstalt Braunschweig [135] (reprinted with permission of the author)

improved the simulation results significantly (Fig. 21). It could be shown, that the parameters precipitation, half-life and sorption strongly affect the calculated transport behaviour [135].

Transport processes in the water-unsaturated zone and in the ground water lead to a propagation of pesticides in space and time because of advection and dispersion. Thus, the maximum concentration of a contaminant is lowered by one to three orders of magnitude depending on the properties of the aquifer. This can be observed in medium-scale and in field experiments. In a 2.5 m deep model aquifer of 100 m length at the Leichtweiss Institute of Hydraulic Engineering, Braunschweig University, the three pesticides lindane, dicrotophos and simazine were injected with initial concentrations of 10 mg/l. The breakthrough

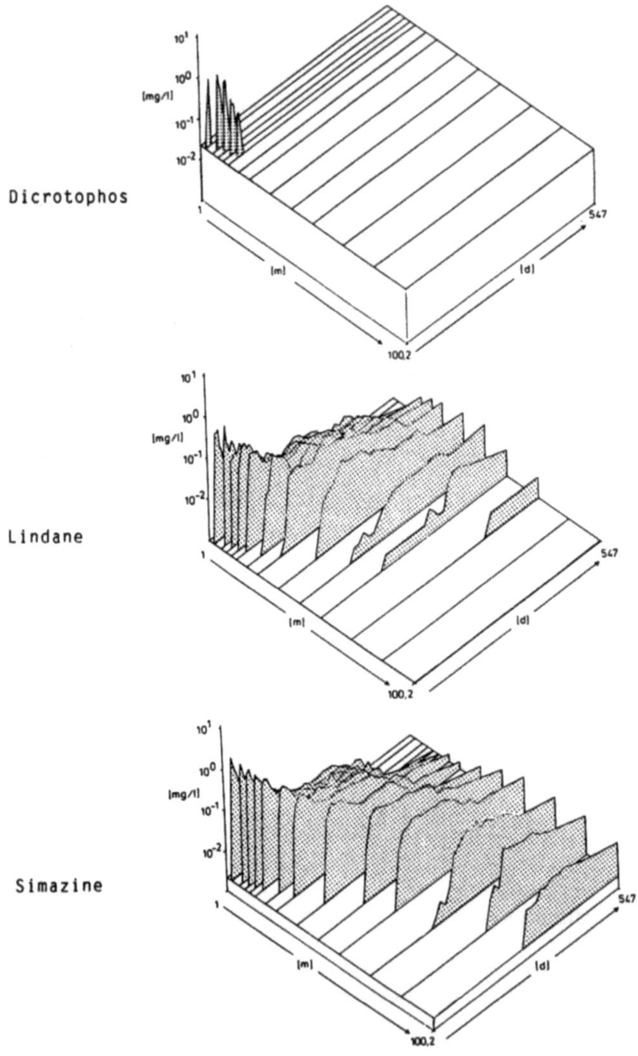

Fig. 22. Distribution of dicrotophos, lindane and simazine in a model aquifer at Braunschweig, FRG [6] (Flow velocity about 1 m/d) (reprinted with permission of Scandinavian University Press, Oslo)

curves were measured at 14 sampling locations. Figure 22 shows that dicrotophos was only detectable at the first five sampling locations. This substance was quickly degraded under its detection limit (19 ppb). Lindane was transported almost through the whole aquifer length; however, its concentration decreased gradually below the detection limit (1 ppb) at the last two sampling locations, presumably due to its adsorption onto the aquifer material. Simazine

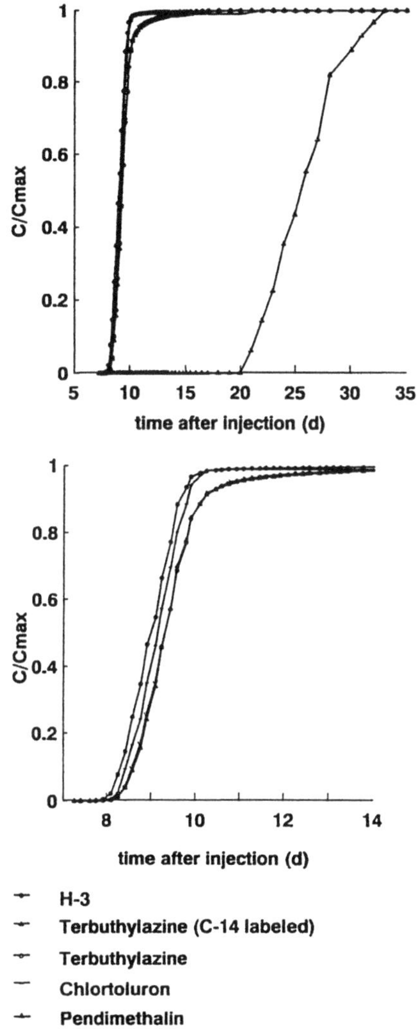

Fig. 23. Cumulative breakthrough-curves of well T4, level 10 at Twin Lake test site, Chalk River, Ontario, Canada [139, 140] (reprinted with permission of A.A. Balkema Publishers, Rotterdam)

was measured all over the aquifer (detection limit 2 ppb) [6]. It was apparently neither sufficiently degraded nor adsorbed.

A field tracer test was carried out with chlortoluron, terbuthylazine and pendimethalin in a glaciofluvial sandy aquifer in Chalk River/Ontario, Canada (Twin Lake test site), where ground water flow velocity is about 1 m/day. The cumulative breakthrough curves (Fig. 23) show that chlortoluron and terbuthylazine were not very much retarded (R_d = 1.0 to 1.2) with respect to the ideal tracer tritium, whereas the retardation of pendimethalin was significantly higher (R_d between 1.7 and 3.0). The latter agent was not detectable 10 m from the injection point within 1 1/2 month. The retardation of the herbicides was generally controlled by the organic carbon content of the local aquifer (0.03–0.08 mass-%) [139, 140].

All investigated herbicides are more persistent in the ground water than in soil. Parallel laboratory experiments at the Institute of Geology and Paleontology, Kiel University, with geochemically similar sediments from the Segeberger Forst test site indicate, that chlortoluron is degraded faster in lime-rich sediments ($t_{1/2} = 92$ to 145 d) than in lime-free sediments ($t_{1/2} = 176$ to 206 d). Pendimethalin was not significantly degraded under ground water conditions. The hydrolytical degradation of terbuthylazine increases with decreasing pH-value ($t_{1/2\mathrm{pH}4.5} = 113$ to 206 d, $t_{1/2\mathrm{pH}7.2} = 217$ to 252 d). The persistence of terbuthylazine in the Twin Lake aquifer (pH 5.6), was significantly higher than in the laboratory investigations [139, 140].

The phenylurea chlortoluron is the most mobile (R_d: 1.00 to 1.15) and persistent ($t_{1/2} = 92$ to 206 d) of the investigated herbicides. This highly soluble pesticide (70 mg/l) may be transported over a long distance without a significant loss of its initial mass due to adsorption or degradation processes and closely followed by terbuthylazine (solubility: 8 mg/l) on a slightly lower level. Pendimethalin (solubility: 330 µg/l) is significantly retarded and – as mentioned above – not detectable after a flow length of 10 m. It may cause deterioration in the ground water quality only very locally [139, 140].

As basic data for the modelling of pesticide transport in the soil, in the water-unsaturated zone and in the aquifer, besides the aquifer parameters the initial concentration, the degradation constant and the retardation factor of the respective pesticide are needed [38, 136, 141–151]. However, the model approach is hampered by the variability of the soil properties. To investigate this problem 200 soil samples were taken from a 130 × 140 m area of arable land used for maize production at the test site Ruhwinkel, near Kiel on a specific day (11-04-1992). Their contents in organic carbon, water and terbuthylazine residues as well as their granulometry were measured. The morphology of the lot is characterised by a N–S directed, S dipping depression and an adjacent small, N–S directed elevation which falls to the W (Fig. 24). The total height difference is 3.6 m [152].

The clay-, silt- and sand contents have mean values of 3.8%, 19.3% and 79.9% with standard deviations of 1.9%, 5.9% and 7.3% and variation coefficients of 50.7%, 30.8% and 9.4%. The organic bound carbon contents vary around a mean value of 1.1 mass-% with a standard deviation of 0.2 mass-% and a variation coefficient of 17.5%. The lowest values were observed in the western part in the relatively high zones (Fig. 25), the highest contents of organic carbon were found at the lowest point of the N–S oriented depression. There is a significant relation between the morphology and the measured contents of organic carbon. In the depression higher values are observed than on the elevated ground. This may be explained by lateral movements of the water including its contents of dissolved and suspended organic substances. The terbuthylazine residues measured in the uppermost 5 cm vary between 11 µg/kg and 179 µg/kg, with an average of 71.1 µg/kg, a standard deviation of 34.6 µg/kg

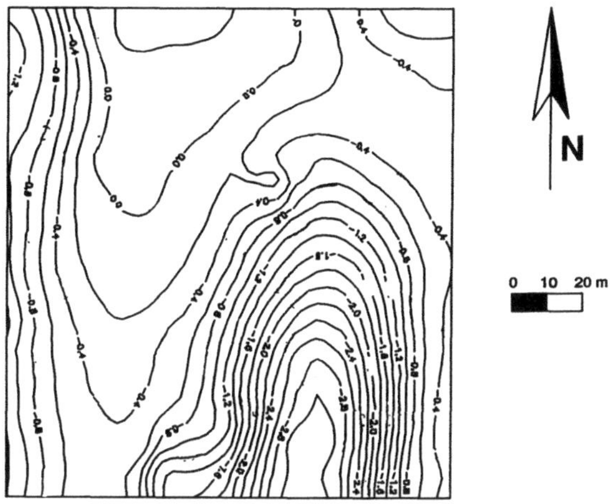

Fig. 24. Morphology of the test site Ruhwinkel [152] (reprinted with permission of the author)

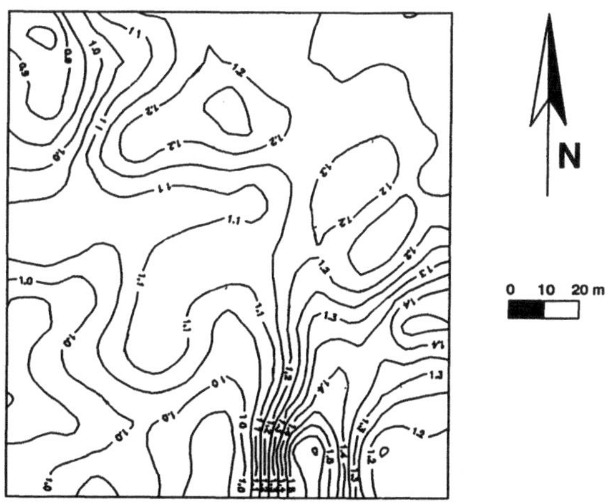

Fig. 25. Contents of organic bound carbon at the test site Ruhwinkel [152] (reprinted with permission of the author)

and a variation-coefficient of 48.6%. The highest terbuthylazine residues are found at the eastern and northern boundary of the test field (Fig. 26). Toward the west the contents decrease following the morphology. There are no significant relations between the terbuthylazine residues and the sedimentological parameters, water content or organic carbon [152]. To assess the effect of these field observation on the model results, a model calculation with PRZM was performed using the mean values of the hydraulic and sorption parameter as well as their standard deviations (field capacity $16.2 \pm 2.1\%$, wilting point

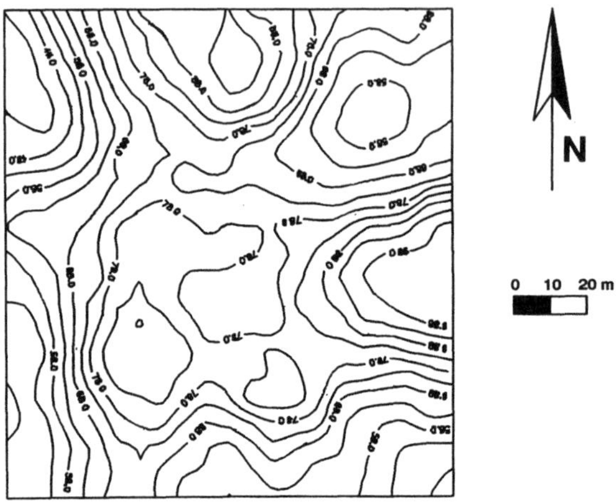

Fig. 26. Contents of terbuthylazine in soil samples at the test site Ruhwinkel [152] (reprinted with permission of the author)

6.9 ± 1.2%, K_d 5.0 ± 2.5 cm^3/g) and assuming two half-lives (70 d, 140 d). If the worst-case standard deviation values (field capacity 14.1%, wilting point 8.1%, K_d-value 2.5 cm^3/g) are taken, the resulting annual leached masses are about 8 or 9 times higher than the masses calculated on the basis of the mean values of the controlling parameters [152]. This illustrates the uncertainty of such a prognostic calculation.

3.7 Volatilization and Gas Exchange

Significant amounts of the pesticide agents spread on soils are lost by vapourisation (volatilization) either directly from the soil surface into the free atmosphere or indirectly by gas exchange between ground air and free atmosphere.

The dispersal of many pesticides into the environment is strongly controlled by volatilization, which cause pesticide losses immediately after their application, often faster than by chemical degradation. In a general way, the total seasonal losses in runoff rarely exceed 5 or 10% of the total applied and the fraction removed by leaching is probably much less. However, volatilization losses up to 80 to 90% have been measured within a few days after application, depending on weather conditions and management practices [153].

The volatilization of residues into the atmosphere involves the evaporation of pesticide molecules into the air from the residues present on soil or plant surfaces, and the dispersion of the resulting vapour into the overlying atmosphere by diffusion and turbulent mixing. The first process represents a phase change into vapour from the liquid or solid state, which can be interpreted in terms of the thermodynamic activity or fugacity of the pesticide agent. The vapour dispersion into the atmosphere is the same mechanism that controls the transfer of water, CO_2, and other gases between soil and plant surfaces and the atmosphere, however the concentration of pesticide vapour in the air more than 2 or 3 m above the ground is virtually zero, thus no significant downward "return flow" occurs [153].

The flow of air over plant and soil surfaces is turbulent except in a 1 mm thick layer in which the flow may be considered as laminar and the dispersal of pesticide vapour can be described by the molecular diffusion coefficient of the vapour. Outside this zone, where the flow in the overlying atmosphere is turbulent, dispersal is to be described in terms of the eddy diffusion coefficient. The dispersion equations for all the environmental phases are additive and fluxes can be described in terms of similar potential gradients throughout the entire environmental system, using the fugacity concept [154, 155]. Here, the vapour pressure of the pesticide in the gas phase is related to the concentration in solution by the use of the Henry equation and to characterize residues adsorbed on soil or plant surfaces by using adsorption isotherms stated in terms of vapour pressures. Thus pesticide volatilization can be simplified to a single flowpath through a series of barriers each with a characteristic, but additive, resistance [153].

The vapour pressures of pesticides range over six orders of magnitude (Table 1). The tabulated values may suggest that the volatilization rates of the compounds are low. However, because the background vapour pressures of pesticides in the air are essentially zero, the residues will evaporate as if they were in a vacuum. Field experiments reveal a wide range of the observed volatilization rates from losses of 70% within a few hours to small losses over weeks or months [153].

In investigations of the transport of vapourized pesticide agents in the subsurface one has to consider the interference of solid material and of water which reduce the fluxes. The exchange between water-unsaturated and saturated zones and the atmosphere has two boundaries which separate the ground water, the ground air and the free atmosphere. The gas transport in the ground air is controlled mainly by diffusion and it is furthermore influenced by thermal and barometric processes [11, 22].

For a porous subsurface material the diffusion rate of a gas in the ground air (and in the ground water) can be described by a modified form of the Fickian transport equations for steady state conditions

$$Q = D^* \cdot \frac{\delta c}{\delta x} \tag{25}$$

and for non-steady state conditions

$$Q = D^* \cdot \frac{\delta^2 c}{\delta x^2} \tag{26}$$

with the amount of gas Q transported per unit of time and area, the gas concentration c, the diffusion distance x and the diffusion coefficient adjusted to the porous medium D^*. D^* is the product of the diffusion coefficient in air (or in water, respectively), the effective porosity n_e and the tortuosity f, which considers the irregularities in the pathways within the pore system. In the water-unsaturated zone D^* is furthermore affected by the water content [156]. In aqueous solutions the diffusion coefficients are generally three orders of magnitude lower than in gas-filled voids.

If the gas transfer at the boundary water – ground air is controlled by diffusion, it can be described by the Fickian equations (25, 26). The equilibrium relation between the concentration of a volatile substance in the gas phase c_g (mg/l air) and in the aquatic phase c_w (mg/l water) is defined by a partition coefficient, which can be experimentally measured or calculated with Eq. (27) [157]:

$$\frac{c_g}{c_w} = \frac{16.04 \cdot P \cdot M}{T \cdot S} \tag{27}$$

with the vapour pressure P (in hPa), the molar mass M (in g/mol), the temperature T (in K) and the solubility in water S (in mg/l).

4 Consequences

The results and hydrologic considerations presented here lead to the following conclusions:

- The protective effect of the subsurface passage is mainly due to dissolution–precipitation, sorption–desorption, degradation, and transport processes. These processes depend on the hydrogeologic, geochemical and biological properties of the respective site and pesticide. The dominating criteria are the degradation and sorption behaviour of the specific substance [158].
- Below the humics-rich, biologically active soil zones the degradation and sorption processes are significantly less effective. The observations indicate, that some agents which have passed the soil zone and reached the ground water, are transported with the same velocity as the water itself [158]. This statement is supported by the results of Enfield et al. [145], Harkin et al. [159], Jones [160], Jones et al. [161], Pestemer and Nordmeier [162], Priddle et al. [163] and Schneider et al. [164].
- The practical application of basic knowledge for the prognostic assessment of the subsurface behaviour of pesticides with respect to a ground water contamination can only exceptionally be successful because important data are missing. The main reason is that the hydraulic and physico-chemical parameters controlling the transport are inhomogeneously distributed in space and – partially – in time. This can be shown by geostatistical studies indicating significant deviations from the model assumptions of horizon-specifically different, however in itself homogeneous physical, chemical and structural

properties. The variances of the controlling parameters, observed in the field, induce non-predictable processes as the above mentioned preferential flow and capillary barrier effects and cause inevitably a diminution of the level of confidence for the prognosis [158].

- The evaluation of the recent knowledge on the subsurface fate of pesticides exhibits the most important gaps in the field of metabolites, which may reach the ground water. There are only a few analytical field data of metabolites available. Albeit the most toxicologically relevant metabolites are known, not all metabolites of all pesticides are identified; thus a thorough toxicological assessment is still not possible [158]. With respect to the necessary, future investigations, it has to be considered that pesticides are usually applied in concentrations of ppm, thus, the concentrations of the metabolites will be extremely small. Their identification and the investigation of their environmental behaviour cannot be achieved by standard methods, but require specific experiments and model approaches [91].
- Pesticides mainly endanger shallow ground water bodies. Especially in dug wells and shallow drilled wells, which use shallow ground water, an increasing deterioration is to be expected with time. The deep wells usually pump their water from unconsolidated and consolidated aquifers with mean residence times of tens to hundreds of years or even more. They will produce at least in a median time range ground water free of pesticides, even when the spectrum of residence times of their pumped water will include successively younger, originally pesticide-containing seepage and ground water [158].
- The prohibition of a pesticide does not remove this agent immediately from the subsurface water. The slow water movement – especially in the water-unsaturated zone – together with the retardation by sorption/desorption processes allow to predict for persistent substances a contamination of ground water on a low trace concentration level over decades [158].
- As a consequence, the possible risk of ground water contamination by pesticides depends on the physico-chemical properties of the pesticide and on the environmental subsurface conditions. Another factor is the type of plant production in which the pesticides are used. Because of crop rotation, the same pesticides are not normally used on the same lots every year. Thus, the substances can be practically totally degraded, until they are applied again on the same lot. The conditions are quite different in special-purpose crops, for example vineyards, nurseries and ornamentals. Here, the same substances are applied every year and an accumulation of residues occurs. Especially on sandy soils, this can lead to leaching of pesticides into lower layers of the unsaturated zone and finally into the aquifer [6].
- In the Federal Republic of Germany, ground water is the most important source of drinking water supply. Therefore a good quality of drinking water can only be maintained if ground water is sufficiently protected against contamination. Ground water protection zones have been established and the guidelines for their use include graded restrictions for the use of pesticides. Those which are not considered as a danger to water are excluded only in

zone 1, an area of about 10–20 m radius around the extraction well. Pesticides which are defined as a danger to drinking water must not be used in zone 2, an area with a radius equivalent to a flow distance of 50 days around the extraction well. The licence given by the Federal Biological Center for Agriculture and Forestry (Biologische Bundesanstalt für Land- und Forstwirtschaft) for each pesticide formulation also includes specific restrictions for their use in catchment areas of waterworks; these have to be printed on the package. The drinking water regulation and the EG guidelines give an upper limit of 0.5 μg/l for total pesticide content in drinking water and of 0.1 μg/l for each individual substance. The possible use of pesticides in the catchment area of any waterworks has recently been discussed in the FRGermany. The conflicting situation between land users and waterworks may lead to stronger restrictions, based on scientific results, in the guidelines for ground water protection zones. Therefore further investigations are necessary to evaluate the wide range of environmental behaviour of pesticides [158].

5 References

1. Matthess G and Ubell K (1983) Allgemeine Hydrogeologie – Grundwasserhaushalt. 438 pp, (Gebr Borntraeger), Berlin- Stuttgart
2. Zullei-Seibert N (1990) Vorkommen und Nachweisbarkeit von Pflanzenbehandlungs- und Schädlingsbekämpfungsmittel-Wirkstoffen in Roh- und Trinkwässern der Bundesrepublik Deutschland. Dortmunder Beitr Wasserforsch 39, 102 pp, Dortmund
3. RIVM and RIZA (1991) Sustainable Use of Groundwater. Problems and threats in the European Communities, 80 pp, (RIVM) Bilthoven, (RIZA) Lelystad, The Netherlands
4. Cohen SZ, Creeger SM, Carsel RF and Enfield CG (1984) Potential pesticide contamination of groundwater from agricultural use. In: Krüger, RF and Sieber, JN (Eds.) ACS Symp Ser 259, Treatment and Disposal of Pesticide Wastes pp 297–325, Washington, DC
5. Cohen SZ, Eiden C and Lorber MN (1986) Monitoring ground water for pesticides. Amer Chem Soc Symp Ser 315, pp 170–196, Washington, DC
6. Matthess G and Isenbeck M, (1987) Pesticide behaviour in Quaternary sediments. Boreas 16, pp 411–418, Oslo
7. Jackson RE, Patterson RJ, Graham BW, Bahr J, Belanger D, Lockwood J and Priddle M (1985) Contaminant hydrology of toxic organic chemicals at a disposal site, Gloucester, Ontario. 1. Chemical concepts and site assessment NHRI Paper 23, IWD sci ser 141, 114 pp National Hydrology Research Institute, Inland Waters Directorate Ottawa, Canada
8. Leuchs W, Plöger E, Friege H, Vogt V and Obermann P (1991) Pestizide in Oberflächengewässern, Teil 2: Belastungsursachen und ihre Wirkungen auf das Grundwasser – Situation im nordwestlichen Münsterland. Vom Wasser, 76, pp 39–50, Weinheim
9. Freeze RA and Cherry JA (1979) Groundwater. 604 pp (Prentice-Hall), Englewood Cliffs, NY
10. Matthess G (1982) The Properties of Groundwater. 406 pp (Wiley) New York etc

11. Matthess G (1990) Die Beschaffenheit des Grundwassers. 2nd ed., 498 pp (Gebr Borntraeger) Berlin-Stuttgart
12. Wohnlich S (1991) Kapillarsperren – Versuche und Modellberechnungen. Schr R Angew Geol Karlsruhe, 15, 127 pp Karlsruhe
13. Eriksson E (1986) Critical loads for acid deposition on groundwater.- Nilsson J Critical Loads for Sulphur and Nitrogen. Nordisk Ministerrad, Miljö Rapp 11, pp 71–86, Stockholm
14. Nace RL (1967) Water resources: a global problem with local roots. Environ Sci Techn 1, pp 1–11, Easton, Pa
15. Parker JM and Foster SSD (1986) Groundwater monitoring for early warning of diffuse pollution. Int Assoc Hydrol Sci Publ 157, pp 37–46, Wallingford, UK
16. Korzun VI (ed, 1978) World water balance and water resources of the Earth. Stud a Rep in Hydrology, 75, 663 pp Paris (UNESCO)
17. Schmitz W (1971) Die Ausbreitung von Spaltprodukten bei Reaktorunfällen in Fließgewässern.- In: Ber Arbeitsgr "Trinkwasser-Kontamination", 15 pp 7 figs (BMI), Bonn
18. Industrie – Verband Agrar (1990) Wirkstoffe in Pflanzenschutz- und Schädlingsbekämpfungsmitteln. Physikalische und toxikologische Daten.- 2nd ed, 478 pp (BLV Publ) München
19. Stumm W and Morgan JJ (1981) Aquatic Chemistry 2nd ed, 780 pp, (Wiley) New York etc
20. MacKay DM (1980) Solubility, partition coefficients, volatility and evaporation rates. Reactions and processes. Handb environm Chem, 2 (A), pp 31–46, (Springer) Berlin
21. Thurman EM (1985) Organic Geochemistry of Natural Waters. 497 pp, (Martinus Nijhoff/Dr W Junk Publ), Dordrecht etc
22. Matthess G (1983) The role of natural organics on water interaction with soil and rock. Int Assoc hydrol Sci publ, 150, pp 11–21, Wallingford, UK
23. Huber StA (1992) Entwicklung und Einsatz eines Analyseverbundsystems zur chromatographischen Charakterisierung natürlicher organischer Wasserinhaltsstoffe PhD. thesis, 88 pp, Karlsruhe
24. Frimmel FH and Christman RF (Eds, 1988) Humic Substances and their Role in the Environment. 271 pp (Wiley), Chichester etc
25. Scheffer F and Schachtschabel P (Eds, 1992) Lehrbuch der Bodenkunde 13th ed, 491 pp (Enke) Stuttgart
26. Schnitzer M and Khan SU (Eds, 1978) Soil Organic Matter. Develop Soil Sci, 8 319 pp (Elsevier), Amsterdam-Oxford-New York
27. Stumm W (1992) Chemistry of the Solid-Water-Interface. 428 pp (Wiley) New York etc
28. Westall JC (1983) Properties of organic compounds in relation to chemical binding. Proc Biofilm Processes in Ground Water Research, Stockholm 1983, pp 65–90, Stockholm
29. Leo A, Hansch C and Elkins D (1971) Partition coefficients and their uses. Chem Rev 71, pp 525–616, Washington, DC
30. Karickhoff SW, Brown OS and Scott TH (1979) Sorption of hydrophobic pollutants on natural sediments. Water Research 13, pp 241–248, Washington, DC
31. Schwarzenbach RP and Westall J (1981) Transport of non-polar organic compounds in a river water-groundwater infiltration system: a systematic approach. Studies in Env Sci 17, pp 569–574, (Elsevier) Amsterdam
32. Lennartz B (1992) Verlagerungsverhalten ausgewählter Pflanzenbehandlungsmittel in wassergesättigten und -ungesättigten Böden. PhD thesis, 112 pp, Kiel
33. Lennartz B and Rambow J (1990) Das Verhalten organischer Chemikalien in wasserteilgesättigten Böden. Internal report Inst Wasserwirtschaft Landschaftsökologie, 51 pp, Kiel

34. Bedbur E (1993) Die Grenzen der Anwendungsmöglichkeit der Freundlich-Adsorptions-Konstanten K_{Fr} und 'N zur Bescheibung der Chemikaliensorption. Meyniana, 45, pp 1–8, Kiel
35. Rambow J (1992) Modellstudie zum Verbleib herbizider Wirkstoffe in wasserteilgesättigten Böden. PhD thesis, 108 pp, Kiel
36. Hansch C and Leo A (1979) Substituent Constants for Correlation Analysis in Chemistry and Biology, 339 pp, (Elsevier), Amsterdam
37. Briggs GG (1973) Molecular structure of herbicides and their sorption by soils. Nature 223, p 1288, London
38. Giger W, Schwarzenbach RP, Hoehn E, Schellenberg K, Schneider JK, Wasmer HR, Westall J and Zobrist J (1983) Das Verhalten organischer Wasserinhaltsstoffe bei der Grundwasserbildung und im Grundwasser. Gas-Wasser-Abwasser 63, pp 517–531, Zürich
39. Matthess G, Isenbeck M, Pekdeger A, Schenk D and Schröter J (1985) Der Stofftransport im Grundwasser und die Wasserschutzgebietsrichtlinie W 101. Umweltbundesamt, Berichte 7/85 181 pp, (E Schmidt) Berlin
40. Chiou CT, Peters LJ and Freed VH (1979) A physical concept of soil-water equilibria for non-ionic organic compounds. Sci 206, pp 831–832, Washington, DC
41. Lambert SM (1968) Omegan, a useful index of soil sorption equilibria. J Agric Food Chem, 16, pp 340–343, Washington, DC
42. Schwarzenbach RP and Giger W (1985) Behaviour and fate of halogenated hydrocarbons in groundwater. In CH Ward, Giger, W and McCarty, PL Ground Water Quality pp 446–471, (Wiley) New York
43. Swanson RA and Dutt GR (1983) Chemistry and physical processes that affect atrazine movement and distribution in soil systems. Soil Sci Soc Amer Proc 37, pp 872–876, Madison, Wis
44. Herklotz K (1985) Sorptions- und Mobilitätsverhalten von ausgewählten Pestiziden in Hausmüll, Böden und Grundwasserleitern. PhD thesis, 174 pp, Hannover
45. Pankow JF and McKenzie SW (1992) Parametrizing the equilibrium distribution of chemicals between the dissolved, solid particulate matter, and colloidal matter compartments in aqueous systems. Environ Sci Technol 25, pp 2046–2053, Washington, DC
46. Eady BJ, Morehead NR and Landrum PF (1990) Three-phase partitioning of hydrophobic organic compounds in Great Lakes waters. Chemosphere 20, pp 161–187, Oxford
47. Chin Y, Weber WJ and Eadie BJ (1990) Estimating the effects of dispersed organic polymers on the sorption of contaminants by natural solids. 2. Sorption in the presence of humic and other natural macromolecules. Environ Sci Technol 24, pp 837–845, Washington, DC
48. Voice TC, Rice CP and Weber WJ (1983) Effect of solids concentration on the sorptive partitioning of hydrophobic pollutants in aquatic systems. Envir Sci Technol 17, pp 513–518, Washington, DC
49. O'Connor DJ and Connolly JP (1980) The effect of concentrations of adsorbing solids on the partition coefficient. Water Res, 14, pp 1517–1523, Oxford
50. Johnson WE, Fendinger NJ and Plimmer JR (1991) Solid- phase extraction of pesticides from water: Possible interferences from dissolved organic material. Anal Chem 63, pp 1510–1513, New York
51. Helling CS (1971) Pesticide mobility in soils II. Application of soil thin-layer chromatography. Soil Sci Soc Amer Proc 35, pp 732–737, Madison, Wis
52. Bengtson G, Enfield, CG and Lindquist R (1987) Macromolecules facilitate the transport of trace organics. Sci Total Environ, 67, pp 159–164, Amsterdam
53. Shih CS and Gloyna EF (1967) Radioactivity transport in water. Mathematical model for the transport of radionuclides. Techn Rep 12 Atomic Energy Comm EHE-04-6702, CRWR 18
54. Matthess G, Pekdeger A and Schröter J (1988) Persistence and transport of bacteria

and viruses in groundwater – a conceptual evaluation. J Contaminant Hydrol 2 pp 171–188, Amsterdam
55. Alexander I and Seiler KP (1985) Feldversuche zur Persistenz und zum Transportverhalten von Bakterien. Versuchsfeld Dornach. Umweltbundesamt, Materialien, 2/85, pp 33–36 (E Schmidt), Berlin
56. Iwasaki T (1937) Some notes on sand filtration. Journ Am Water Works Assoc 29, pp 1591–1609, Baltimore, Md
57. Yao K-M, Habibian MT and O'Melia CR (1971) Water and waste water filtration: concepts and applications.- Environ Sci a Techn, 5 (11), pp 1105–1112, Chapel Hill
58. Matthess G, Bedbur E, Gundermann KO, Loof M and Peters D (1991) Vergleichende Untersuchungen zum Filtrationsverhalten von Bakterien und organischen Partikeln in Porengrundwasserleitern. Zbl Hygiene 191, pp 53–97, 347–395, Stuttgart-New York
59. Betz V, Fischer G, Gorbauch H, Rump HH, Scholz B, Schneider W, Golwer A, Matthess G, Pekdeger A, Herklotz H and Pestemer W (1983) Zum Verhalten von Umweltchemikalien in künstlichen Grundwassergerinnen. Fresenius Z analyt Chemie, 314, pp 464–474, Berlin
60. Kustermann H (1962) Der Einfluß der Organismen in Kiesschotterböden der Flüsse auf die Wasserversickerung.- Gas – Wasser – Abwasser, Monatsbull Schweiz Ver Gas- u Wasserfachmänner, 42, pp 25–34, 54–63, 82–83, Zürich
61. Stundl K (1965) Die Auswirkung land- und forstwirtschaftlicher Kulturmaßnahmen auf die Bodenorganismen und das Grundwasser. Wasser und Abwasser, pp 139–150, Wien
62. Schmidt KH (1963) Die Abbauleistung der Bakterienflora bei der Langsamsandfiltration und ihre Beeinflussung durch die Rohwasserqualität und andere Umwelteinflüsse. Veröff Inst Wasserforsch (5), 1, 170 pp, Dortmund
63. Hurle K and Walker A (1980) Persistence and its prediction. In RJ Hance (ed) Interactions between Herbicides and the Soil, pp 83–122, (Academic Press) London etc
64. Hurle K (1973) Lag-Phase Studium mit DNOC und 2,4-D im Boden. Proc Europ Weed Res Counc Symp Herbicides and the Soil, pp 151–162, Oxford
65. Hance RJ and McKone CE (1971) Effect of the concentration on the decomposition rates in soil of atracine, linuron and picloram. Pest Sci, 3, pp 31–34, London
66. Hamaker JW (1972) Decomposition: quantitative aspects. In Goring, CAJ and Hamaker, JW (eds) Organic Chemicals in the Soil Environment, pp 253–340, (Dekker), New York
67. Scharpenseel HW, Stephan S, Theng B, Kruse E and Lay A (1977) Infiltration and Translokation von polychlorierten Biphenylen in natürlich gelagerten Bodenprofilen; biotischer und abiotischer Abbau. 1. Adsorption und Einbau polychlorierter Biphenyle (PCB) im Boden. Z Pflanzenernähr Bodenk 140, pp 285–301, Weinheim
68. Choudhry GG, Roof AAM and Hutzinger O (1979) Mechanisms in sensitized photochemistry of environmental chemicals. Toxicol Environ Chem 2, pp 259–302, New York
69. Choudhry GG and Webster GRB (1985) Protocol guidelines for the investigations of photochemical fate of pesticides in water, air and soils. Residue Reviews, 96, pp 79–136
70. Howard PH, Saxena J and Sikka H (1978) Determining the fate of chemicals. Environ Sci Technol 12, pp 398–409, Washington, DC
71. Choudhry GG (1984) Humic Substances. Part II. Photophysical, photochemical and free radical characteristics. Toxicol Environ Chem 4, pp 261–288, New York
72. Lemaire J, Campbell I, Hulpke H, Guth JA, Merz W, Philip J and von Waldow C (1982): An assessment of test methods for photodegradation of chemicals in the environment. Chemosphere 11, pp 119–164, Oxford

73. Guth JA (1981) Experimental approaches to studying the fate of pesticides in soil. In: Hutson, DH and Roberts TR (Eds) Progress in Pesticide Biochemistry, Chap 2, pp 84–114, (Wiley), New York
74. Choudhry GG, Roof AAM and Hutzinger O (1982) Photochemistry of halogenated benzene derivatives. Part 2. Photoreactions of α-substituted *p*-chlorotoluenes. J Chem Soc Perkin Trans I, pp 2957–2961, London
75. Roof AAM (1982) Aquatic Photochemistry In Hutzinger O (ed) The Handbook of Environmental Chemistry. Vol 2, Part B, pp 43–72, (Springer) Berlin-New York
76. Cooper WJ and Zika RG (1983) Photochemical formation of hydrogen peroxide in surface and ground waters exposed to sunlight. Sci 220, pp 711–712, Washington, DC
77. Draper WM and Crosby DG (1983) The photochemical generation of hydrogen peroxide in natural waters. Arch Environ Contam Toxicol 12, pp 121–126, New York
78. Zafiriou OC, Joussot-Dubien J, Zepp RG and Zika RG (1984) Photochemistry of natural waters: Many compounds and environments are affected by sunlight-induced photochemistry. Environ Sci Technol, 18, pp A358–A371, Washington, DC
79. Mansour M, Feicht E and Meallier P (1989) Improvement of the photostability of selected substances in aqueous medium. Toxicol Environ Chem 20/21, pp 139–147, New York
80. Draper WM and Crosby DG (1984) Solar photooxidation of pesticides in dilute hydrogen peroxide. J Agric Food Chem 32, pp 231–237, Washington, DC
81. Mansour M, Hustert K, Moza PN and Kettrup A (1989): Photochemischer Abbau von chlorierten aromatischen Verbindungen durch Sauerstoffspezies in aquatischen Systemen. VDI-Berichte, 745, pp 927–936, Düsseldorf
82. Smith JH, Mabey WR, Bohonos N, Holt BR, Lee SS, Chou TW, Bomberg DC and Mill T (1977) Environmental pathways of selected chemicals in freshwater systems. I: Background and experimental procedures. US EPA-600/7-77-113, Athens, GA
83. Choudhry GG, van den Broeck JA and Hutzinger, O (1983) Formation of polychlorodibenzofurans (PCDFs) by the photolysis of polychlorobenzenes in aqueous acetonitrile containing phenols. Chemosphere, 12, pp 487–492, Oxford
84. Armstrong DE and Konard JG (1974) Non-biological degradation of pesticides.- In WD Guenzi et al (eds) Pesticides in Soil and Water, pp 123–131, Madison, Wis
85. Voudrias EA and Reinhard M (1986) Abiotic reactions at mineral surfaces. In JA Davis and Hayes KF (Eds.) Geochemical Processes at Mineral Surfaces, Amer Chem Soc Symp Ser, 323, pp 462–486, Washington, DC
86. Helling CS, Kearney PC and Alexander M (1971) Behaviour of pesticides in soils. Adv Agron, 23, pp 147–240, New York-London
87. Fritz R and Spiteller M (1993) Metabolismusforschung für den Umweltschutz.- Labor-Praxis, Sonderh Labor 2000-Umwelt im Blickpunkt, pp 20–31, (Vogel) Wüzburg
88. Zepp RG and Cline DM (1977) Rates of direct photolysis in aquatic environment. Environ Sci Technol, 11, pp 359–366, Washington, DC
89. O'Brein RD (1960) Toxic phosphorus esters. 434 pp, (Academic Press) New York
90. Alexander M (1969) Microbial degradation and biological effects of pesticides in soil. Natural resources research 9, pp 209–240 (UNESCO) Paris
91. Häfner M (1989) Wichtige Aspekte zum Schutz des Grundwassers vor Pflanzenschutzmittel-Rückständen – dargestellt am Beispiel der Chlortriazine Atrazin, Simazin und Terbuthylazin. In Milde G and Müller-Wegener U (eds) Pflanzenschutzmittel und Grundwasser-Bestandsaufnahme, Verhinderungs- und Sanierungsstrategien, pp 261–294, (Fischer) Stuttgart-New York
92. Alexander M (1977) Introduction to Soil Microbiology. 467 pp, (Wiley), New York
93. Wallnöfer PR and Engelhardt G (1989) Microbial degradation of pesticides. In Chem Plant Protection 2, pp 1–115 (Springer) Berlin-Heidelberg

94. Zoeteman BCJ, Greef Ede and Brinkman FJJ (1981) Persistency of organic contaminants in groundwater, lessons from soil pollution incidents in the Netherlands. Studies in Env Sci 17, pp 465–480, Amsterdam (Elsevier)
95. Dictor MC, Soulas G, Tagaki K, Anderson JPE and Lewis F (1992) Subsoil microbiology: microbial activity and potential for pesticide degradation in the unsaturated zone. In Anderson JPE, Arnold DJ, Lewis F and Thorstenson L (eds) Proc int symp on Environmental Aspects of Pesticide Microbiology, pp 284–290 (Dpt Microbiol Swed Univ Agric Sci) Uppsala
96. Simon L, Spiteller M, Haisch A and Wallnöfer PR (1992) Influence of soil properties on the degradation of the nematocide fenamiphos. Soil Biol Biochem 24, pp 769–773
97. Scholz K, Fritz R, Hellpointner E and Spiteller M (1992) The lysimeter facility at the Crop Protection Research Center, Bayer AG Monheim, Germany. BCPC Mono 53, pp 55–64
98. Althaus H, Jung KD, Matthess G and A Pekdeger (1982) Lebensdauer von Bakterien und Viren in Grundwasserleitern. Umweltbundesamt, Materialien 1/82 190 pp (Schmidt), Berlin
99. Chapelle FH (1992) Ground-Water Microbiology and Geochemistry. 424 pp, (Wiley) New York etc
100. Fliersmans CB and Hazen TC (1991) Microbiology of the deep Subsurface. Proc Ist Int Symp 7 Chaps
101. Matthess G, Frimmel F, Hirsch P, Schulz HD and Usdowski HE (1992) Progress in Hydrogeochemistry – Organics – Carbonate Systems – Silicate Systems – Microbiology – Models.- 544 pp, (Springer), Berlin etc
102. Simon L, Spiteller M and Wallnöfer PR (1992) Metabolism of fenamiphos in 16 soils originating from different geographic areas. J agricult a food Chem, 40, pp 312–317, Washington, DC
103. Kuznetsov SJ (1962) Geologic activity of mikroorganisms. Transact, Institute Microbiology 9, 12 pp New York
104. Hirsch P and Rades–Rohkohl E (1983) Microbial diversity in a groundwater aquifer in Northern Germany. Develop industr Microbiol, 24, pp 183–200, Washington, DC
105. Wilson JT and McNabb JF (1983) Enumeration and characterization of bacteria indigenous to a shallow water-table aquifer. Groundwater 21, pp 134–142, Worthington, Ohio
106. Wuhrmann K (1980) Mikrobiologische Aspekte der Gewinnung und Aufbereitung und Verteilung von Trink- und Brauchwasser. Swiss Food 2, pp 15–23, Zürich
107. McCarty PL (1983) Biofilm transformations of trace organic compounds in groundwater. Proc Biofilm Processes in Ground Water Res, Stockholm 1983, pp 91–112, Stockholm
108. McCarty PL, Reinhard M and Rittmann BE (1981) Trace organics in groundwater. Environ Sci Techn, 15, pp 40–51, Washington, DC
109. Kaufman DD, Kearney PC, Von Endt, DW and Miller DE (1970) Methylcarbonate inhibition of phenyl-carbamate metabolism in soil. J Agr Food Chem, 18, pp 513–519, Washington DC
110. Hance RJ (Ed, 1980) Interactions between Herbicides and the Soil. 349 pp, (Economic Press) London etc
111. Helweg A (1975) Degradation of ^{14}C-labelled maleic hydrazide in soil as influenced by sterilization, concentration and pretreatment. Weed Research 15, pp 53–58, Oxford
112. Simon L, Spiteller M, Haisch A and Wallnöafer, PR (1990) Metabolismus von Pflanzenschutzmitteln in verschiedenen Böden – eine Literaturstudie. Pflanzenschutz-Nachr Bayer 43, pp 111–139
113. Hurle K (1981) Untersuchungen zum Abbau von Herbiziden in Böden. Acta Phytomedica, 8, 120 pp, Berlin-Hamburg

114. Kearney PhC (1982) IUPAC Pesticide commission report. J Assoc off anal Chem, 65, pp 1030–1032, Washington, DC
115. Wheatley GA, Suett DL and Hardman JA (1972) Some factors influencing the persistence of chlorfenvinphos in soil.- In AS Tahori (Ed.) Fate of Pesticides in the Environment, Proc 2nd IUPAC Congr Pesticide Chemistry, 6, pp 77–85, (Gordon a Breach) New York
116. Eichelberger JW and Lichtenstein JJ (1971) Persistence of pesticides in river water. Environ Sci Technol 5, pp 541–544, Washington, DC
117. Walker A (1973) Use of a simulation model to predict herbicide persistence. J Environ Qual 3, pp 396–401, Madison, Wis
118. Walker A (1978) Simulation of the persistence of eight soil- applied pesticides. Weed Research 18, pp 305–313, Oxford
119. Bear J and Verrujt A (1987) Modelling Groundwater Flow and Pollution. 414 pp Dordrecht, Boston, Lancaster, Tokyo
120. Bear J (1972) Dynamic of Fluids in Porous Media, 764 pp, (Elsevier), New York
121. Behrens H and Seiler KP (1982) Beziehungen zwischen der Dispersivität und der Länge des Fließweges des Grundwassers nach Geländeversuchen in den glazio-fluviatilen Kiesen des oberen Loisachtales (Oberau) und von Dornach (Münchener Schotter Ebene). In Beiträge über hydrologische Tracermethoden und ihre Anwendung, GSF-Bericht R290, Neuherberg
122. Gelhar LW and Axness CL (1983) Three-dimensional stochastic analysis of macrodispersion in aquifers. Water Resourc Res 19, pp 161–180, Washington, DC
123. Gelhar LW, Gutjahr AL and Naff RL (1979) Stochastic analysis of macro dispersion in a stratified aquifer. Water Resourc Res 15, pp 1387–1397, Washington, DC
124. Pickens JF and Grisak GE (1981) Scale-dependent dispersion in a stratified granular aquifer. Water Resourc Res 17 (4), pp 1191–1211, Washington, DC
125. Schroeter J (1983) Der Einfluß der Textur- und Struktureigenschaften poröser Medien auf die Dispersivität. PhD thesis, 152, pp, Kiel
126. Schroeter J (1984) Mikro- und Makrodispersivität poröser Grundwasserleiter. Meyniana, 36, pp 1–34, Kiel
127. Bertsch W (1978) Die Koeffizienten der longitudinalen Dispersion – ein Literaturüberblick.- Deutsche gewässerkundliche Mitt, 22 (2) pp 37–46, Koblenz
128. Hull DE (1962) Dispersion and persistence of tracers in river flow measurements. Int J Appl Rad Isot, 13, pp 63–73, New York
129. Van Genuchten MTH and Wierenga PJ (1976) Mass transfer studies in sorbing porous media I. Analytical solutions. J Soil Sci Soc Amer 41, pp 272–278, Madison, Wis
130. Van Genuchten MTH (1981) Non-equilibrium transport parameters from miscible displacement experiments. Res Rep 119, US Salinity Lab Riverside, Ca
131. Parker H and Van Genuchten MTH (1984) Determining transport parameters from laboratory and field tracer experiments.- Bull 84–3, Virginia Agr Exp Stat, Blacksburg, Va
132. Nicholls PH, Walker A and Baker RJ (1982) Measurement and simulation of the movement and degradation of atrazine and metribuzine in a fallow soil. Pesticide Sci 2 (22), pp 123–137, London
133. Wagnet RJ and Hutson JL (1989) LEACHM: A process based model of water and solute movement, transformations, plant uptake and chemical reactions in the unsaturated zone. Vol 2, Version 2, 148 pp, (Centre Environm Res, Cornell Univ), Ithaca, NY
134. Carsel RF, Smith CN, Mulkey LA, Dean JD and Jobise PP (1984) User's manual of the pesticide root zone model (PRZM): Release 1. EPA-600/3-84-109 US EPA, Athens, Ga
135. Dibbern H (1992) Zur Simulation des Ausbreitungsverhaltens der Pflanzenschutzmittel Atrazin, Chlortoluron, Isoproturon, Lindan und Terbuthylazin im Boden und Grundwasser. PhD thesis, Ber Rep Geol Paleont Inst 49 102 pp, Kiel

136. Boesten JJTI (1986) Behaviour of Herbicides in Soil Simulation and experimental Assessment PhD thesis, 263 pp, Wageningen
137. Knisel WG, Leonhard RA and Davis FM (1989) GLEAMS User Manual. 35 pp, (Southern Watershed Res Lab), Tifton, Georgia
138. Bonazountas M, Wagner J and Little AD (1984) SESOIL: A seasonal soil compartment model. Off Toxic Substances, EPA, Washington, DC
139. Reichling J (1991) Mobilität und Persistenz der Herbizide Chlortoluron, Terbuthylazin und Pendimethalin in einem porösen Grundwasserleiter.- PhD thesis, 93 pp, Kiel
140. Reichling J, Bedbur E, Champ D and Matthess G (1992) Mobility and persistence of three herbicides under groundwater conditions. In Hötzl H and Werner A (Eds) Tracer Hydrology, pp 165–171, (Balkema) Rotterdam
141. Dean JD, Strecker EW, Salhotra AM and Mulkey LA (1987) Exposure assessment for the pesticide aldicarb in Florida, USA.- In Van Duijvenbooden, W and Van Waegeningh HG (Eds) Vulnerability of Soil and Groundwater to Pollutants. Proc and Inf 38, pp 983–993, Den Haag (TNO)
142. Dean JD, Huyakorn PS, Donigian AS, Voos KA, Schanz RW, Meeks YJ and Carsel RF (1989a) Risk of unsaturated/saturated transport and transformation of chemical concentrations (RUSTIC) Vol I, Theory and code verification EPA/600/3-89/048a, 203 pp, US EPA, Athens, Ga
143. Dean JD, Huyakorn PS, Donigian AS, Voos KA, Schanz RW, Meeks YJ & Carsel RF (1989b) Risk of unsaturated/saturated transport and transformation of chemical concentrations (RUSTIC) Vol II, User's manual.- EPA/600/3-89/048b, 340 pp US EPA, Athens, Ga
144. Enfield CG, Carsel RF, Cohen SZ, Phan T and Walters DM (1982) Approximating pollutant transport to groundwater. Ground Water 20, pp 711–721, Worthington, Ohio
145. Enfield CG, Carsel RF and Phan T (1981) Comparison of a one-dimensional, steady state hydraulic model with a two-dimensional, transient hydraulic model for aldicarb transport through soil. In van Duijvenbooden W, Glasbergen P and van Lelyfeld H (Eds) Stud Environ Sci 17, pp 507–510, (Elsevier) Amsterdam
146. Jury WA, Spencer WF & Farmer WJ (1983) Behaviour assessment model for trace organics in soil: model description. J Environ Qual 12, pp 558–564, Madison, Wis
147. MacKay DM, Roberts PV and Cherry JA (1985) Transport of organic contaminants in ground water.- Environ Sci Technol 19, pp 384–392, Washington, DC
148. Oliver GR and Laskowsky DA (1986) Development of environmental scenarios for modeling the fate of agricultural chemicals in soil. J Environ Qual 12, pp 553–558, Madison, Wis
149. Timmermans GHF, Hendriks CA, Van't Klooster HA, Uffink GJM and Van der Gaast NG (1986) FLODIN, a computer program for prediction and description of the transport of hydrophobic organic pollutants in soil. H_2O (19) 1986, 16, pp 374–379, Rotterdam
150. Van Genuchten MTH, Davidson JM and Wierenga PJ (1974) An evaluation of kinetic and equilibrium equations for the prediction of pesticide movement through porous media. J Soil Sci Soc Amer 38, pp 29–34, Madison, Wis
151. Velikov BL and Al Mohammed Y (1987) Laboratory investigations on the migration of pesticides in porous medium. In Van Duijvenbooden W and Van Waegeningh HG (Eds) Vulnerability of soil and groundwater to pollutants. Proc and Inf 38, pp 849–861, Den Haag
152. Gass M (1993) Feld- und Modelluntersuchungen zum Transportverhalten von Terbuthylazin im Wasser-ungesättigten Bereich glazifluvialer Ablagerungen. PhD thesis, 86 pp, Kiel
153. Taylor AW and Spencer WF (1990) Volatilization and vapor transport processes. In Cheng HH (Ed) Pesticides in the Soil Environment: Processes, Impacts and Modeling, SSSA book ser 2, pp 213–269, Madison, Wis

154. Lewis GN (1901) The law of physico-chemical change. Proc Amer Acad, 37, p 49; Das Gesetz physiko-chemischer Vorgänge. Z phys Chem, 38 pp 205–226, Leipzig
155. MacKay DM and Paterson S (1981) Calculating fugacity. Environ Sci Technol 5, pp 1006–1014, Washington, DC
156. Albertsen M (1977) Labor- und Feldversuche zum Gasaustausch zwischen Grundwasser und Atmosphäre über natürlichen und verunreinigten Grundwässern PhD thesis, 145 pp, Kiel
157. Dilling WL (1977) Interphase transfer processes. II. Evaporation rates of chloro methanes, ethane, ethylenes, propanes, and propylenes from dilute aqueous solutions – Comparison with theoretical predictions. Environ Sci Techn, 11, pp 405–410, Washington, DC
158. Matthess G, Bedbur E, Dibbern H, Hurle K and Pestemer W (1993) Hydrochemische Bewertung und modellhafte Auswertung des Transportverhaltens von PSM im Untergrund. DVGW Schr (Wasser), 109, pp 105–110 Eschborn
159. Harkin JM, Jones FA, Fathulla RN, Dzantork EK and Kroll DG (1986) Fate of aldicarb in Wisconsin groundwater. In Evaluation of Pesticides in Groundwater, ACS Symp Ser 315, pp 219–255, Washington, DC
160. Jones RL (1986) Field, laboratory and modelling studies on the degradation of aldicarb residues in soil and ground water. In Evaluation of Pesticides in Groundwater, ACS Symp Ser 315, pp 197–218, Washington, DC
161. Jones RL, Hornsby AG, Rao PSC and Andersen MP (1987) Movement and degradation of aldicarb residues in the saturated zone under citrus groves on the Florida ridge. J Contam Hydrol, pp 265–287, New York
162. Pestemer W and Nordmeyer H (1990) Modelluntersuchungen mit ausgewählten Pflanzenschutzmitteln im Bodenprofil im Hinblick auf die Beurteilung einer Grundwasserbelastung. Mitt Biol Bundesanstalt 259, 80 pp, Braunschweig
163. Priddle MW, Jackson RE, Novakowski KS, Denhoed S, Graham BW, Patterson RJ, Chaput D and Jardine D (1987) Migration and fate of aldicarb in the sandstone aquifer of Prince Edward Island. Water Poll Res J Canada, 22, pp 173–185, Burlington, Ont
164. Schneider AD, Wiese AF and Jones OR (1977) Movement of three herbicides in a fine sand aquifer. Agron J 69, pp 432–436, Madison, Wis

Pesticides in Soil: A German Approach of Predicting their Movement into Ground and Surface Water

M. Häfner

Landesanstalt für Pflanzenschutz, Reinsburgstr. 107, D-70197 Stuttgart, FRG

In using chemical pesticides, water protection is an essential priority. The quantification of the potential ground water pollution from using pesticides in the Federal Republic of Germany is included as early as in the registration procedure. Of central importance are model tests in the laboratory using soil columns, lysimeter tests, and calculations with the help of computer-aided simulation models. In addition, results of other techniques and methodical tests are taken to quantify the potential risk of ground water pollution by pesticides within the registration procedure. In this context it can be concluded from the results collected so far that possible danger to ground water from applying pesticides can be predicted with the help of model tests in the laboratory using soil columns and with controlled drainage water tests.

Chemistry of Plant Protection Vol. 9

1 Introduction

Economic production in the cultivation of vegetables and fruits as well as in agriculture cannot be achieved without chemical pesticides. The use of pesticides not only guarantees nourishment but also contributes to maintaining the high quality standards of food from plants in industrialized countries. At the same time, pesticides have an important function in preserving the profitability of the production units and with this, their ability to compete.

However, the use of pesticides is controversial. There have been heated discussions about the question of possible damage to health and environmental pollution resulting from the use of pesticides. Since the beginning of the 1980s, the headlines have been filled more and more with the pollution of ground and surface water by pesticides.

The European Community also contributed to the discussion about pesticide residues in ground and drinking water in 1980. Its guideline "About the Quality of Drinking-Water for Human Use" [1] lists the maximum amounts of pesticides in drinking water. The limit for an active ingredient of a pesticide including its main toxic metabolites for the single substance is 0.1 μg per litre and the sum of those substances the limit is 0.5 μg per litre. The translation of these into national law was carried out in the Federal Republic of Germany in the directive for drinking water from May 22nd, 1986 [2]. At this time, the revision of analytical methods was not yet complete, so the limiting values were not enforced and did not take effect until October 1st, 1989. These legal limits for pesticides in the Federal Republic of Germany were also included in the new version of the directive for drinking water from December 1990 [3].

With regard to the directive on drinking water of 1986, many samples of ground and drinking water were analysed for residues of pesticides in the Federal Republic of Germany by laboratories from universities, governmental agencies, industry and in private and communal laboratories. As a result, a great amount of data accumulated in the middle/at the end of the 1980s which contained a lot of positive results for pesticides.

Organizations from the field of water supply and environmental protection collected the positive results of these laboratories in lists without interpreting the individual results and without commenting on the analytical method that was used. These lists – continually brought up to date and with new reports of positive results of residues of pesticides being added – were widely distributed and the public was under the impression that any pesticide could be found in any concentration when analysing drinking water and ground water. Table 1 shows some of the residues of pesticides that were displayed in such lists as positive results.

The published positive results about pesticides in ground and drinking water instigated a number of activities concerning pesticides. In this connection, some research work was carried out and the Federal Republic of Germany revised and completed the registration procedures for pesticides. The Landesanstalt für

Table 1. Examples for positive results of pesticides in ground and drinking water, summarized in a list of the Bundesverband der deutschen Gas- und Wasserwirtschaft (Federal Association of the German Gas and Water Industry) [4]

Atrazine	MCPA
Azinphos-ethyl	Mecoprop
Bentazone	Metazachlor
Bromacil	Methabenzthiazuron
Chlorfenvinphos	Metobromuron
Chloridazon	Metolachlor
Chlortoluron	Metoxuron
Clopyralid	Pendimethalin
2,4-D	Propazine
Dimethoat	Pyrdiate
Dinoseb acetate	Simazine
Isoproturon	2,4,5-T
Lindane	Terbuthylazine

Pflanzenschutz Stuttgart (State Institute for Plant Protection) carried out a research project from 1987–1992 which was sponsored by the Bundesministerium für Forschung und Technologie (Federal Ministry for Research and Technology). It was the aim of this research project to evaluate soil as a "filter and buffer" for pesticides and such to make clear statements about the danger to ground water through these substances.

2 Methods for Evaluating the Mobility of Pesticides in Soils

The degree of mobility of a pesticide in soil and with this the potential risk of ground water pollution is a result of the chemical structure of the pesticide as well as the local structure of the ground water region.

The potential risk of ground water pollution by pesticides is evaluated by six methods described in Table 2. The most important authorization procedures for pesticides in the Federal Republic of Germany are soil column tests and recently lysimeter tests. In addition, models like the programme "Pesticide Root Zone Model (PRZM)" are used to classify the danger to ground water. Simultaneously, physico-chemical characteristics of these substances such as water solubility, hydrolysis constants and K_{oc} values as well as biochemical data – for instance the decomposition speed of pesticides in different soils – are considered in the final evaluation of the potential risk of ground water pollution by pesticides.

With regard to the results of the extensive studies, pesticides which show a potential of risk to ground water pollution have a water protection regulation imposed on them in the Federal Republic of Germany. The water protection

Table 2. Evaluation models on mobility of pesticides in soils

1. Soil column tests based on the guidelines for the official test of pesticides part IV 4-2 of the Federal Biological Research Centre for Agriculture and Forestry (leaching behaviour of plant protection products) [5].
2. Lysimeter studies [6].
3. Calculation models such as the programme "Pesticide Root Zone Model (PRZM)" [7] or "Seasonal Soil Compartment Model (SESOIL)" [8].
4. Analyses of drainage water.
5. Core drilling and installation of variously developed water-gauges. The soil cores are analysed for pesticides. In addition, monitoring is carried out using water taken at specific intervals.
6. Ground water monitoring.

regulation restricts the use of these pesticides in catchment areas of water works taking groundwater and springwater, mineral springs and drinking water reservoirs as well as in other ground water sensitive districts. The Federal Biological Research Centre for Agriculture and Forestry (BBA) in Braunschweig differentiates between W1 and W2 regulations [9] and combines the latter with legally defined water protection areas according to the guidelines of the Deutschen Vereins des Gas- und Wasserfaches e.V. (DVGW) (German Association of the Gas and Water Industry) [10].

The criteria for a legal definition according to the guidelines of the DVGW as well as the definitions of W1 and W2 regulations are shown in Fig. 1. The Federal Republic of Germany has authorized 216 active ingredients for pesticides with 33 having been made subject to a water protection regulation by March 11, 1993. Those active ingredients with a water protection regulation are shown in Table 3 [11].

The great number of authorized or formerly authorized pesticides, including their metabolites, as well as the complexity of the soil system caused by several reciprocally influencing parameters, demand systematic planning and mode of working for evaluating the mobility of pesticides in soils. Because pesticides are transported with the water flow through the "solid state" soil during which reactions between pesticides and parts of the soil may occur, mobility studies of pesticides in soils are treated as a chromatographic system where solid particles are the stationary phase and running water the mobile one.

Each soil constituent shows different sorption activity. There is high activity with clay, fine silt and humus-particles in the soil. At the same time, the eluting power of the water may vary depending on its salt content. A rule of thumb is that polarity and with this the eluting power increases with increasing salt content.

The balance of contact surface reactions in soils can be described by the classical laws of adsorption, the Freundlich adsorption isotherm, the Langmuir adsorption isotherm and the Brunauer/Emmett/Teller equation.

Since the 1960s, the complex correlations of movements of pesticides in soils have been investigated more and more. In 1970, G.W. Bailey and J.L. White

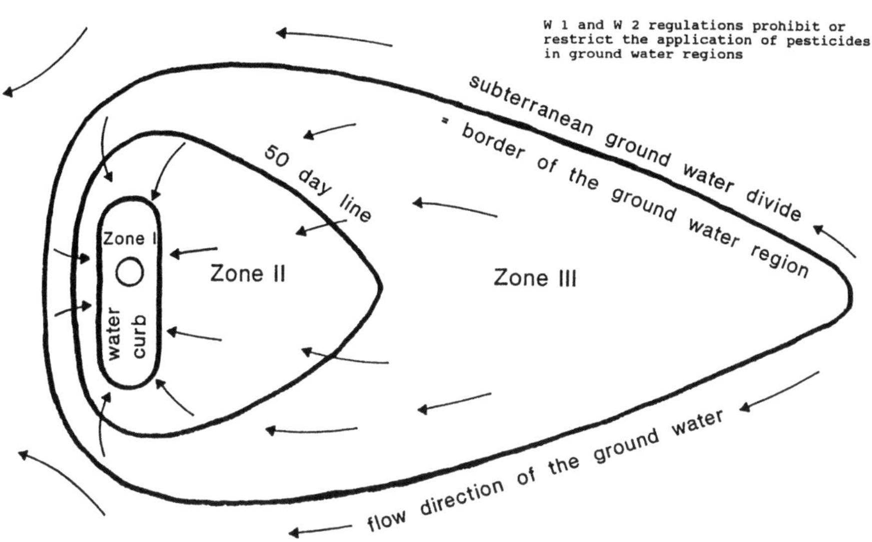

Fig. 1. Ground plan of a water protection area. The establishment of ground water protection areas is a precondition for the legislator to be able to take action and provide protection for groundwater in the Federal Republic of Germany. In establishing a protected area, it is separated into three zones according to the guidelines of the German Association of the Gas and Water Industry. Zone I (= curb district) encloses the immediate surroundings of the spring and is usually fenced in. The adjoining Zone II (= narrow protection zone) encloses the area from which the ground water needs 50 days to reach the spring curb. Zone III (= extended protection zone) encloses the area between the 50 day line and subterranean ground water divide. When Zone III is very large it is often divided into subdistricts IIIa and IIIb.

The BBA can prohibit and restrict the use of pesticides in ground water protected regions by the following water protection regulations W 1 and W 2:

W 1 (does not refer to the guidelines of the German Association of the Gas and Water Industry for the establishment of water protection areas) = not applied in the catchment areas of ground and spring water supply plants, mineral springs and reservoirs and other ground water sensitive areas.

W 2 (refers to the guidelines of the German Association of the Gas and Water Industry for the establishment of water protection areas) = application in ground water regions of ground water and spring water supply plants only in areas where the flow of the water takes longer than 50 days to reach the water curb according to the authorized water office; that means that in water protection areas that are established according to these guidelines application may take place only in zone III. No application in ground water regions of reservoirs.

gave a summary of the literature of mobility studies of pesticides in soils. Their literature catalogue shows that by that time most studies were on s-triazines, herbicidal active ingredients up to now [12]. In the same year, J.B. Weber reported in detail about adsorption mechanisms of the s-triazines on clay particles of soil [13], M.H.B. Hayes summarized the results of research on

Table 3. List of the pesticides classified as ground water pollutants in the Federal Republic of Germany up to March 11, 1993 and so a water proctection regulation imposed

active ingredient of pesticide*	water protection W1	regulation W2
Aluminium phosphide	+	
Aldicarb	+	
Amitrole	+	
Bacillus thuringiensis		+
Benazolin	+	
Bendiocarb	+	
Bentazone as salt	+	
Butocarboxim	+	
Carbofuran	+	
Calcium carbide	+	
Calcium phosphide	+	
Clopyralid	+	
2,4-D	+	
Dicamba as salt	+	
Didecyldimethyl ammonium chloride	+	
Ethiofencarb	+	
Ethoprophos	+	
Fluazifop	+	
Fluroxypyr	+	
Haloxyfop	+	
8-Hydroxichinolin	+	
Linuron	+	
Metalaxyl	+	
Metam as salt	+	
Metazachlor	+	
Methamidophos	+	
Metobromuron	+	
Metribuzin	+	
Nitrothal-isopropyl	+	
Oxadixyl	+	
Sethoxydim	+	
Triclopyr	+	

* pesticides with a risk of ground water pollution, that were not authorized up to March 11, 1993 in the Fedral Republic of Germany, are not included in this list. Examples of these active ingredients are atrazine and simazine

adsorption of triazine herbicides onto the organic parts of soil [14] and C.S. Helling produced a short survey on the mobility of s-triazines in soils [15]. A more up-to-date summary on adsorption processes concerning pesticides has been given by W. Ziechmann and U. Müller-Wegener [16].

Modern mobility studies of pesticides in soils are based on theoretical speculations where the kind and strength of interactions of these substances with the colloidal substances of the soil play a central part. These interactions may be the result of physical or chemical adsorption where at the same time several adsorption mechanisms may be concerned adding their effects. As for the adsorption phenomenon, van der Waals forces, hydrogen bonds, charge transfer, ligand replacement, hydrophobic binding, ionic bonding or ion exchange have to be discussed. Consequently the kind and strength of the interactions are determined by the physico-chemical parameters of the pesticides and in the final analysis are given by the chemical structure of these substances. The potential risk of ground water pollution of pesticides including their metabolites can therefore be theoretically quantified by the physico-chemical characteristics of these substances and in the first instance the multiple complex factors of soil science may be set aside in mobility studies of pesticides. Essential hints about the mobility of pesticides in soils result from easily carried out soil column tests in the laboratory which will be described and discussed in the following. In this context, the reader is referred to the guidelines for the official test of pesticides part IV 4-2 of the Federal Biological Research Centre for Agriculture and Forestry [5] and to constructive publications [17, 18, 19, 20, 21, 22, 23, 24, 25, 26, 27, 28].

2.1 Model Experiments in the Laboratory Using Soil Columns

The test setup for the simple to use soil column tests in the laboratory may be seen in Fig. 2. Preferredly, standard soils are used as column filling material. The soil characterization is shown in Table 4.

The soil column is produced by filling the glass column with air-dried soil. It is then vibrated and finally saturated with water. After applying the pesticide, the soil is irrigated with fully deionized water and the percolates are analysed for pesticide residues. When working out the elution curves of the pesticides, it must be irrigated for a longer time and the percolates must be collected and analysed in fractions. After the end of the irrigation the column soil may be analysed further for pesticides in fractions to get more results about their mobility. This last method is extremely significant if a pesticide is only partially or not eluted at all after a longer period of irrigation.

Figures 3-6 show the elution curves that were determined with laboratory columns for the s-triazines – atrazine, simazine, terbuthylazine and terbutryn including desethylated metabolites, using sand with low amounts of humus as the standard soil. Figures 7–9 show the elution curves of metolachlor (Fig. 7), the phenyl urea herbicides – isoproturon, chlortoluron and metobromuron (Fig. 8) and the total herbicides – bromacil and hexazinone (Fig. 9) compared with the elution curve of atrazine that was always applied with the previously mentioned active ingredients to the low humus sand in the laboratory columns. The results of percolation trials of the percolates in soil columns of low humus sand are shown in Fig. 10 for pesticides containing nitro groups such as nitralin, pendimethalin, quintozene and trifluralin and in Fig. 11, for the halogen-

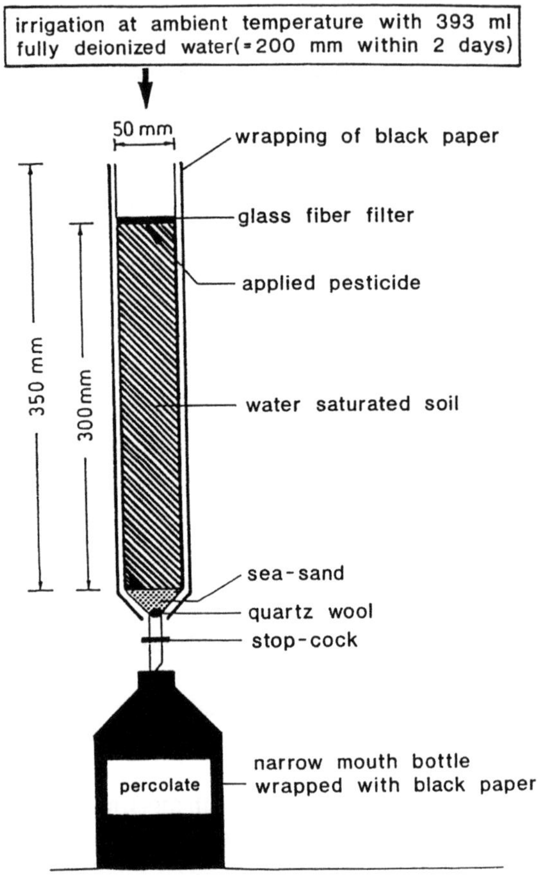

Fig. 2. Trial assembly for testing the leaching behavior of pesticides. The tests are carried out under "worst-case" conditions. The results obtained from analyzing the percolate in fractions render the determination of elution curves for pesticides and their metabolites possible

Table 4. Characterization of standard soils used for soil column tests

Standard soil	Org. C (%)	Particles < 20 µm* (%)	pH
low humus sand	0.25–0.75	< 10	5, 5–7, 5
high humus sand	2–3	10–15	5, 5–7, 5
low humus sandy loam	0.5–1.5	20–30	5, 5–7, 5

* parts < 20 µm = poor clay, fine and medium silt

containing isocyclic hydrocarbons – 1,3,5-trichlorobenzene and hexachlorobenzene.

Figure 12 shows the elution curves, that were analysed with soil columns of varying height. The heights of the soil cores were 5, 10, 30, 50 and 200 cm. Atrazine was used as the model substance for examining the interdependence of

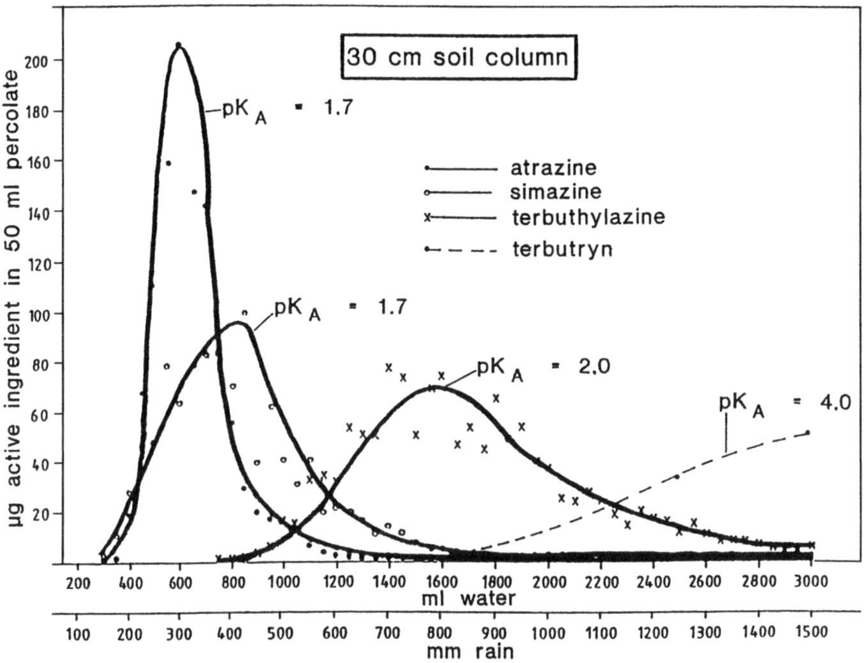

Fig. 3. Atrazine, simazine, terbuthylazine and terbutryn elution curves for low humus sand. In the soil column test the dose was ten times higher than used in agriculture or for railway systems

the retention times and the slopes of the elution curves from the height of the soil layer. Finally, Fig. 13 demonstrates the elution curves of atrazine obtained in tests to monitor the reproducibility of the soil column tests.

In the past, there was controversial discussion about the acceptability of evidence of the results of the soil column tests considering the potential risk of ground water pollution by pesticides and their metabolites. It was finally admitted by all sides that the results of the easy to handle laboratory model tests gave significant impulses to quantify the risk of ground water pollution by pesticides. In the following, some aspects and forecasts are shown that were obtained from the laboratory column tests shown in Figs. 3 – 13 [17, 18, 19, 29].

1. Figure 3 shows that s-triazines – especially atrazine and simazine – have a relatively high mobility. A more detailed analysis of the elution curves of the s-triazines leads to the conclusion that the degree of mobility depends on the pkA-value of those substances and decreases with increasing pkA-value. So ionic bonding and ion exchange as consequence of protonizing s-triazine molecules play an essential part in adsorption processes of s-triazines in soils.

2. Due to *N*-desethylation being connected with a lowering of the pkA-value, the *N*-desethylated metabolites are more mobile than the corresponding

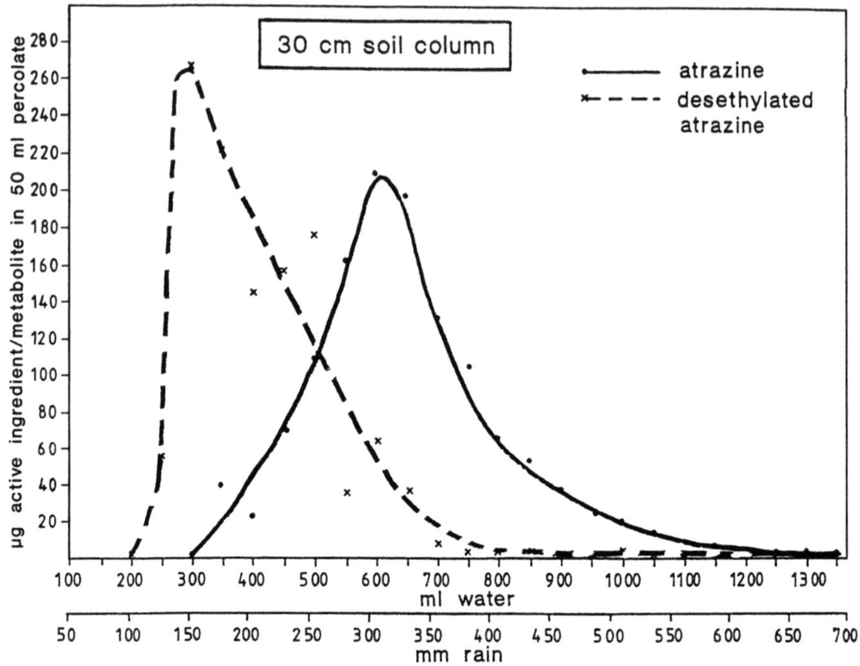

Fig. 4. Elution curves of atrazine and its desethyl metabolite in low humus sand. Ten times as much atrazine was applied to the soil column than is normally used in maize. The amount of desethylated atrazine used is equivalent to that of atrazine

active ingredients. These theoretical reflections are confirmed by experiments. Figures 4–6 show that the desethylated metabolites of atrazine, simazine and terbuthylazine are more mobile than the respective active ingredient, as expected. In particular, it must be remarked that the desethylated metabolite of the moderately mobile terbuthylazine has a similarly high mobility in soil to atrazine. The degrees of mobility of the s-triazines and their desethylated metabolites graduate as follows:

terbutryn < terbuthylazine < simazine ≤ atrazine ≤ desethylated terbuthylazine < desethylated atrazine ≤ desethylated simazine

3. Figure 7 shows that the active ingredient metolachlor is like atrazine, i.e. relatively mobile in soils. Metolachlor is as a herbicidal active ingredient contained together with terbuthylazine in the commercial product Gardoprim plus.

4. In the 1960s, the percolation behavior of the phenyl urea herbicides was tested. Hance [30] found out that urea derivatives have a varying high mobility and the tendency to adsorb depending on their substituents. Mobility increases

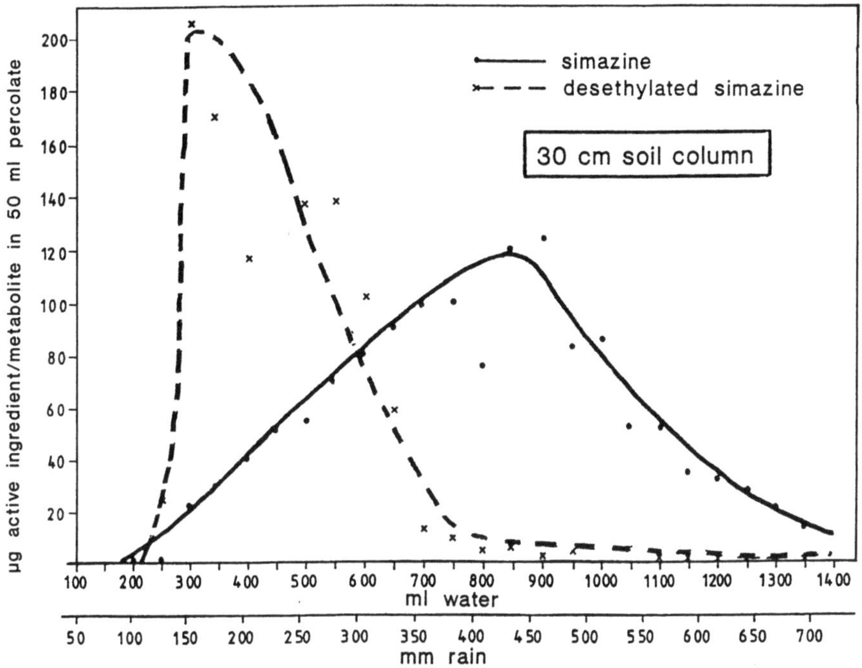

Fig. 5. Elution curves of simazine and its desethyl metabolite in low humus sand. Ten times as much simazine was applied to the soil column than is used in normal agricultural practice in maize. The amount of desethylated simazine used is equivalent to that of simazine

as follows:

urea < fenuron < methyl urea < phenyl urea < monuron ≤ monolinuron < diuron ≤ linuron < neburon < chloroxuron

The results of a soil column test in Fig. 8 show a graduated mobility for the phenyl urea herbicides isoproturon, chlortoluron and metobromuron that decreases from isoproturon to chlortoluron to metobromuron. A reason for the graduation of the mobility behavior of the phenyl urea herbicides may be effects of the functional groups. They may give the herbicidal molecules a substantial electrical dipole moment. The strength of this dipole moment is given by what kind of groups they are and what position they occupy within the molecule. A rule of thumb is that with increasing polar character of the phenyl urea herbicides or pesticides the adsorption bondings of these substances to colloidal soil particles get stronger.

5. The dipole moments of pesticides can be determined with optical and mathematical methods. A fast mathematical quantification of the dipole moments can be, to put in the amounts of the individual atomic bond of Table 5.

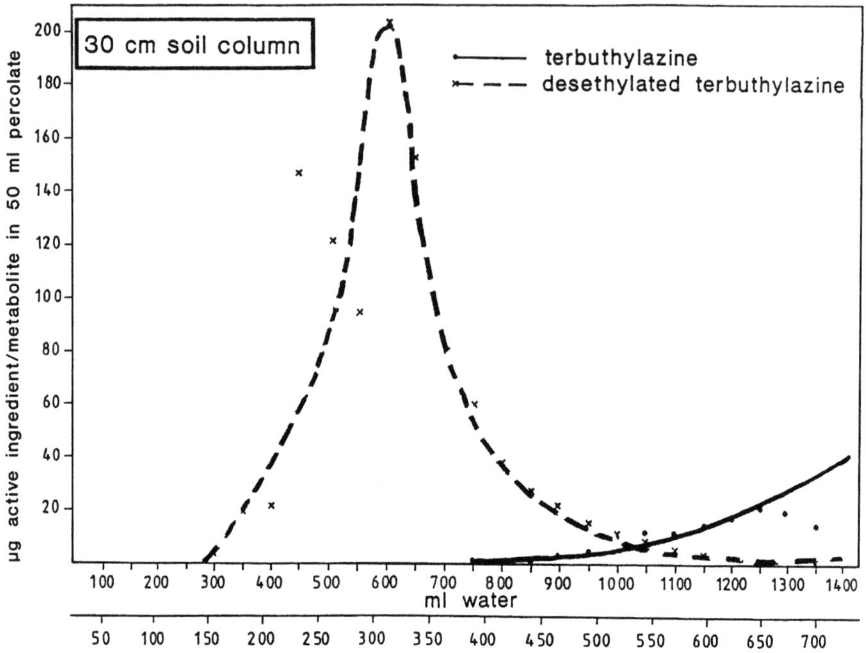

Fig. 6. Elution curves of terbuthylazine and its desethyl metabolite in low humus sand. Ten times as much terbuthylazine was applied to the soil column than is used in normal agricultural practice in maize. The amount of desethylated terbuthylazine used is equivalent to that of terbuthylazine

Methodical analyses at the aromatic ringsystem show that halogen-, cyano-, nitro- and carbonylgroups in group 1 of Table 5 act as electron acceptors to the aromatic systems and so are negatively loaded. Methyl-, amino- and dimethylaminogroups of group 2 are electron donors and so form the positive dipole end. In benzenes with several functional groups, the dipole moment is obtained with the help of vectorial addition, where the direction of the individual moment must be considered.

In the hypothetical critical case that the bonding forces of polar pesticides to colloidal soil components are based exclusively on the dipole strengths, the retention times of the elution curves agree with the strength of the dipole moments and the higher the dipole moment of a pesticide the later it is eluted.

High dipole moments may be characteristic of – according to Table 5 – for example, pesticides containing nitro groups. It is therefore of general interest to study the leaching behaviour of these substances. Experiments in the laboratory using soil columns were carried out with the pesticides containing nitro groups – nitralin, pendimethalin, quintozene and trifluralin. In this context, Fig. 10 shows

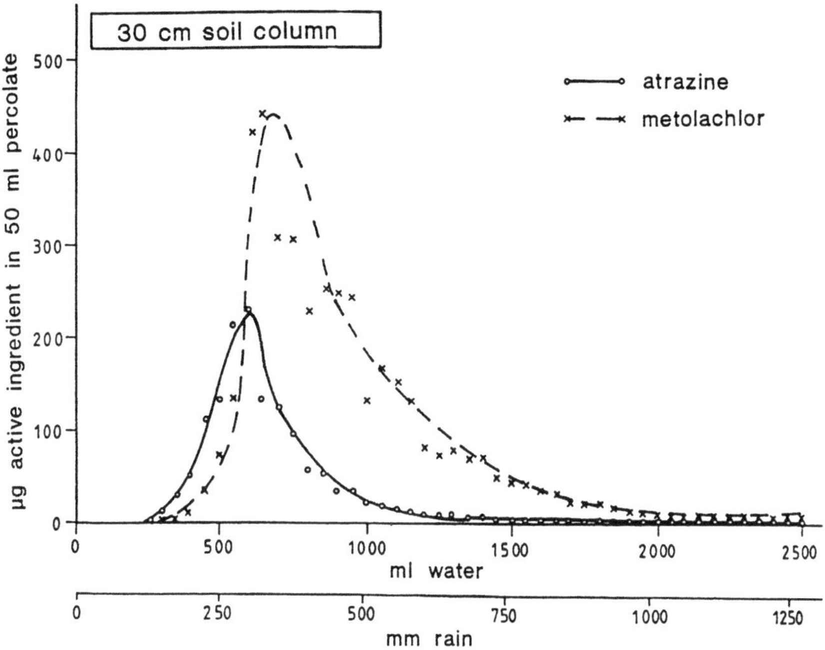

Fig. 7. Elution curves of metolachlor and atrazine in low humus sand. In the soil column test a dose a ten times higher was used than in normal agricultural practice

Table 5. Increments for the amounts of individual atomic bonds with regard to the dipole moments of pesticides

	group 1					
	-F	-Cl	-Br	-C(=O)-CH_3	-CN	-NO_2
alkyl	1.89	2.01	2.00	2.75	3.60	3.25
phenyl	1.48	1.60	1.57	2.96	4.05	4.01
	group 2					
	-CH_3		-NH_2		-N (CH_3)$_2$	
alkyl	0		1.32		0.86	
phenyl	0.35		1.52		1.61	

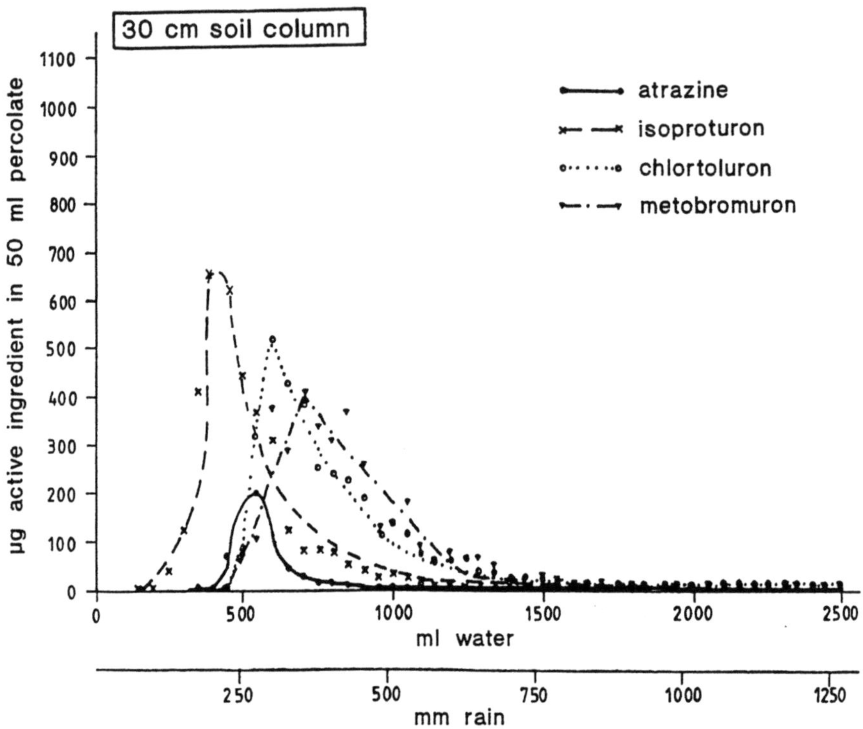

Fig. 8. Elution curves of isoproturon, chlortoluron, metobromuron and atrazine in low humus sand. In the soil column test a dose was used ten times higher than in normal agricultural practice

that these substances are still totally adsorbed to the upper 10 cm soil core of a soil column with low humus sand even after irrigation with 2,500 ml water (≙ 1,273 mm rain). The lower quantities of active ingredients found in the soil columns compared with the amount of active ingredient used can be explained by the recovery rates which were between 75–100% for the pesticides containing nitro groups. Besides the pesticides containing nitro groups – nitralin, pendimethalin, quintozene and trifluralin – it can be deducted that as a consequence of functional groups the number of immobile pesticides in soils is much higher than so far assumed.

6. Further examples for nearly immobile substances are halogenated benzenes. Figure 11 shows that 1,3,5-trichlorobenzene and hexachlorobenzene – formerly used worldwide as a seed disinfectant – are also fully adsorbed to the upper 10 cm soil layer of a low humus sand in a laboratory column even after irrigation with 2,500 ml water (≙ 1,273 mm rain). Under the aspect of the physico-chemical characteristics of pesticides it can therefore in this context be summarized that besides dipole strengths other physical forces as charge transfer are to be considered.

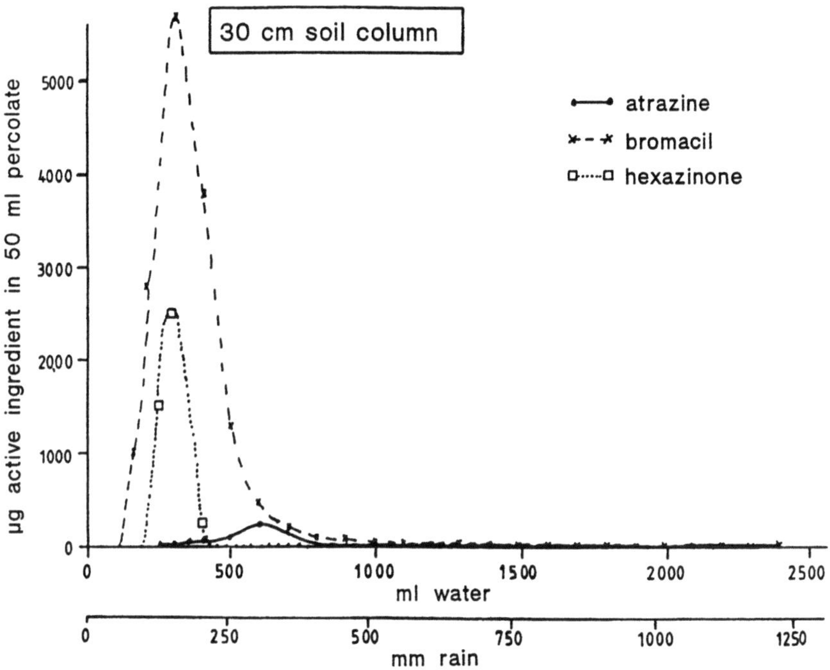

Fig. 9. Elution curves of bromacil, hexazinone and atrazine in low humus sand. In the soil column test a dose was used 5 to 10 times higher than in normal agricultural practice or in railway systems.

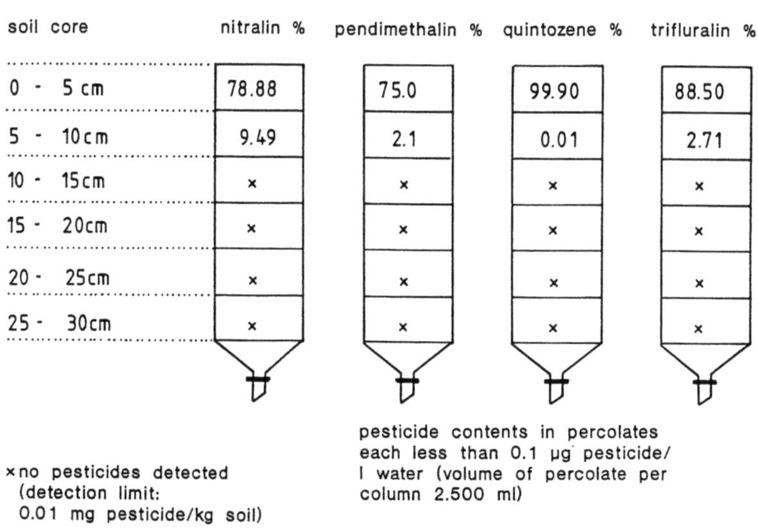

Fig. 10. Results of soil column tests carried out with the nitro-group-containing pesticides nitralin, pendimethalin, quintozene and trifluralin. Each time, low humus sand was irrigated with 2500 ml water (≙ 1273 mm rain). The percolate and- after irrigation- the soil column were analyzed for pesticides in fractions in each case

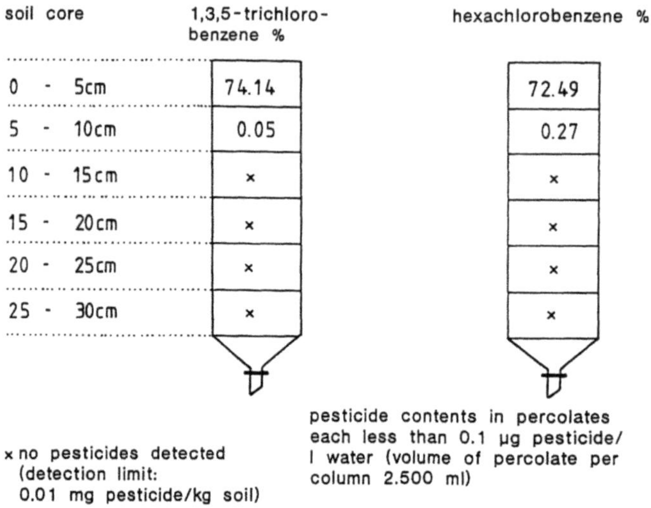

Fig. 11. Results of soil column tests carried out with the halogen-containing isocyclic hydrocarbons 1,3,5-trichlorobenzene and hexachlorobenzene. In both cases, low humus soil was irrigated with 2500 ml water (≙1273 mm rain). The percolate and- after irrigation- the soil column were analysed for pesticides in fractions both times

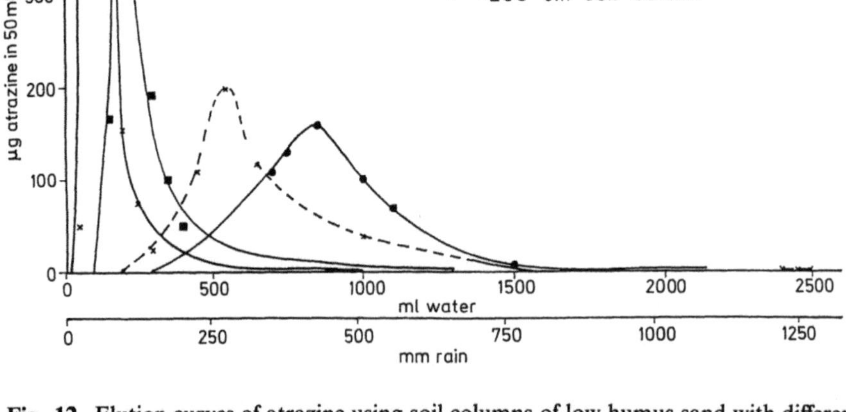

Fig. 12. Elution curves of atrazine using soil columns of low humus sand with different heights. The amount of atrazine applied was identical in each of the five tests

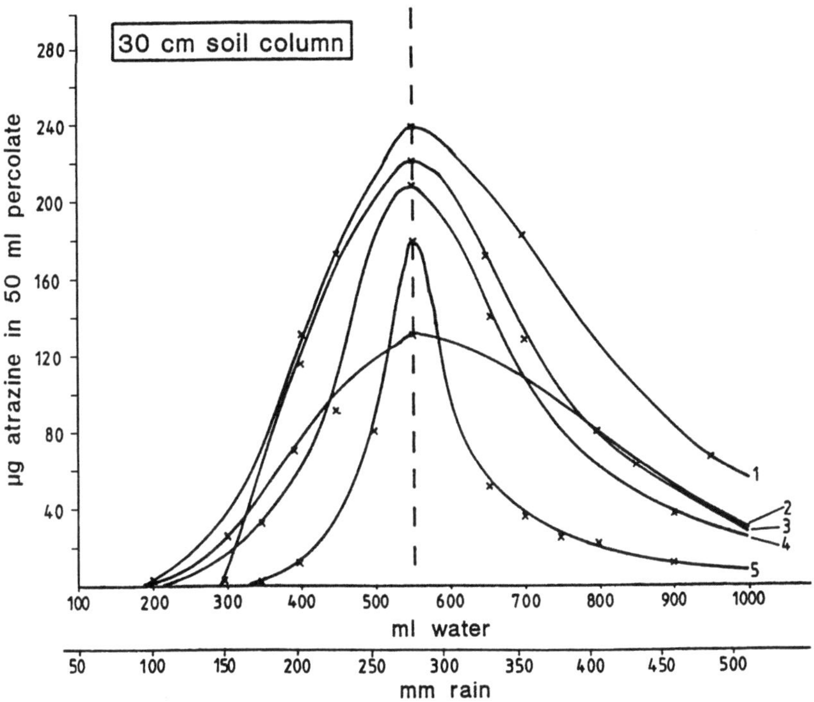

Fig. 13. Elution curves of atrazine that were obtained on examining the reproducibility of soil column tests with five soil columns filled with low humus sand. Immediately before irrigation, each soil column had the same amount of atrazine applied

7. The percolation trials in the laboratory are "worst-case-tests" – from the point of view of applying pesticides to agricultural soils – because of the high amounts of water used for irrigation as well as the use of low humus sand. In applying herbicides to railway systems with their ballast being free from clay, silt and humus, the conclusions drawn from the mobility behavior of these substances lose validity. The situation is aggravated by using the relatively highly mobile herbicides atrazine, simazine, bromacil and hexazinone extensively during the last few years. The results of the mobility study for the total herbicides bromacil and hexazinone to low humus sand are shown in Fig. 9. Atrazine was used as the reference substance within the mobility study.

8. As a consequence of the elution curves for atrazine in soil columns with different heights shown in Fig. 12, it can be said that the retention times and the band width of the elution curves correspond with the height of the soil columns and the breakthrough of mobile pesticides is delayed more the higher the soil layer is. At the same time, the elution curves get broader and flatter with increasing height of the soil layer. The mobility studies using soil columns with different heights make it clear that the "chromatographic concept" is essential for evaluating the potential risk of ground water pollution by the pesticides.

Mobility studies can resort to the models of chromatography and with the help of these theories the quantities or concentrations of pesticides can be described depending on place and time in soils. The beginning of the atrazine elution curve in a 2 m soil column after irrigation with about 2500 ml water is shown in Fig. 12 and it can be seen that the concentration of atrazine is below 0.1 μg atrazine per litre of ground water after passing through several metres of good filtering surface soil.

9. The results of reproducibility tests of the laboratory mobility studies are shown in Fig. 13. According to the elution curves shown, which were determined using five different soil columns of low humus sand, the soil column tests provide reproducible information on mobility of pesticides in soils. It is striking, in this context, that the retention times characterized by the peak maximum show small differences. As a consequence of the reproducibility, from the results that are gained with the model studies in the laboratory, general rules about the mobility of pesticides in soils and a graduated range of the pesticides according to their degree of mobility can be drawn.

2.2 Calculations with the Help of Computer-Aided Simulation Models to Quantify the Potential Risk of Ground Water Pollution by Pesticides

The first attempts to use mathematical calculations with the help of computer aided simulation models to quantify the potential risk of ground water pollution go back to the 1970s. On the basis of their results obtained from adsorption studies of pesticides in soils, many authors developed simple forecast models about the leaching behavior of these substances with regard to various environmental conditions [31, 32, 33]. The continuing improvements of these models ended in computer programs for calculating and showing the translocation of pesticides into the subsoil depending on the specific characteristics of those substances as well as environmental data. The programs vary greatly because of the many parameters which influence the transport of pesticides into the subsoil and the selection and the importance laid on specific parameters.

The Pesticide Root Zone Model (PRZM) and the Seasonal soil Compartment Model (SESOIL) that were published in the middle of the 1980s [7, 8] found wide acceptance and since then have been the basis for many authors to quantify the potential risk of ground water pollution by pesticides. For example M.N. Lorber and C.K. Offutt debated the dislocation of aldicarb under various meteorological and soil conditions with the help of the Pesticide Root Zone Model [34] and M. Klein used a modified version of the computer program, Seasonal Soil Compartment Model [35], in translocation studies on the active ingredients cloethocarb and bentazone. For the registration procedure of the Federal Republic of Germany calculations with the help of computer aided simulation models as PRZM and SESOIL to quantify the potential risk of ground water pollution by pesticides became of significant importance as long ago as the end of the 1980s.

The first publications on calculations of the translocation of pesticides into the subsoil were full of enthusiasm gradually waned. Nowadays, the possibilities and limits of computer simulation models are evaluated more critically. The inputs of those computer simulation models were often too sweeping and proved to be a handicap in trying to optimize the output. It must be said, however, that the computer simulation models led to a better and deeper understanding of the translocation of these substances by systematically linking soil parameters and the hydrostatic behavior of water in soils to physico-chemical data and metabolism kinetics of pesticides. Figure 14 shows schematically the important parameters the Pesticide Root Zone Computer Model uses for the calculation of the translocation of pesticides into the subsoil. So mobility studies considered hydrological components such as evaporation and transpiration and from the soil parameters, for example, the pore volume with the help of computer programmes. The computer programmes are being continually changed and corrected to reflect new practical experience and knowledge about the behavior of pesticides in soils.

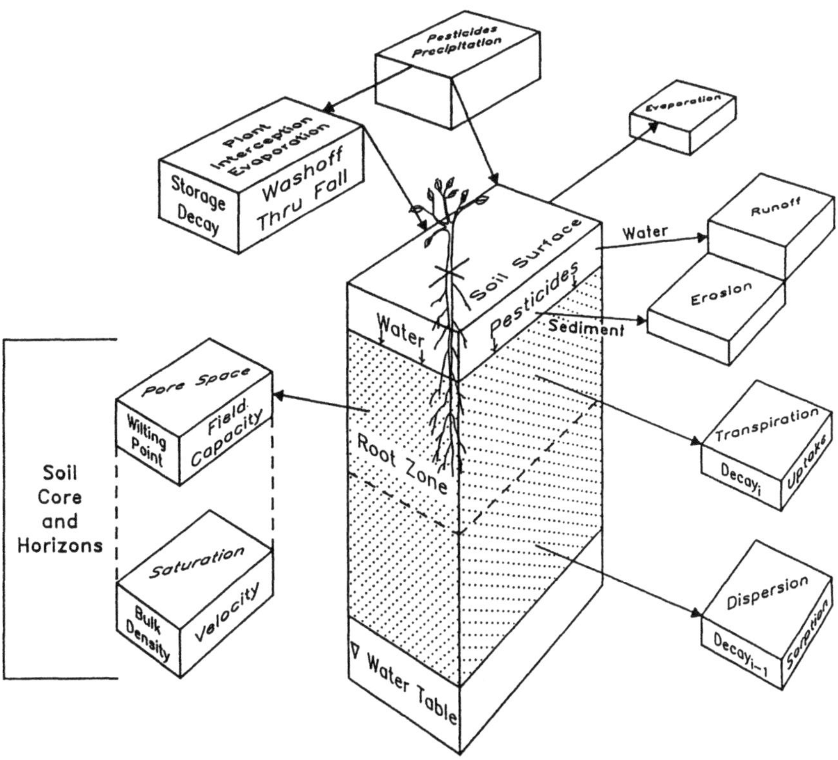

Fig. 14. Parameters that are used with computer aided simulation models based on the Pesticide Root Zone Model (PRZM) to quantify the potential risk of ground water pollution by pesticides (according to [7])

2.3 Lysimeter Tests

Lysimeter tests are a connecting link between model trials with the help of soil columns in the laboratory and the real situation in ground water regions. Lysimeter studies can be important in evaluating the potential risk of ground water pollution by pesticides in the registration procedure of the Federal Republic of Germany in the following cases:

1. Calculations with the help of computer aided simulation models as for example SESOIL (Seasonal Soil Compartment Model) or PRZM (Pesticide Root Zone Model) show that an active ingredient may occur in the water of the new formation of ground water with a concentration of > 0.1 µg/l and so entry of this substance into the ground water cannot be excluded.
2. Leaching tests according to the BBA-guideline IV 4-2 using soil columns in the laboratory lead to the following results:

a. The concentration of an active ingredient or metabolite that is persistent according to the BBA guideline IV 4-1 [36] (DT 90 > 100 days) exceeds 5% in the percolate and there are no results of "aged-leaching" tests. The aged-leaching tests are a variant of the leaching tests and according to BBA guideline 4-2 are carried out with active ingredients or their metabolites that were aged before application to the soil column. Descriptions of the soil for the aging of the active ingredients or metabolites and its duration are part of the guideline 4-2.
b. The concentration of an active ingredient or metabolite that is classified as persistent according to the BBA guideline IV 4-1 is greater than 2% in the percolate of aged-leaching tests.

Lysimeters for mobility studies are containers above or buried in the ground which are filled with soil in disturbed or undisturbed packings. The construction of lysimeters varies considerably and depends on its purpose. Primarily, lysimeters were developed purely for soil science investigations but more and more they came to be used in other sectors, e.g. to determine the evapotranspiration or outcome of the movement processes of environmentally important substances in soils. Fundamentally, we can differentiate between weighable and non-weighable lysimeters where the expensive weighable lysimeters register the change of weight in addition to the leaching within the soil monoliths and so the outcome of the processes in soils can be determined – the leaching behavior as well as the changing of soil water capacity.

Mobility studies of pesticides in soils generally use non-weighable lysimeters. Figure 15 shows the scheme of a widely used model of a non-weighable lysimeter which is buried and was developed to examine the nutrient content of soils and can also be used for the analysis of the percolate for pesticide residues. The lysimeters, buried in agricultural areas, differ in filling and size, where 1 m^2 for the catching area and 1.5 m for the height of the container are normal sizes. The distance of the upper container rim above the ground varies with the lysimeters where the depth of burial is determined by the aim of the soil science examination.

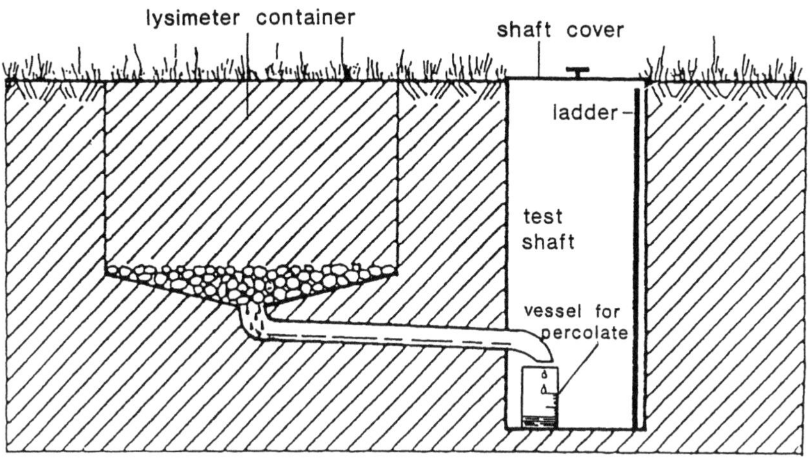

Fig. 15. Scheme of a non-weighable, buried lysimeter with its upper container rim ending level with the field

The differences of the lysimeters in installation, filling, form, size of the reception area and the height affect the percolation results of pesticides. At the same time, agricultural practices at the location of the lysimeters and meteorological conditions affect the tendency of leaching of these substances and determine kind and extent of the monitoring procedures during the duration of test for a correct interpretation of the results. So the use of lysimeters for quantifying the potential risk of ground water pollution was constantly a subject of technical discussions in the past.

As a result of thorough consultation, rules for lysimeter studies in the Federal Republic of Germany were established and are contained in the BBA guideline part IV 4-3 "Lysimeter investigations of the displacement of plant protection products into the subsoil". According to this, investigations on the mobility of pesticides in registration tests must use lysimeters with undisturbed soil cores of 1 m to 1.30 m in height and with at least a 0.5 m^2, if possible 1 m^2 catching area are to be used. As a soil material, loamy or silky sand with a low humus content should be used where the sum of clay and silk should amount to a maximum of 30% – clay below 10% – over the total profile and the content of organic bound carbon should not exceed 1.5%. The resulting percolate has to be collected via an outlet valve at the base of the lysimeter and is then analysed. The guideline leaves the question unanswered whether to use lysimeters above or in the ground and non-weighable lysimeters may also be used in mobility studies for registration procedures.

In lysimeter studies for registration procedures the radioactive marked unformulated – alternatively the formulated – active ingredient is normally applied commonly with the highest dosage envisaged at the scheduled time. The farming procedures in the vicinity of the lysimeter such as plowing, harrowing, sowing or planting, fertilizer application and further plant protection activities are orientated on normal agricultural practice.

According to guideline IV 4-3, the procedures include collecting the percolate at appropriate time intervals, i.e. at least once a month, determination of the amount of percolate and analyses of the active ingredient and its metabolites. During the test period, those climatic data which are relevant to distribution and decomposition such as temperature and rainfall distribution are to be collected and the total amount of percolate is to be determined. To find how much of the active ingredient remains, the soil core should be divided into 10 cm-layers that are analysed for the not-extractable radioactivity as well as the active ingredient and its metabolites.

In the past, a succession of authors have investigated the mobility of pesticides and their metabolites with the help of lysimeter studies. In the Federal Republic of Germany, the team led by F. Führ was involved at an early stage in the development and settling questions of methodology for the lysimeter technique and carried out numerous investigations into the displacement of pesticides into the subsoil using lysimeters. Recent investigations by this team concern the herbicidal active ingredients pyridate, clopyralid, fluroxypyr, 2,4-Dichlorprop-P and chloridazon. The results of these lysimeter experiments were so significant that the precautional restriction against use in water protection areas for these pesticides could be withdrawn by the registration authority lately [37, 38, 39]. Extensive data on lysimeter studies were also presented by H.J. Jarczyk [40, 41]. The results of more than 3,000 individual analyses, involving 12, 081 litres of percolate from lysimeter monoliths, are summarized in Table 6 [42]. In these studies, Jarczyk simulated applications of pesticides with lysimeter monoliths in agricultural practice, whereas lysimeter IV of Table 6 shows extremely unfavourable conditions because the field lay fallow for the duration of the test.

Lysimeter experiments carried out under open land conditions give similar results to those on agricultural land. Therefore farming methods and atmospheric conditions can influence the leaching behaviour of pesticides as in the agricultural areas. The real chromatographic system soil/water of the agriculturally used soils is reflected better in lysimeter soil cores than in laboratory columns. Therefore a comparison of the results of lysimeter studies and of the results of the laboratory studies with the help of soil columns concerning displacement of pesticides into the subsoil is of great interest. The results of investigations by F. Führ show that the rules about the degree of mobility of pesticides and their metabolites worked out with the help of laboratory columns are still valid under open land conditions. Lysimeter studies show, as do the laboratory model studies with the help of soil columns, that the desethylated metabolite of atrazine is more mobile than atrazine and the desethylated metabolite of terbuthylazine is more mobile than terbuthylazine.

Table 6. Summary of results from more than 3,000 individual analyses, involving 12,081 litres of percolate from lysimeter monoliths by Jarczyk in the period 1977–1986 (October). The analyses given are average values throughout a year (mg/l).

	1977–1982		1983		1984		1985		1986 (09/30)		
lysimeter	II	IV	II	IV	II	IV	II	IV	II	IV	HPLC-practical detection limit from 1.1.85 (mg/l)
percolate water (litres)	1735	3625	506	721	66	1186	145	1160	239	559	
amitrole	–	n.d.	–	n.d.	–	–	–	n.d.	–	n.d.	0.01
azinphosethyl	n.d.	n.d.	n.d.	n.d.	*n.d.	*n.d.	n.d.	n.d.	n.d.	n.d.	0.0001
ethiofencarb	–	1979–1980 0,00006	n.d.	n.d.	***n.d.	***n.d.	–	–	–	–	–
metamitron	n.d.	n.d.	n.d.	n.d.	**n.d.	**n.d.	n.d.	n.d.	n.d.	n.d.	0.0001
methabenz-thiazuron	n.d.	n.d.	n.d.	n.d.	**n.d.	**0.0064	n.d.	0.0013	n.d.	0.0007	0.0001
oxidemethon-methyl	n.d.	n.d.	n.d.	n.d.	***n.d.	***n.d.	n.d.	n.d.	n.d.	n.d.	0.001
triadimefon	n.d.	n.d.	n.d.	n.d.	*n.d.	*n.d.	n.d.	n.d.	n.d.	n.d.	0.0002
triadimenol	n.d.	n.d.	*n.d.	n.d.	*n.d.	*n.d.	n.d.	n.d.	n.d.	n.d.	0.0005
trichlorphon	n.d.	n.d.	n.d.	n.d.	***n.d.	***n.d.	n.d.	n.d.	n.d.	n.d. (mg/l)	0.0005

* detection limit 0.005 mg/l
** detection limit 0.002 mg/l
*** detection limit 0.01 mg/l (TL-method)
n.d. = not detected

2.4 Investigations of Drainage and Ground Water to Quantify the Local Risk of Ground Water Pollution by Pesticides

The potential risk of ground water pollution by using a pesticide results from two factors – the specific potential risk of pollution posed by the pesticide and local potential risk of ground water pollution of the catchment area. While the specific potential risk of ground water pollution by a pesticide is decided by the physico-chemical characteristics of this substance, the local potential of risk of ground water pollution by pesticides is affected by the characteristics of the surface, the layer of the soil and the environmental influences acting on it.

The lysimeter studies and the calculations made with the help of computer-aided simulation models to quantify the potential risk of ground water pollution in using pesticides are intended to determine the specific potential of risk of ground water pollution by the pesticide combined with the local potential risk of ground water pollution by pesticides. Therefore, results produced are dependent on both potential risks. In connecting the effects of the specific potential risk of ground water pollution by a pesticide and the effects of the local potential risk of ground water pollution by pesticides in a single value may lead to complications in interpreting the results and give false conclusions because of the many different influencing factors. It therefore seems logical to examine the local potential of risk of ground water pollution by pesticides separately from the specific potential risk of ground water pollution by pesticides and to work out criteria for the quantification of the local potential risk of ground water pollution by pesticides. The independent view of the specific elements of risk of ground water pollution which can be found out by standardized soil column tests in the laboratory (as described in Sect. 2.1) and the criteria for the local potential risk of ground water pollution by pesticides shows the effects of individual elements of risk and allows valid conclusions about the local potential risk of ground water pollution from using pesticides. The following text demonstrates significant aspects that play a role in quantifying the local potential risk of ground water pollution.

In evaluating the surface soil with regard to local potential risk of ground water pollution by pesticides, the main consideration is the run off of pesticides. How this happens and its extent depends on the geomorphology of the soil surface. In general, with regard to a mobile pesticide, the part, that is not evaporated, decomposed or stored in soils for example as a bound residue, runs off with the water. The run-off happens in various regions above and below the soil surface. The total run-off consists of three components

a. Run-off above the soil surface

b. Underground run-off through the surface soil without reaching the deeper ground water. If there are stagnation layers, ground water may accumulate near the surface and may reach the light of day as spring water. Another part of the underground running water follows the hydraulic incline to run, for example, into rivers (draining water).

c. Subsoil run-off through the surface soil till it reaches the deeper ground water. The run-off may continue from the ground water and also run, for example, into rivers.

A closer look at the components of run-off shows, that a mobile pesticide that exists on the soil may reach the water circulation and so the pollution of the surface water and ground water with pesticides may not be discussed in isolation. One must also consider at the same time that with pollution of ground water with pesticides, these substances can run off from the ground water to the surface water. The part of the pesticides that runs off from the ground water into rivers may be monitored whereas that part that runs into other areas cannot be checked. The run-off may occur in the reverse direction as well. For example, the river water may enter ground water near rivers as bank filtrate. As a consequence, ground water pollution can be caused by pesticides in river water, their journey may be several hundred of kilometres long. Figure 16 shows as a scheme the different components of run-off for a mobile and a persistent pesticide that exists on the surface soil. Figure 17 shows schematically the run-off events that happen to water and the pesticides it contains in different components of run-off.

The lysimeters whose soil monoliths are tightly enclosed by containers can inevitably not register these components of run-off and so they do not show the real run-off in a surface soil. The percolate collected at the base of the lysimeter shows a different composition from the percolate from a comparable natural section of surface soil and contains parts of the run-off on the soil surface, the draining water and the percolate that contribute to the new formation of ground water. The allocation of the total run-off of a lysimeter into the single components can only be done with the use of correction factors.

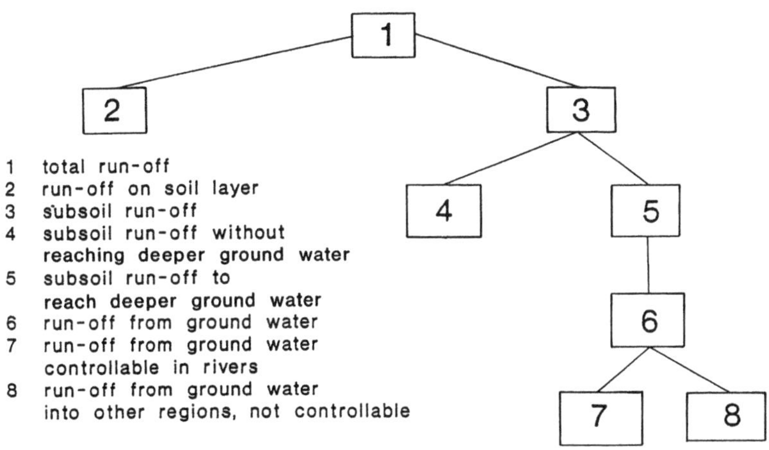

Fig. 16. Scheme of the run-off components for mobile and persistent pesticides that are present on the surface soil. These substances run off with the water

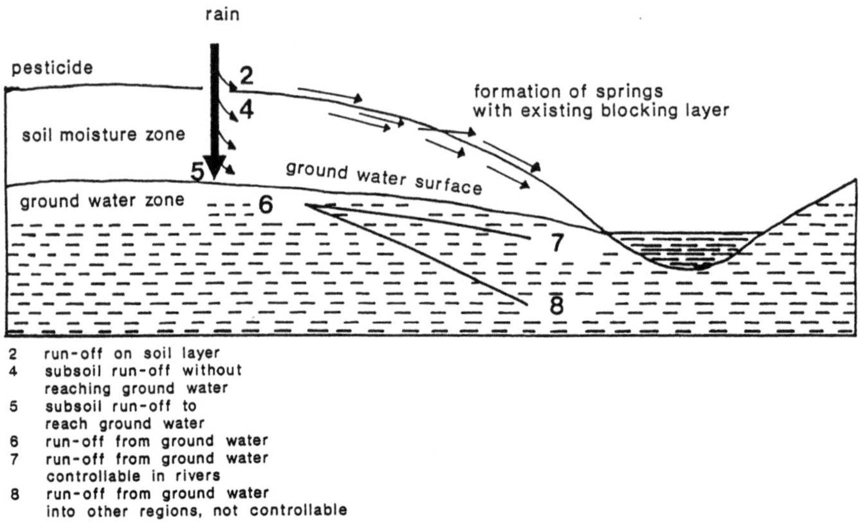

Fig. 17. Scheme of the run-off events that may have an effect on a pesticide on the surface soil. The run-off events connect ground and surface water. Pollution of the surface water with pesticides may result from the direct entry into the surface as well as from the indirect entry of these substances with the ground and draining water

The real evaluation of the influence of the surface soil to the run-off of pesticides is obtained from analysing the water near the surface and the deeper lying ground water. The draining water should be included in these investigations.

The draining of the soil by underground conduits serves to take up and lead off the surplus soil water. This is a widely used method to promote the structure formation of the soil. H.J. Collins estimates, that in the Federal Republic of Germany about 1.4 million ha out of the total agricultural area of 17.1 million ha are drained by subsoil conduits [43]. Therefore, the drained areas are estimated to be 8% of the total and this must be taken into consideration in the run-off processes of pesticides.

When laying the drainage pipes, for example, outfall conditions and soil permeability are considered. Shallow draining systems 0.6 to 1.0 m deep are selected for heavy slightly permeable soil or unfavourable outfall; medium depth drainage systems 1.0 to 1.2 m deep for medium heavy, permeable soils; deep drainage systems 1.2 to 1.5 m deep for deep, medium heavy soils; and finally, very deep drainage systems 1.5 to 1.8 m deep when deep rooted hops are grown.

Samples of draining water for investigating residues of pesticides can be taken from the outlets of the draining pipes before the water runs into the water course. The depth of the surface soil is included in the examinations and depends on the bedding of the drainage pipe system and lies between 0.6 and 1.8 m.

Since the beginning of the 1970s, many authors have reported on pesticide residues in drainage water [44, 45, 46, 47, 48, 49]. The results of these investigations – made with different drainage systems, different surface soils and environmental influences – are in the same sequence as the degrees of mobility of pesticides determined with the help of soil columns in the laboratory and confirm in principle the validity of the scale of the degrees of mobility based on the physico-chemical characteristics. The differences in the contents of the drain water found out with a pesticide in various test series – can be ascribed to the influence of the drainage system used, the state and structure of the surface soil and the environmental influences on the surface soil during the duration of the test. For instance, Table 7 shows the amounts of atrazine and desethylated

Table 7. Atrazine and desethylated atrazine content in drain water from a maize field that was repeatedly treated with atrazine*

Sample taken on	atrazine content in μg/l	desethylated atrazine content in μg/l
May 26nd, 1973**	0.33	0.70
June 6th, 1973***		
June 8th, 1973	0.29	0.60
June 9th, 1973	0.24	0.56
June 12th, 1973	1.49	0.40
June 15th, 1973	0.62	0.85
June 20th, 1973	0.63	1.11
June 27th, 1973	0.52	0.82
July 4th, 1973	0.40	1.08
July 18th, 1973	0.32	0.90
May 5th, 1974	0.51	0.36
April 15th, 1974	0.23	0.57
April 29th, 1974	0.13	0.67
May 13th, 1974	0.21	1.16
June 3rd, 1974	0.12	0.58
June 12th, 1974***		
June 20th, 1974	0.06	0.47
July 1st, 1974	0.15	0.77
July 10th, 1974	1.09	1.10
July 19th, 1974	0.82	0.74
July 30th, 1974	10.82	7.71
September 3rd, 1974	0.41	0.47
November 23rd, 1974	0.41	1.00
December 8th, 1974	1.78	2.15
December 22nd, 1974	0.24	1.05

* Size of the trial area: 1.75 ha
Soil: sandy loam
draining depth: 1.20 m
draining incline: 0.25%
** The maize field was treated with atrazine containing pesticides several times before test start
*** Date of atrazine application (Atrex 90 W); 2.8 kg/ha

atrazine contained in the drain water tests of a maize field after preceeding applications of atrazine [44]. Figure 18 shows the terbuthylazine and desethylated terbuthylazine content in drain water from a maize field after application of terbuthylazine for the first time on this area [49]. As expected, the results indicate a higher mobility of the metabolites desethylated atrazine and desethylated terbuthylazine than of the corresponding active ingredient.

The run-off of mobile active ingredients of pesticides and their metabolites with the drain water after a single application of the active ingredients may last several years. It can be concluded from the results of drain water trials that applications of pesticides in spring may lead to pollution of the water of the short-term new formation of ground water in late spring/early summer as well as to the water of the long-term new formation of the ground water in autumn/winter. The degree of pollution caused by mobile pesticides to the water of the long-term new formation of the ground water depends on the speed of their decomposition in the surface soil and the time of application of these ingredients. Because of autumn and winter application of pesticides with high dosages possibly leading to high pesticide contents in soils in the period of long-term new formation of ground water, these applications appear to be problematic from the view of ground water protection [50].

A review of the extensive data material about pesticide content in draining water shows no such rigorous connections between percolation volume and the

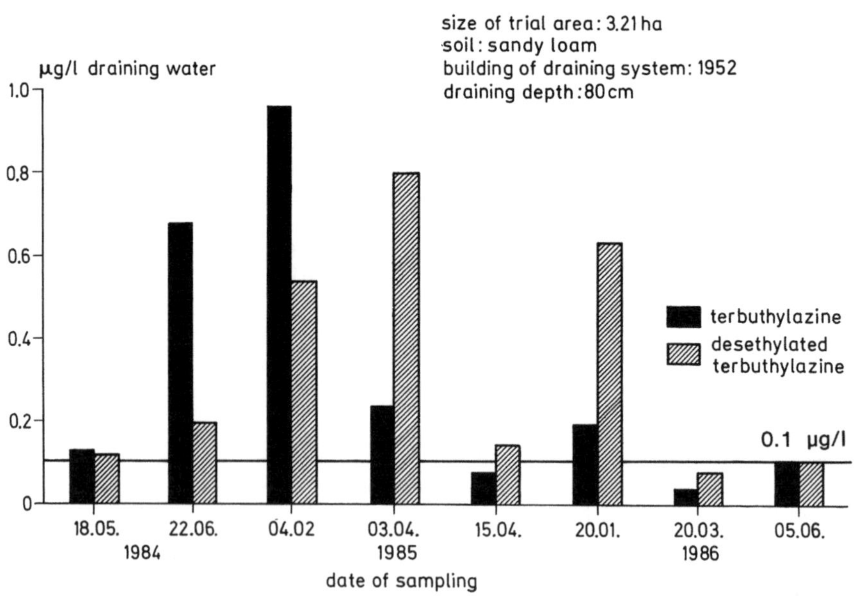

Fig. 18. Terbuthylazine and desethylated terbuthylazine content of the drainage water of a maize field that had been applied on April 27th, 1984 with 21 of the terbuthylazine containing pesticide Gardoprim 500 liquid (content of active ingredient: 490 g/l)

transport of substances as shown by the soil tests in the laboratory. The elution curves of pesticides in drain water cannot be described using the Gaussian curve that is obtained in the ideal case after application of a pesticide to the soil surface and a subsequent linear, not ideal adsorption chromatographic process, but at first sight shows an irregular form with sporadic extreme ups and downs.

With an exact analysis with a great number of single values about the pesticide contents of the drainage water over the trial period, larger parts of the peaks of the elution curves may be fitted better to the passage diagram of a front chromatography the longer the time period since the pesticide application.

At the same time, the reasons for extreme values for the pesticide contents of drain water can be described including soil parameters of the surface soil and meteorological conditions. So general statements can be made about the elution curves of pesticides in drain water. The factual details may be seen in Figs. 19 a–d.

Primarily, the elution curves of a linear, not ideal adsorption chromatographic process (19a) and an ideal front chromatographic process (19b) are shown schematically. Figure 19c in contrast, shows as an example the terbuthylazine and desethylated terbuthylazine contents of samples of drain water of a trial area to which a terbuthylazine containing pesticide was applied at the end of April 1986. It should be noted that the rain volume on May 8th, 1986 was 24 mm and on May 7th, 1986 was 22 mm. So the terbuthylazine content of the drain water sample of May 9th, 1986 was more than 10 μg/litre draining water as a consequence of the preceeding intense rain.

If there are no abnormal amounts of precipitation, the content of the active ingredient in drainage water after an application of a pesticide containing atrazine or terbuthylazine over the period of one year after application – as Fig. 19d shows – is usually between 0.1 and 1 μg per litre of drainage water and afterwards dwindles away. The following rule of thumb can be deduced from the many results of drainage water analyses: About three years after the application of the previously mentioned pesticides there is no more run-off of atrazine or terbuthylazine with the drainage water [51].

The reasons for the front chromatographic phenomenon in run off of the pesticides with the drainage water and for sporadically appearing irregularities are due to the porous ground mass of the soil and to the rainfall. Important keywords in this context are: pore volume, pore size and pore size distribution. It is persumed that heavy rains occuring soon after application of a pesticide facilitate the transport of these substances to the macropores of the soil and as a consequence of the fast uncontrolled run-off there is a relatively high pesticide content in the percolate.

With an increasing retention period of a mobile pesticide in soil, mainly the water motion in soil and with it the hydromechanical dispersion, the molecular diffusion and sorption processes cause a continuously increasing volume in which the pesticide is found and an equal distribution of it in the surface soil. Here a fluctuation between ascending and descending and more or less inclined chromatography occurs which leads to a band broadening of the pesticides in

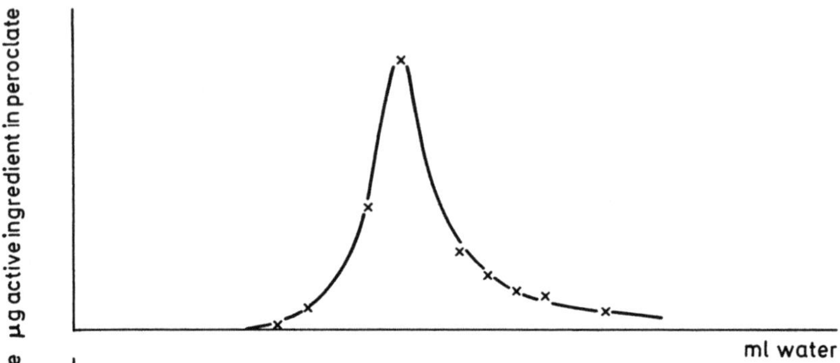
µg active ingredient in peroclate
ml water

µg active ingredient in peroclate
ml water

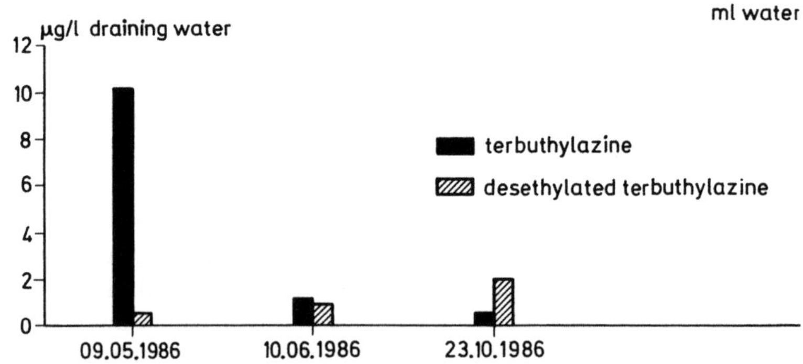
µg/l draining water
12
10
8
6
4
2
0
terbuthylazine
desethylated terbuthylazine
09.05.1986
10.06.1986
23.10.1986

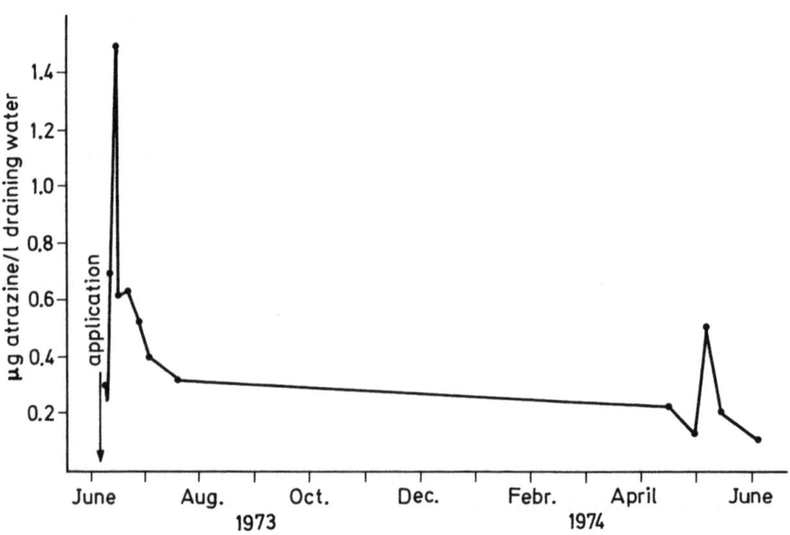
µg atrazine/l draining water
1.4
1.2
1.0
0.8
0.6
0.4
0.2
application
June
Aug.
Oct.
Dec.
Febr.
April
June
1973
1974

the percolate. This is because of the ever changing flow direction of the water that acts as the "mobile phase". The chromatographic events within the real soil system are so much more complex and multi-layered – depending on the constantly changing flow direction of the water – than those described in Sect. 2.1 with the soil column tests in the laboratory that are based exclusively on continuously descending chromatography.

Furthermore, the previously mentioned molecular diffusion and the hydromechanical dispersion essentially lead to a levelling out of the bands of mobile pesticides and their metabolites in drainage water. The slower the water moves in the surface soil layer or the longer the flow distance was at constant speed, the greater becomes the importance of the levelling out of the bands caused by diffusion.

The extent of the diffusion is determined by the depth of the surface soil, the characteristics of the soil and the soil structure, as well as pore volume, pore size and pore size distribution. It can therefore be concluded that the elution bands of pesticides in drainage water are flatter, the deeper the drainage pipes are laid and the more fine-grained the soil is. As a consequence, when we know the depths of drainage pipes and the grain-size distribution or the granulometric curve of the fine soil, qualitative statements can be made about the course of the elution curves of the pesticides in drainage water. The statements can be further differentiated if we consider the secondary pore system of the soils. Aggregate packages in which water flows slower than in the coarse passages in the soil lead to further levelling out of the bands in the run-off of pesticides. The following observation may be due to diffusion – with a drying up or a sparse flow of drainage water, a lower concentration of pesticides is found that fast flowing drainage water. The running or sparsely flowing drainage water consists mainly of water from the fine pore system of the soil, whereas the fast running drainage water is fed by the water from the coarse pore system and as a consequence of the quick run-off of the pesticides from soil may be highly polluted.

The time factor favors the exchange between bodies of water in the soil containing different amounts of molecular disperse residues or fine colloidal residues of the pesticides. As a consequence of the concentration-equalizing effects of molecular diffusion, the hydromechanical dispersion and the sorption events, as well as the distinct interdependence of these effects on the soil pore-system together with the meteorological events after the application of a pesticide, a relatively constant pesticide run-off with the drainage water may result. Furthermore, the pesticide elution curves may show several peak values in contrast to the elution curves of classical soil column tests in the laboratory,

Fig. 19. Scheme of the elution curves of a linear, not ideal adsorption chromatographic process (19a) and an ideal, front chromatographic process (19b). Compared with this are the terbuthylazine and desethylated terbuthylazine contents of drainage water samples after heavy rain (19c) and the content of atrazine in drainage water for a period of one year after the last atrazine application (19d) [44, 51]

obtained after a linear, not ideal adsorption chromatographic process. The peak values in permeable soils follow the rule of thumb: The stronger the leaching rates of rain are and the closer the corresponding leaching event is to the time of pesticide application, the higher the pesticide content in drainage water may be in the short-term. At the same time, the peak values may show so called "advancing effects". In other words: The first maximum of a pesticide elution curve, for example, may occur much earlier in natural soils, because of a mainly undisturbed run-off of the pesticides in macropores, than may result with elution curves of soil columns used in the laboratory.

Taking into consideration the many results from drainage water investigations for pesticides [for example 44, 45, 46, 47, 48, 49, 50, 51] with regard to their possible pollution of the ground water, the following statements can be deduced about the specific degree of risk to ground water by pesticides on the one hand and the local potential risk of ground water pollution on the other:

a. Ground water pollution by pesticide residues is almost exclusively caused by *mobile* pesticides resp. *mobile* metabolites of these substances. The basis of this statement lies in the results of soil column tests in the laboratory and in draining water studies, where pesticides classified as immobile on laboratory tests were not detected as pollutants in the drainage water. Pesticides, irreversibly attached to colloidal soil particles, are not worth mentioning in the run-off from the fine-pores nor in macro-pores of natural soils. So with an application of immobile pesticides in agricultural soils there is no potential risk of ground water pollution by these substances, independent of what the local risks are or how high they are. The result is that with the help of model tests using soil columns as described in Sect. 2.1 it can be decided which pesticides may be ground water pollutants.

b. As a consequence of the model tests with soil columns in the laboratory, those pesticides which were classified as being mobile can be separated into two groups according to the results of drainage water studies of the potential risk of ground water pollution.

The first major group includes those mobile pesticides that are microbially or chemically decomposed by hydrolysis so fast that they do not occur as pollution factors in drainage water and in the water involved in the new formation of ground water. As a consequence of their rapid decomposition in surface soils, there is no risk of ground water pollution after an application of these pesticides to agricultural areas, no matter what the local risks are or how high they are. Certainly with locally occurring, extremely unfavorable factors – such as exceptionally heavy rain immediately after the application of those substances to shallow layers of permeable soil – as risk to the ground water cannot be excluded. For unstable pesticides in ground water, the ground water pollution is temporary and not so long lasting as with relatively mobile and persistent active ingredients of pesticides. These relatively mobile and persistent pesticides can be lumped together under the aspect of potential risk of ground water pollution and so form a second main group within the mobile pesticides.

A closer look shows that all mobile pesticides can be integrated into the second major group being slowly decomposed and still present as residues in the surface layer at the beginning of the short-term new formation of ground water and at the beginning of the long-term new formation of the ground water. A consequence of results of analyses of drainage water for residues of pesticides is that, above all, active ingredients of the herbicidal chlorotriazines and their desethylated metabolites are pollution factors for the ground water.

Regarding the results from the analyses of drainage water with inclusion of the annual sales of the herbicidal chlorotriazines – atrazine, simazine, terbuthylazine and propazine – it was estimated for the Federal Republic of Germany that about 90% of the ground water pollution was caused by chlorotriazines and their desethylated metabolites. Inevitably it must be assumed that atrazine and especially its desethylated metabolite are the leading pollutants of the ground water by pesticides. Atrazine was used as the standard herbicide in maize cultures until the end of the 1980s. According to the results of the drainage water and the soil column tests, endangerment to ground water can be the effect of just a few pesticides. In this connection, the results of model tests in the laboratory using soil columns, the total herbicides bromacil and hexazinone – used to keep rail track free of weeds – should be discussed as ground water pollutants, too [17, 18, 19, 29, 45, 49, 50, 51, 52, 53, 54, 55, 56, 57, 58, 59].

The molecular structures of the ground water endangering chlorotriazines – atrazine, propazine, simazine and terbuthylazine – as well as their desethylated metabolites are shown in Fig. 20. This figure shows, furthermore, that from the decomposition of the different chlorotriazines by microbial desalkylation, identical metabolites may result and so, for example, desisopropylated atrazine which has the same molecular structure as desethylated simazine and des-*tert*.-butyl-terbuthylazine can be formed by the decomposition of atrazine, simazine and terbuthylazine.

A complete desalkylation of the before mentioned chlorotriazines leads to 2-chloro-4,6-diaminotriazine which is a major metabolite of these substances. As a consequence of the chemically identical metabolites of the triazines, the desisopropylated atrazine content determined in soils after an application of atrazine, simazine and terbuthylazine to agricultural soils often cannot be apportioned to the applied active ingredients with certainty. As a general principle, it should be noted that a isopropyl group is microbially decomposed slower than an ethyl group and so desisopropylated atrazine is a subordinate metabolite of atrazine.

The chemical decomposition of the chlorotriazines for example by hydrolysis, may lead – as Fig. 20 also shows – to the formation of hydroxytriazines that do not occur as polluting factors in ground water according to the results up till now.

c. With the mobile and relatively persistent residues of pesticides, the local risks of endangering factors determine the existing potential risk of ground water pollution. High local endangering factors mainly ensue from shallow surface layers of soil, permeable soils low on humus, and heavy rain immediately

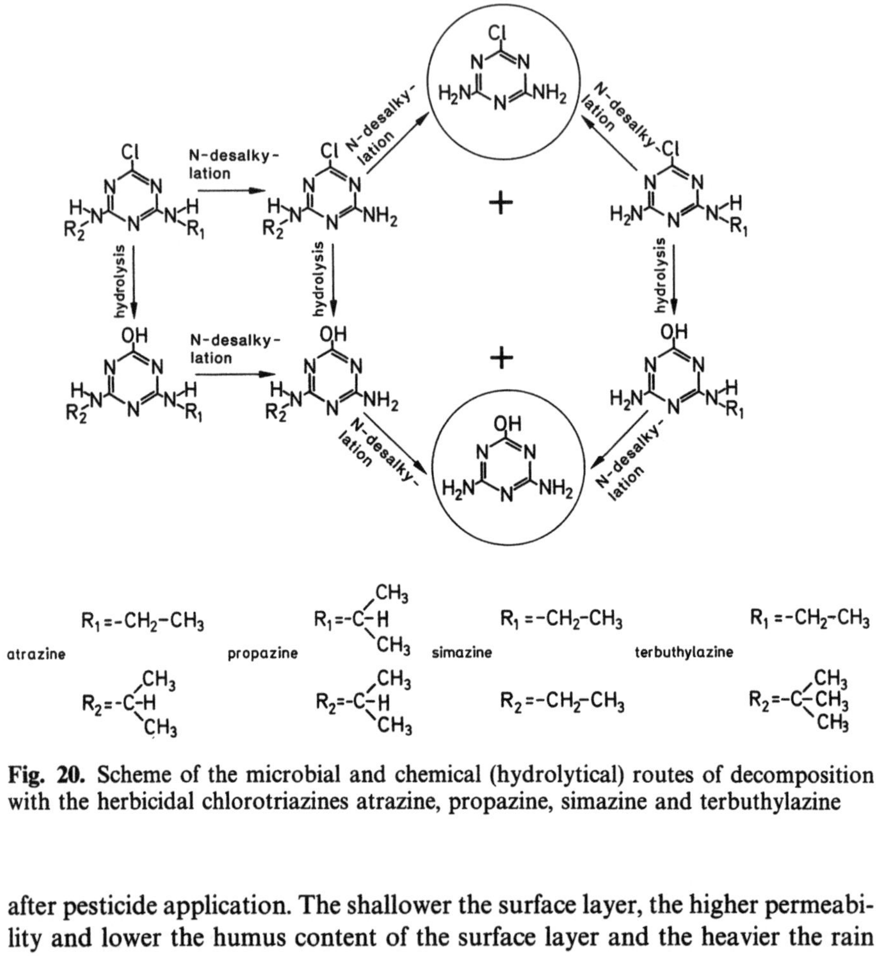

Fig. 20. Scheme of the microbial and chemical (hydrolytical) routes of decomposition with the herbicidal chlorotriazines atrazine, propazine, simazine and terbuthylazine

after pesticide application. The shallower the surface layer, the higher permeability and lower the humus content of the surface layer and the heavier the rain immediately after pesticide application are, the higher the degree of danger to the ground water is. In this context, the term "permeability of soils" is defined as pore volume, pore space and pore space distribution as well as the grain-size distribution curves and granulometric curves of the fine soils where in applying pesticides, the local potential of risk of ground water pollution rises with an increasing proportion of macropores and increasing grain size of the soil particles.

The danger of polluting ground water with mobile and persistent pesticides, when the results of the drainage water investigations are analysed is reduced conclusively in areas, where the previously mentioned criteria of endangering factors are fulfilled. For example, with a loess soil several metres thick rich in humus, the EG limit of 0.1 µg pesticide/litre drinking water cannot be exceeded even after repeated applications of atrazine because of decomposition and dilution by diffusion, mechanical dispersion and sorption. As a consequence, it was also deduced in the middle of the 1980s that more than 90% of water works have no problems with the EG limit for pesticides and with the remaining approx. 10% of water works, nearly 90% of the ground water pollution – as explained in item b – results from chlorotriazine pesticides.

d. Finally, the results of drainage water analyses related to the amount of atrazine applied in the water catchment area, allow the time atrazine and desethylated atrazine appear as well as their quantities in ground water to be estimated. According to this, with high ground-water-endangering factors, atrazine and its desethylated metabolite can be measured in ground water the faster, the less water there is which is unpolluted as ground water store, the larger the atrazine applied areas are in the water catchment area and the higher the dosage of atrazine applied.

As an example, a simplified calculation shows with an average desethylated atrazine-drainage water content of 1.0 μg desethylated atrazine/litre drainage water and the atrazine applied area were 10% of the water catchment area, the water of the shallow new formation of ground water contains on average 0.1 μg desethylated atrazine/litre water. The application of four times as much atrazine/ha and atrazine-applied areas of 50%, leads to a rise in the desethylated atrazine content to 2 μg desethylated atrazine/litre water. Analogously, the average pesticide content in the water of the shallow new formation of ground water can be estimated, for example, even after determining higher pesticide contents in drainage water as a consequence of shallow or highly permeable surface layers of the soil. According to the results of drainage water analyses, it must be assumed at the same time that after years of atrazine application, the yearly concentration profiles of atrazine as well as of desethylated atrazine are nearly constant and extreme values in those concentration profiles appear faster and higher, the shallower the ground water is.

After atrazine application has stopped, the ground water pollution with atrazine and desethylated atrazine declines more, the less the ground water supply is contaminated with pesticides of residues and the faster the water of the subsequent new ground water formation reaches the ground water supply. As a consequence, the potential risk of ground water pollution in applying mobile and relatively persistent pesticides can be quantified according to the local risk factors for ground water pollution.

In the past, the use of drainage systems to evaluate a local potential of risk of ground water pollution was the subject of many discussions. The discussions often became heated over the question of how far draining systems reflect the real run-off events in the surface layers of the soil.

In this context, for example the specific influence of the drain depth, the draining incline the drainage distance and the age of the drainage system on the results of investigations were debated and with it their relevance to quantifying the local potential risk of ground water pollution.

A closer examination shows that the pipes drain off the mobile water in the pore space free of tension and so accelerate natural drainage. A consequence is that the results of drain water analyses can be treated as if they were real ground water pollution situations. However, in individual cases, differences in degree of the influence of the drains must be considered. According to this, the draining effect causes the maxima of the pesticide elution curves to appear earlier than in a system without drainage. Furthermore the draining effects will lead to higher maxima and to less band levelling-out.

The flow to the drainage pipe can be mathematically formulated as one, two or three dimensional water movement according to the laws of Poiseuille, Darcy, Bernoulli, Richards and Laplace. The total flow is divided into vertical, horizontal and radial flow and described as flow system with streamlines (Ψ) and equipotential lines (ρ). According to the distance of the drainage pipe from the blocking layer below it, which influences the vertical, horizontal and radial resistance, three cases result with stationary streaming state:

1. The drainage pipe is directly above or only a small distance above the blocking layer. In this case, the horizontal resistance predominates.
2. The drainage pipe is more than 1/4 of the spacing between the drainage pipes and the blocking layer. In this case, the radial resistance predominates.
3. The drainage pipe is nearer than 1/4 of the spacing between the drainage pipes and the blocking layer. In this case, horizontal and radial resistance work together.

Figure 21 shows vertical, horizontal and radial streaming to a drainage pipe.

The mathematical consideration of the streams to a drainage pipe enables to work out the specific streaming conditions of a drainage system against a run-off where there are no drains and to draw more conclusions about the relevance of results of drainage water analyses or to get corresponding correction factors. From the middle of the 1980s, many reports about pesticide contents in ground water appeared, so a direct comparison of results of drainage water analyses and

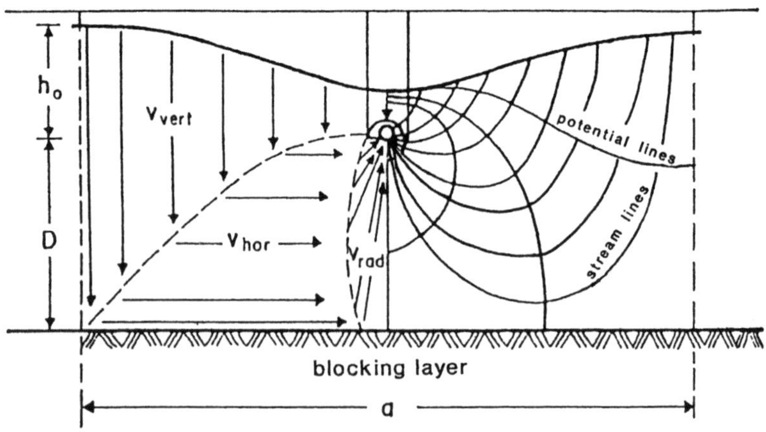

Fig. 21. Scheme of the flow pattern of a drainage pipe. The total flow is separated into vertical, horizontal and radial flow to the drainage pipe

results of ground water analyses is possible. Thereby an elegant method offers itself to confirm the statement of the rules obtained with the help of drainage systems about the potential risk of ground water pollution by pesticides. At the same time, the aspects and forecasts about the specific factors of ground water pollution by pesticides with the use of simple model tests in laboratory using soil columns according to the BBA guideline for the official test of pesticides part IV 4-2 can be critically evaluated.

3 Evaluation of Theoretical Prognostic Methods for Quantifying the Specific Potentials of Risk of Ground Water Pollution by Pesticides as well as the Local Factors of Risks of Ground Water Pollution Associated with Their Use

In evaluating the prognostic methods with results of analyses that were determined under actual conditions, it should be noted that the analysis methods used in the middle of the 1980s to monitor ground water for pesticides have since been improved. In the course of improving these methods, it was realized that many of the earlier methods of the ground water monitoring gave incorrect results that irritated public [19].

The revision of the ground water monitoring/analytical methods was completed in the Federal Republic of Germany at the end of the 1980s/at the beginning of the 1990s. With the new numerical data, the statements based on theoretical prognostic methods, for example, the model tests in laboratory using soil columns, may be directly correlated with the results of the ground water monitoring.

Out of 161, 466 test results that were reported to the Federal Environmental Agency by the states and the Water Supply Industry 147, 440 (91, 3%) were negative results. At the same time, the fraction of the herbicidal chlorotriazines including their metabolites with test results above 0.1 μg pesticide residue per litre water is 81% [60]. Leader in the number of complaints in the Federal Republic of Germany is the earlier highly popular atrazine and its desethylated metabolite. As a consequence of their pollution of ground and drinking water the application of atrazine- and simazine-containing pesticides in water protection areas was prohibited [61]. Furthermore, in 1990, in BadenWürttemberg, the application of terbuthylazine containing pesticides, that was planned as a replacement herbicide for atrazine, was prohibited in water protection areas as well [62]. In 1991 the application of atrazine-containing pesticides was totally forbidden [63].

Moreover, it can be deduced from the analyses of ground water monitoring, that endangerment by pesticides is as a rule confined to hydrogeologically unfavourable areas. When the limits for pesticides are exceeded, most are in the range between > 0.1 to 0.5, in the range between > 0.5 and 1.0 and in individual cases, in the range between > 1.0 and 3.0 μg pesticide per litre of ground water. Higher pesticide content in ground water as a consequence of agricultural use can be excluded according to the results of the ground water monitoring. At the same time, the chlorotriazine contents of the ground water show a certain temporal constancy with a tendency to decrease. The tendency to decrease is a direct consequence of the restriction or the prohibition of chlorotriazine-containing pesticides in the Federal Republic of Germany. Figure 22 shows that well 1 of commune A has provided drinking water with a atrazine and desethylated atrazine content below the limit since 1991. The atrazine content of well 1 of commune B, in contrast, decreased from 2.51 μg per litre in 1989 to 0.39 μg per litre in 1993 (Fig. 23). At the same time, the terbuthylazine and desethylated terbuthylazine content, which, in 1989, was still above the limiting value for pesticides sank below the limit for pesticides as a consequence of the prohibition of terbuthylazine-containing pesticides in water protection areas. Similar tendencies are shown by the atrazine and desethylated atrazine content and the terbuthylazine content of spring 1 of commune B (Fig. 24) as well as spring 1 of commune C [51] (see Fig. 25).

The results of the Federal Republic of Germany's ground water monitoring affirm impressively the statements of the theoretical prognosis systems based on model studies in the laboratory using soil columns, draining water tests, lysimeter tests and using computer-aided simulation models to quantify the potential risk of ground water pollution by pesticides.

As a summary of the extensive results of the Federal Republic of Germany's ground water monitoring program, it can be concluded that model tests in the laboratory using soil columns (based on the guideline for the official test of pesticides part IV 4-2 of the Federal Biological Research Centre for Agriculture

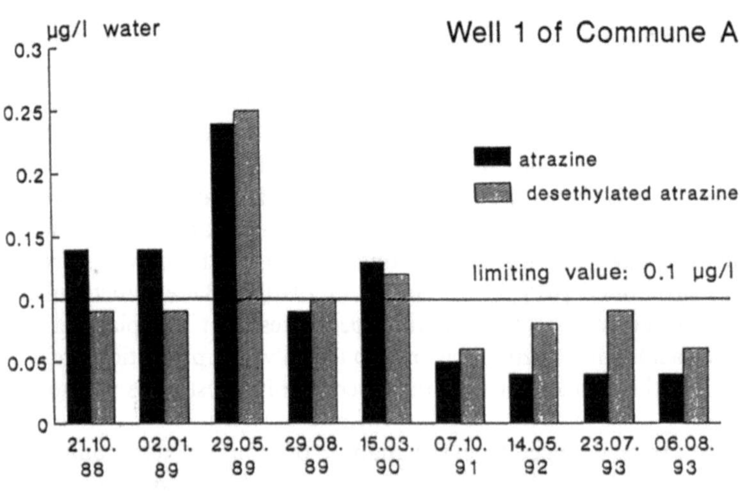

Fig. 22. Course of the atrazine and desethylated atrazine contents in well 1 of commune A from 1988–1993

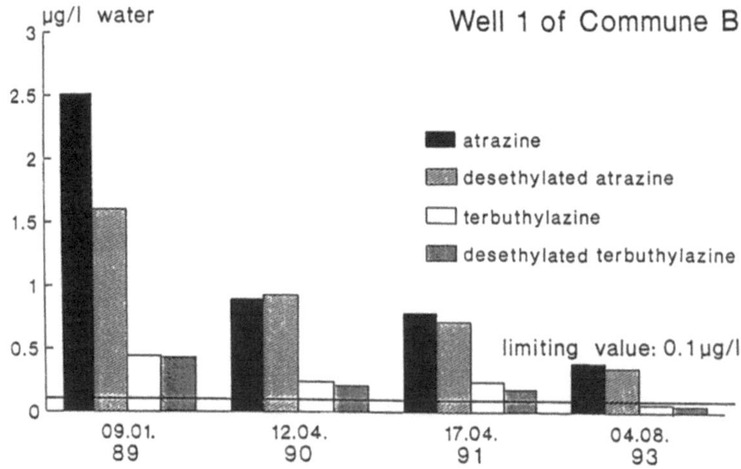

Fig. 23. Course of the atrazine, desethylated atrazine, terbuthylazine and desethylated terbuthylazine in well 1 of commune B from 1989–1993

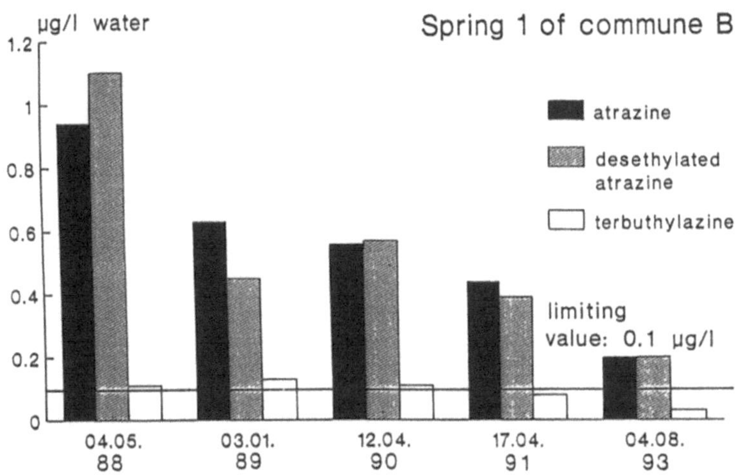

Fig. 24. Course of the atrazine, desethylated atrazine and terbuthylazine contents in the spring of commune B from 1988–1993

and Forestery) as well as controlled drainage water tests allow us to make a fast and reliable evaluation of the potential risk of ground water pollution by pesticides as well as the local risk factors of ground water pollution. It can be concluded with the help of soil column tests in the laboratory which pesticides are mobile in surface soil layers or what their degree of mobility is. So the amount of ground water pollution and the course of tendency of pesticides in

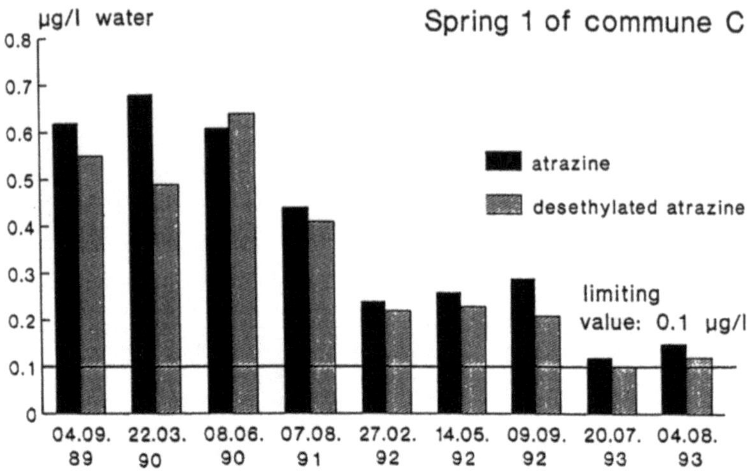

Fig. 25. Course of the atrazine and desethylated atrazine contents in spring 1 of commune C from 1989–1993

ground water can be definitely predicted for the different water regions with simple drainwater investigation. Possible ground water endangerment from using pesticides can be shown before their application with the help of those simple prognostic methods.

4. References

1. Anonym (1980) Richtlinie des Rates vom 15. Juli 1980 über die Qualität von Wasser für den menschlichen Gebrauch (80/778/EWG) Nr. L 229, p 11
2. Anonym (1986) Verordnung über Trinkwasser und über Wasser für Lebensmittelbetriebe (Trinkwasserverordnung-TrinkwV) vom 22. Mai 1986, BGB1.I, p 760
3. Anonym (1990) Bekanntmachung der Neufassung der Trinkwasserverordnung vom 5. Dez. 1990, BGBl. I, p 2612
4. Anonym (1988) Wasserwerke fordern Schutz vor Pestiziden. PresseInformation Bundesverband der deutschen Gas- und Wasserwirtschaft e.V.
5. Biologische Bundesanstalt für Land- und Forstwirtschaft Deutschland (ed) (1986) Versickerungsverhalten von Pflanzenschutzmitteln. Richtlinien für die amtliche Prüfung von Pflanzenschutzmitteln Teil IV 4-2
6. Biologische Bundesanstalt für Land- und Forstwirtschaft Deutschland (ed) (1990) Lysimeteruntersuchungen zur Verlagerung von Pflanzenschutzmitteln in den Untergrund. Richtlinien für die Prüfung von Pflanzenschutzmitteln im Zulassungsverfahren Teil IV 4-3
7. Carsel RF, Smith CN, Mulkey LA, Dean JD, Jowise P (1984) Users Manual for the Pesticide Root Zone Model (PRZM), Release 1. EPA Athens
8. Bonazountas M, Wagner J (1984) SESOIL: A seasonal soil compartment model. Office of Toxic Substances, US Environmental Protection Agency, Washington, DC

9. Biologische Bundesanstalt für Land- und Forstwirtschaft Deutschland (ed) (1993) Pflanzenschutzmittel-Verzeichnis 1993 (einschließlich Wachstumsregler)
10. Anonym (1975) Richtlinien für Trinkwasserschutzgebiete. I. Teil. Schutzgebiete für Grundwasser. DVGW-Regelwerk. Technische Regeln, Arbeitsblatt W 101 p 1–12
11. Biologische Bundesanstalt für Land- und Forstwirtschaft Braunschweig (ed) (1993) Bekanntmachung Nr. 55 vom 17.03.1993 mit der 122. Bekanntmachung über die Zulassung von Pflanzenschutzmitteln vom 11.03.1993
12. Bailey GW, White JL (1970) In: Residue Reviews. Springer Berlin Heidelberg New York, p 29
13. Weber JB (1970) in: Residue Reviews, Springer Berlin Heidelberg New York, p 93
14. Hayes MHB (1970) In: Residue Reviews. Springer, Berlin Heidelberg New York, p 131
15. Helling CS (1970) in: Residue Reviews, Springer Berlin Heidelberg New York, p 175
16. Ziechmann W, Müller-Wegener U (1990) Bodenchemie. BI Wissenschaftsverlag Mannheim
17. Häfner M (1989) In: Pflanzenschutzmittel und Grundwasser. Schriftenreihe des Vereins für Wasser-, Boden- und Lufthygiene. Gustav Fischer. Stuttgart, p 261
18. Häfner M (1990) In: Zeitschrift für Pflanzenkrankheiten und Pflanzenschutz. Eugen Ulmer, Stuttgart, p 427
19. Häfner M (1991) In: Gewässerschutz und Wasseraufbereitung- Grundlagen einer zuverlässigen Trinkwasserversorgung. Kommissionsverlag R. Oldenbourg, München, pp 165–200
20. Herzel F (1987) Nachrichtenbl. Deut. Pflanzenschutzd. 39: 97
21. Herzel F, Schmidt G (1987) Bundesgesundhbl. 30 Nr. 11, 396
22. Herzel F, Schmidt G (1988) Nachrichtenbl. Deut. Pflanzenschutzd. 40: 25
23. Herzel F, Schmidt G (1989) Nachrichtenbl. Deut. Pflanzenschutzd. 41: 91
24. Nordmeyer H, Pestemer W (1989) Nachrichtenbl. Deut. Pflanzenschutzd. 41: 193
25. Herzel F, Schmidt G (1990) Nachrichtenbl. Deut. Pflanzenschutzd. 42: 53
26. Herzel F, Schmidt G (1991) Nachrichtenbl. Deut. Pflanzenschutzd. 43: 13
27. Bunte D, Pekrun S, Utermann J, Nordmeyer H, Pestemer W (1991) Nachrichtenbl. Deut. Pflanzenschutzd. 43: 17
28. Bunte D, Pestemer W (1991) Nachrichtenbl. Deut. Pflanzenschutzd. 43: 238
29. Häfner M. Unveröffentlichte Untersuchungsergebnisse zum Versickerungsverhalten von Pflanzenschutzmitteln auf der Grundlage der Richtlinien für die amtliche Prüfung von Pflanzenschutzmitteln Teil IV 4-2
30. Hance RJ (1965) Weed Research 5: 98
31. Davidson JM, Brusewitz GH, Baker DR, Wood AL (1975) Use of Soil Parameters for Describing Pesticide Movement Through Soils, U.S. Environmental Protection Agency, Athens, GA. Report No. EPA-660/2-75-009
32. Enfield CG, Carsel RF, Cohen SZ, Phan T, Walters DM (1983) Ground Water 20: 711
33. Gureghian AB, Ward DS, Cleary RW (1979) J. Hydrol. 41: 253
34. Lorber MN, Offutt CK (1986) In: Garner WY, Honeycutt RC, Nigg HN (eds) Evaluation of Pesticides in Ground Water. Developed from a symposium sponsored by the Division of Pesticide Chemistry at the 189th Meeting of the American Chemical Society, Miami Beach, Florida, April 28– May 3, 1985. American Chemical Society, Washington, DC, p 342
35. Klein M (1991) Pestic. Sci. 31: 389
36. Biologische Bundesanstalt für Land- und Forstwirtschaft Deutschland (ed) (1986) Verbleib von Pflanzenschutzmitteln im Boden- Abbau, Umwandlung und Metabolismus. Richtlinien für die amtliche Prüfung von Pflanzenschutzmitteln Teil IV 4-1
37. Brumhard B, Führ F (1992) Lysimeter Studies of the Fate of Pesticides in Soil. BCPC Monograph No. 53, p 103
38. Brumhard B, Snel M, Baloch R, Führ F (1992) Z. PflKrkh. PflSchutz, XII, 449

39. Führ F (1992) In: IRA Ergebnisbericht 1992, Institut für Radioagronomie des Forschungszentrums Jülich GmbH
40. Jarczyk HJ (1983) Untersuchungen zum Sickerverhalten von Pflanzenschutzmitteln in einer Monolith-Lysimeter-Anlage. Vortrag 95. VDLUFA-Kongreß Kiel, 19. –24.9.1983. Landwirtsch. Forsch., Kongreßband, p 273
41. Jarczyk HJ (1983) Investigation on the leaching behaviour of herbicides under natural agronomic conditions of lysimeter monoliths. Vortrag, Assoc. Appl. Biol., Europ. Weed Res. Soc., Soc., Chem. Ind., Oxford, St. Chatherine's College, 12.–15. Dez. 1983, Proc., Aspects of Appl. Biol. 4, p 441
42. Jarczyk HJ (1987) Pflanzenschutz-Nachrichten Bayer 40: 49
43. Collins HJ (1986) Zeitschrift für Kulturtechnik und Flurbereinigung. Journal of Rural Engineering and Development Vol 27: 269
44. Muir DC, Baker BE (1976) J. Agric. Food Chem., 24: 122
45. Meinert G, Häfner M (1987) in: Milde G, Friesel P (eds) Grundwasserbeeinflussung durch Pflanzenschutzmittel. Schr. -ReiheWaBoLu 68, Gustav Fischer, Stuttgart, p 51
46. Hurle K, Gießl H, Kirchhoff J (1987) In: Milde G, Friesel P (eds) Grundwasserbeeinflussung durch Pflanzenschutzmittel. Schr. -Reihe WaBoLu 68, Gustav Fischer, Stuttgart, p 169
47. Felding G (1992) Pestic. Sci. 35: 271
48. Southwick LM, Willis GH, Selim HM (1992) J. Agric. Food Chem 40: 1264
49. Häfner M (1988) In: Cavalloro R, El Titi A. (eds) Weed Control in Vegetable Production. Proceedings of a Meeting of the EC Experts' Group/Stuttgart 28–31 October 1986. A.A. Balkema/Rotterdam/Brookfield, p 35
50. Häfner M (1986) In: 45. Deutsche Pflanzenschutztagung in Kiel 6. –10. Oktober 1986. Mitteilungen aus der Biologischen Bundesanstalt für Land- und Forstwirtschaft Berlin Dahlem, Heft 232. Kommissionsverlag Paul Parey, Berlin und Hamburg, p 300
51. Häfner M, Unveröffentlichte Ergebnisse von Drän-, Grund- und Oberflächenwasseruntersuchungen
52. Häfner M (1989) In: Riewenherm/Lieth/Stegmann (eds) Verhandl. d. GfÖ Bd. XIX. I, Göttingen, p 162
53. Häfner M (1989) In: Riewenherm/Lieth/Stegmann (eds) Verhandl. d. GfÖ Bd. XIX.I, Göttingen, p 291
54. Häfner M (1990) In: Statusseminar Bodenbelastung und Wasserhaushalt, Bundesministerium für Forschung und Technologie, Projektträger Biologie, Energie, Ökologie Forschungszentrum Jülich GmbH
55. Häfner M (1990) TASPO magazin p 29
56. Häfner M (1991) Pflanzenschutz-Praxis 4, p 36
57. Häfner M (1990) Mitteilungen aus der Biologischen Bundesanstalt für Land- und Forstwirtschaft Berlin-Dahlem, Heft 266, p 94
58. Häfner M (1987–92) Jahresberichte, Bundesministerium für Forschung and Technologie, Projektträger Biologie, Energie, Ökologie Forschungszentrum Jülich GmbH
59. Häfner M (1993) In: Abschlußbericht, Bundesministerium für Forschung und Technologie, Projektträger Biologie, Energie, Ökologie Forschungszentrum Jülich GmbH
60. Anonym (1992) Pflanzenschutzmittelbefunde im Wasser nach Mitteilungen der Länder und Wasserversorgungsunternehmen, Umweltbundesamt
61. Anonym (1988) Verordnung über Anwendungsverbote für Pflanzenschutzmittel (Pflanzenschutz-Anwendungsverordnung) Vom 27. Juli 1988, BGbl. I p 1196
62. Anonym (1990) Verordnung des Umweltministeriums zur Änderung der Schutzgebiets- und Ausgleichsverordnung Vom 22. März 1990, GBl. Nr. 11, p 166
63. Anonym (1991) Erste Verordnung zur Änderung der Pflanzenschutz-Anwendungsverordnung Vom 22. März 1991, BGBl. I, p 796

Subject Index

Previously Published Volumes in Chemistry of Plant Protection

Vol. 1: Sterol Biosynthesis, Inhibitors and Anti-Feeding Compounds

T. Kato
Sterol-Biosynthesis in Fungi, a Target for Broad Spectrum Fungicides

D. W. Krämer
Chemistry of Sterol Biosynthesis Inhibiting Fungicides

K.-H. Kuck, H. Scheinpflug
Biology of Sterol-Biosynthesis Inhibiting Fungicides

D. M. Norris
Anti-Feeding Compounds

Vol. 2: Degradation of Pesticides, Desiccation and Defoliation, ACh-Receptors as Targets

P. R. Wallnöfer, G. Engelhardt
Microbial Degradation of Pesticides

U. Schoknecht, D. Otto
Enzymes Involved in the Metabolism of Organophosphorus, Carbamate and Pyrethroid Insecticides

R. Richter, D. Otto, H.-J. Mengs
Insecticide Compounds, Acting on the Acetylcholine Receptor of the Insect Nervous System

H. Bermann, D. Martin
Chemical Manipulation of Desiccation and Defoliation and Essential Aspects for the Application and Development of New Chemical Compounds in the Future

Vol. 3: Pyrethroid Residues, Immunoassays for Low Molecular Weight Compounds

W. Blaß
Residues of Pyrethroid Insecticides and Their Analytical Determination

E. W. Weiler
Immunoassay for the Quantitation of Low Molecular Weight Compounds in Plants

Vol. 4

K. Naumann
Synthetic Pyrethroid Insecticides: Structures and Properties

Vol. 5

K. Naumann
Synthetic Pyrethroid Insecticides: Chemistry and Patents

Vol. 6: Controlled Release, Biochemical Effects of Pesticides, Inhibition of Plant Pathogenic Fungi

M. Bahadir, G. Pfister
Controlled Release Formulations of Pesticides

M. Eto
Biochemical Mechanisms of Insecticidal Activities

M. A. Q. Khan
Biochemical Effects of Pesticides on Mammals

P. Böger, G. Sandmann
Modern Hibicides Affecting Typical Plant Processures

H. Buchenauer
Physiological Reactions in the Inhibition of Plant Pathogenic Fungi

Vol. 7: Herbicide Resistance - Bressinosteroids, Gibberellins, Plant Growth Regulators

D. Gross
Plant Growth Regulatory Substances Both of Microbial and Plant Origin

G. Adam, B. Voigt, M. Lischewski
Synthesis and Labelling of Gibberellins

V. Marquardt, G. Adam
Recent Advances in Brassinosteroid Research

K. C. Vaughn, S. O. Duke
Biochemical Basis of Herbicide Resistance

Vol. 8: Terrestrial Behavior of Pesticides

I. Scheunert
Physical and Physico-Chemical Processes Governing the Residue Behavior of Pesticides in Terrestrial Ecosystems

I. Scheunert
Transformation and Degradation of Pesticides in Soil

I. Scheunert
Fate of Pesticides in Plants and in Soil Fauna

H. Parlar
Mechanism for the Behavior of Pesticides on Surfaces

Herbicide Resistance – Brassinosteroids, Gibberellins, Plant Growth Regulators

With contributions by **G. Adam**, **S.O. Duke**, **D. Gross**, **M. Lischewski**, **V. Marquardt**, **K.C. Vaughn**, **B. Voigt**

1991. XI, 176 pp. 10 figs., 4 tabs. (Chemistry of Plant Protection, Vol. 7)
ISBN 3-540-54197-7

D. Gross, Halle/FRG: *Plant Growth Regulatory Substances Both of Microbial and Plant Origin*.- G. Adam, B. Voigt, M. Lischewski, Halle/FRG: *Synthesis and Labelling of Gibberellins*.- G. Adam, V. Marquardt, Halle/FRG: *Recent Advances in Brassinosteroid Research*.- K.C. Vaughn, S.O. Duke, Stoneville/MS: *Biochemical Basis of Herbicide Resistance*.

Terrestrial Behavior of Pesticides

With contributions by **I. Scheunert**, **H. Parlar**

1992. XI, 144 pp. 31 figs., 27 tabs. (Chemistry of Plant Protection, Vol. 8)
ISBN 3-540-54238-8

Contents: I. Scheunert, Neuherberg, FRG: *Physical and Physico-Chemical Processes Governing the Residue Behavior of Pesticides in Terrestrial Ecosystems.- Transformation and Degradation of Pesticides in Soil.- Fate of Pesticides in Plants and in Soil Fauna.*- H. Parlar, Kassel, FRG: *Mechanism for the Behavior of Pesticides on Surfaces*.

MIX
Papier aus verantwortungsvollen Quellen
Paper from responsible sources
FSC® C105338

If you have any concerns about our products,
you can contact us on
ProductSafety@springernature.com

In case Publisher is established outside the EU,
the EU authorized representative is:
Springer Nature Customer Service Center GmbH
Europaplatz 3, 69115 Heidelberg, Germany

Printed by Libri Plureos GmbH
in Hamburg, Germany